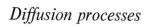

Diffusion processes

Diffusion processes

Proceedings of the Thomas Graham Memorial Symposium
University of Strathclyde

VOLUME 1

Edited by

J. N. SHERWOOD
A. V. CHADWICK
W. M. MUIR
F. L. SWINTON

Department of Pure and Applied Chemistry
and the Bioengineering Unit
University of Strathclyde

GORDON AND BREACH SCIENCE PUBLISHERS
London　　　　　　　New York　　　　　　Paris

Library of Congress catalog card number 76-125012. ISBN 0 677 14820 8. All rights reserved. No part of this book may be reproduced or utilized in any form or by any means, electronic or mechanical, including photocopying, recording, or by any information storage and retrieval system, without permission in writing from the publishers. Printed in east Germany

Preface

September 16th, 1969, marked the centenary of the death of Thomas Graham, F.R.S. As will be seen from the short biographical note, in 1830 Graham was appointed the first independent Professor of Chemistry in Anderson's University, Glasgow; the teaching of the subject having been previously associated with the Chair of Natural Philosophy. Anderson's University which had been founded as a technological university thirty-four years earlier under the name of Anderson's Institution was to regain this designation in 1964 as the University of Strathclyde. During his association with the University, Graham published the famous series of papers: "On the Law of the Diffusion in Gases". The work and ideas reported in these papers led to the establishment of the "Laws of Diffusion" which are now associated with his name. By this and his subsequent researches Graham laid the foundations of the scientific study of diffusion in all its aspects and although he achieved eminence in other branches of chemistry it is for this that he is chiefly remembered. On this centenary it seemed appropriate to commemorate Graham's contributions to physical science by holding an international symposium on diffusion processes in the building which now bears his name at the University of Strathclyde.

The symposium attracted a wide spectrum of papers on all aspects of diffusion processes in solids, liquids and gases and the technological and biological applications of diffusion. Those who attended the symposium were unanimous in their declarations that they had benefited from attending and discussing lectures which covered wider aspects of the subject than were usually presented at symposia on their specialised fields. As a consequence of this general feeling we have decided to publish the proceedings of the symposium as a further commemoration to the man whose work initiated the study of this diverse subject.

University of Strathclyde

A. V. CHADWICK
W. M. MUIR
J. N. SHERWOOD
F. L. SWINTON

v

Thomas Graham 1805–1869
Portrait in oils by John Graham-Gilbert 1837

[Photograph by R.Cowper]

Biographical note

Thomas Graham, M.A., D.C.L., F.R.S.
1805–1869

1805	Born Glasgow, 20th December.
1826	Graduated M.A. Glasgow University.
1826	Published his first scientific paper "On the absorption of gases by liquids".
1829	Lecturer on Chemistry at Mechanics' Institution, Glasgow.
1830	Appointed to first Independent Chair of Chemistry, Anderson's University.
1834	Keith Medal of Royal Society of Edinburgh for his paper "On the law of diffusion of gases".
1836	F.R.S.
1837	Appointed Professor of Chemistry at University College, London.
1838	Royal Medal of the Royal Society for his paper "Inquiry respecting the constitution of salts, oxalates, nitrates, phosphates, sulphates, and chlorides".
1840	Elected first President of the Chemical Society.
1842	Published text-book "Elements of Chemistry".
1846	Elected first President of the Cavendish Society.
1850	2nd Royal Medal of the Royal Society for his papers "On the motion of gases".
1853	D.C.L., Oxford.
1855	Appointed Warden and Master Worker of the Mint.
1855	Chevalier of the Legion of Honour.
1862	Copley Medal of the Royal Society for his papers on osmotic force, and diffusion applied to analysis.
1868	Paper "On the occlusion of hydrogen gas by metals".
1869	Died, London, 16th September.

Contents of Volume 1

Section 4   Diffusion in metals

Contents of Volume 2

Diffusion in gases

E. A. MASON

Brown University, Providence, Rhode Island

ABSTRACT

Serious study of gaseous diffusion dates from Thomas Graham's work, beginning in 1829. A review is given of the development of the subject up to the present, with some indications of the current areas of greatest activity and of greatest ignorance. The subject can be roughly divided into two main aspects, phenomenological and molecular. In the phenomenological aspect one seeks to describe how gases mix and flow, and what the variables are that influence this behaviour; ultimately such a description would take the form of some differential equations containing coefficients to be found by experiment. The task of the molecular aspect is to relate these coefficients to atomic and molecular properties. Graham, in his studies of diffusion, effusion, and transpiration, had the genius to delineate clearly these three main mechanisms of gas transport, which usually occur in combination. This is an especially remarkable feat because the kinetic theory of gases as we know it had not yet been born.

It is curious that most of the interesting surprises in modern times have been on the experimental and phenomenological rather than the molecular side, which has developed instead into a tool for studying intermolecular forces. Some of the phenomena, such as diffusion in the transition region and the diffusion baroeffect, go straight back to Graham's work, and one of the strangest stories has been the repeated and unrecognized rediscovery since 1953 of Graham's law of diffusion.

1 HISTORICAL BACKGROUND

Our purpose is to give, in rather broad terms, a review of the development of the subject of diffusion in gases up to the present, with some indication of the current areas of greatest activity and of greatest ignorance. This seems especially appropriate for a Thomas Graham Memorial Symposium, for the

3

subject can be said to date from Thomas Graham's work, beginning in 1829. The first sentence of his first paper on diffusion[1] still seems, even after 140 years, to be a fair statement of the situation at that time:

"Fruitful as the miscibility of the gases has been in interesting speculations, the experimental information we possess on the subject amounts to little more than the well established fact, that gases of a different nature, when brought into contact, do not arrange themselves according to their density, the heaviest undermost, and the lightest uppermost, but they spontaneously diffuse, mutually and equably, through each other, and so remain in an intimate state of mixture for any length of time."

Graham then proceeded to obtain the first quantitative results on gaseous diffusion by studying the loss of various gases out of a closed vessel into the surrounding air through a small tube. He also made a few observations on diffusion through a small tube connecting two gas reservoirs. He did not obtain any diffusion coefficients from these measurements. Indeed the term "diffusion coefficient" did not even exist, for the mathematical statement of the law of diffusion, analogous to Fourier's law of heat conduction, was not given until 26 years afterwards by Adolph Fick[2] who was a demonstrator of Anatomy in Zurich. Perhaps Graham would not have calculated diffusion coefficients even if the mathematical formulation had existed; apparently it was not customary in those days for chemists to solve partial differential equations, that being the province of theoretical physicists. At any rate, it was Maxwell[3] who finally extracted diffusion coefficients from Graham's observations. Those from the 1829 paper turned out to be much too low, although of the correct order of magnitude. But those obtained in a later experiment[4] with a closed vertical tube gave a diffusion coefficient for CO_2-air which is only 5–6% lower than the best modern value. This seems to be the first accurate diffusion coefficient obtained.

Thus Graham not only performed the first quantitative diffusion experiments, but also obtained the first accurate diffusion coefficient (by modern standards). Both of these efforts were characterized by ingenious experimentation. The methods used have survived, with modifications, to the present time. For instance, the 1829 method was used, under the name of "capillary-leak", to obtain the first high-temperature diffusion coefficients by Klibanova *et al.* in 1942.[5] The 1863 closed-tube technique was slightly modified a few years later by Loschmidt[6,7], who made an extensive series of measurements on ten gas pairs and achieved a level of accuracy of about 2%, an accuracy rarely exceeded, and often not reached, even today. This is a comment both on Loschmidt's skill as an experimenter and on the difficulty of measuring diffusion coefficients with high accuracy.

Despite these two notable "firsts" in gaseous diffusion, the connection of Graham's name with diffusion comes from an entirely different line of work. This work has been enshrined in learned treatises and textbooks (almost invariably either erroneously or incompletely) as "Graham's Law of Diffusion", and it has been repeatedly misunderstood, forgotten, rediscovered, and misunderstood again.[8,9] There are some things about it that we don't fully understand yet. This work is worth discussing in a little detail, for much effort and confusion in recent years might have been avoided if it had been remembered and understood.

Graham's investigations were reported in a paper read before the Royal Society of Edinburgh in 1831, and later published in three sections.[10] He took a calibrated glass tube with a porous plate at one end and with the other end immersed in a vessel of water (or mercury). The gas to be investigated was added to the tube by displacement of water, and its volume noted. As the gas diffused out and the air diffused in through the porous plate, the water level tended to rise or fall, depending on whether the gas was lighter or heavier than air. Since a change in the water level would have caused a pressure difference across the porous plate, Graham kept the pressure uniform by flowing water into or out of the outer vessel to keep the outer water level the same as the one inside the tube. After some time all the gas had diffused out and been replaced by the air that had diffused in. Graham noticed that the ratio of the volumes of gas diffused out and air diffused in was equal to the square root of the inverse ratio of their molecular weights.

It should be noted that these experiments were *not* in the free-molecule regime; the free-molecule result is called *effusion*, which was also discovered by Graham, but much later.[11] It should also be noted that no diffusion coefficients were obtainable from these experiments. They can be obtained from the rate of change of the water level[12], but the differential equation involved is complicated enough to have given trouble perhaps even to Maxwell.

The importance of Graham's law of diffusion is obviously not what it says about diffusion coefficients, for it says nothing, but is its bearing on the coupling between diffusion and flow. The crucial difference between these experiments and diffusion experiments in a closed tube is the uniform total pressure of the former. A small pressure gradient has to be present in the closed-tube experiment in order to keep the fluxes of the two diffusing gases equal and opposite, which they must be in order that the pressure does not continuously increase on one side of the tube and decrease on the other side. This pressure gradient is almost immeasurably small, except in capillary tubes, and was not detected by direct measurement until comparatively recent times[13,14,15]. Because of this confusion over the crucial importance of uniform total pressure, and probably confusion with the law of effusion, Graham's law of diffusion was relegated to the scrap heap of results which are of some historical interest, but which are only crude approximations. In fact, however, the law is very precise, the level of accuracy being about 1% in most cases.

It was finally rediscovered independently in 1953 as a new experimental result[16,17], but took another 14 years to be recognized as Graham's Law[8].

Graham also studied the *effusion* of gases through apertures in thin plates[11] and the *transpiration* of gases through long capillary tubes.[11,18] At first sight these would seem to have little relation to diffusion, but the three phenomena usually occur in some sort of combination in an actual experiment, and have to be unscrambled before the experiment can be properly understood. Graham was perfectly clear that diffusion, effusion, and transpiration were three different processes. These are, in fact, the three main modes of gas transport, and a complete phenomenological theory can be based on their recognition as independent processes.[19] It is a remarkable feat of insight to delineate clearly these three mechanisms, especially since the kinetic theory of gases as we now know it did not yet exist. Graham's work seemed to have considerable influence on his contemporaries; it is a pity that he has been so badly served in modern times that his original law of diffusion had to be rediscovered experimentally 120 years later.

The remainder of this review is devoted to experimental techniques, and to the phenomenological description and the kinetic-molecular theory of diffusion. Logically, the phenomenology should come first, but it seems appropriate to discuss experimentation instead, in the spirit that Graham was a master of ingenious experimentation.

2   EXPERIMENTAL TECHNIQUES

The classic techniques for measurements of gaseous diffusion coefficients are the closed-tube and the evaporation methods. With a few minor exceptions, all diffusion coefficients measured prior to World War II were obtained by one of these two methods, and covered only a very limited temperature range. During and after World War II there was a great revival of interest in diffusion, inspired in part by interest in isotope separation and by problems involving high temperatures in flames. The ready availability of isotopic tracers was also a great stimulus, and made so-called self-diffusion coefficients experimentally accessible. It was also gradually recognized that diffusion coefficients could be a good source of information on forces between unlike molecules, provided they could be measured over a substantial temperature range, and this was a further stimulus. Important new work was done with two-bulb, point-source, gas-chromatography, and diffusion-bridge apparatuses, as well as with dissociated gases. A host of miscellaneous methods, often showing great ingenuity, have been tried out but not used extensively. The existence of an accurate kinetic-molecular theory has also made possible several indirect methods, in which diffusion coefficients are deduced from measurements of other properties that at first sight seem to have no particular connection with diffusion.

A few comments, not at all comprehensive or complete, follow. When a series of papers has appeared from one laboratory, only the most recent reference is usually given, from which earlier papers can be traced.

A   Closed Tube

This method is usually associated with the name of Loschmidt[6,7] who is surely entitled to major credit for its development and exploitation. However, as already remarked, the invention of the method should really be credited to Graham[4]. Loschmidt covered a maximum temperature range of about $-20°$ to $20°C$. During the 1880's von Obermayer[20] studied many systems up to a maximum temperature of about $60°C$. However, many of his results are, for some unknown reason, systematically low by about 5%. In the early 1900's a series of investigations was carried out at the University of Halle to test the composition dependence of the diffusion coefficient, a crucial theoretical point; this work is summarized by Lonius[21]. Boardman and Wild[22] and Coward and Georgeson[23] devised a rotating-plate version of the apparatus, which has been much copied. Since the war this type of apparatus has been adapted for continuous composition analysis by means of radioactive tracers[24] and optical techniques[25,26]. The lowest temperature attained is $-78°$ and the highest is $200°C$. Reproducibility is often better than 1%, but absolute accuracies are no better than 2%; a major effort involving careful variation of many experimental parameters would probably be necessary to achieve much improvement.

B Evaporation Method

The rate of evaporation of a liquid placed in the bottom of a long tube is controlled by the diffusion of the vapor through the surrounding gas, and can be used to measure the vapor-gas diffusion coefficient[27]. The method has been very widely used, but is restricted to volatile liquids and limited temperature ranges. The accuracy achieved has been generally disappointing.

C Two-Bulb Apparatus

The two-bulb configuration was first used by Graham[1] and its mathematical analysis was given by Maxwell[28]. It was reinvented by Ney and Armistead[29], who measured the self-diffusion coefficient of UF_6. It is easily adapted to cover large temperature ranges, and with moderate care gives a level of accuracy about the same as the closed-tube apparatus. It has been widely used in recent years. By meticulous attention to details, van Heijningen *et al.*[30] have apparently obtained the best absolute accuracy (1%) to date, over a maximum temperature range form 65° to 400°K.

D Point Source

This flow method was devised by Walker and Westenberg[31] to measure diffusion at high temperatures. One component is introduced through a fine hypodermic tube (the point source) into a slowly-flowing stream of the other component, and spreads by diffusion as it moves downstream. The diffusion coefficient can be determined by sampling the gas with a fine probe and measuring its composition at different downstream positions. Temperatures up to 1200°K could be obtained without difficulty. Temperatures up to 1800°K have been obtained by using the burned gas from a flat flame stabilized above a porous plate as the flowing stream[32]. Reproducibility is 1–2%, and accuracy probably better than 5%.

E Gas Chromatography

This is another flow method, in which a pulse of one component is injected into a stream of the other component flowing through a long tube (i.e., an unpacked gas chromatograph column).The diffusion coefficient can be determined by measuring the amount of spreading of the gas pulse as it emerges from the end of the column. Such apparatus is readily adapted to routine operation[33], and to fairly wide temperature ranges[34]. Reproducibility and accuracy are comparable to that of the point-source technique.

F   Diffusion Bridge

This is also a flow method, but is really Graham's classic work all over again, because it is usually operated at uniform total pressure. Gas streams A and B flow across opposite faces of a porous septum or ends of a capillary tube, and the emerging streams are analyzed for how much B diffused into the A stream and *vice versa*. The flow rates can be controlled by valves in the lines, and adjusted to produce any desired pressure difference across the septum or capillary. This technique was first used by Wicke and Kallenbach[35] and was used to rediscover Graham's Law of Diffusion[16,17]. It has been widely used to study gaseous diffusion in porous media; almost everyone who uses it sooner or later rediscovers, but does not recognize, Graham's Law of Diffusion[9]. It has been used only once with a capillary, to obtain absolute values of diffusion coefficients down to very low temperatures[36]. Since it is a steady-state method without moving mechanical parts, it is readily adaptable to operation over wide temperature ranges.

G   Dissociated Gases

Direct measurements of the diffusion of highly reactive species such as free radicals and valence-unsaturated atoms are very difficult, but are needed for basic understanding of many phenomena in chemical reactions and at high temperatures. A variety of direct techniques have been used to measure the diffusion of H, N, and O atoms in different gases[37,38,39,40,41]. As might be expected, scatter and consistency are not too good (10% or more in many cases). The best results, in fact, have been obtained by indirect methods such as mixture viscosities[42] and combined molecular-beam scattering and semi-empirical quantal calculations[43].

H   Miscellaneous Methods

It is futile, as well as boring, to enumerate all the miscellaneous methods by which a diffusion coefficient has been determined at some time or other. Here are a few that are worth noting, for both general applicability and experimental ingenuity.

The interdiffusion of ortho and para hydrogen was measured down to 20°K by an ingenious steady-state flow system, in which one component diffuses upstream against the second flowing component[44]. The composition at one or more upstream points can be used to determine the diffusion coefficient.

The "capillary leak" leak method has already been mentioned as an adaptation of Graham's 1829 method. It gave results of low accuracy, but should be capable of much improvement.

When two gases interdiffuse, a small transient temperature gradient is set up; this is called the Dufour effect or the diffusion thermoeffect. The asymptotic time decay of this temperature difference can be used to determine the diffusion coefficient[45,46]. The inverse of the Dufour effect is thermal diffusion, in which an imposed temperature gradient causes a composition difference to be set up. The speed with which this happens can also be used to determine the diffusion coefficient[47].

In solids, the net drift of inert markers placed near a diffusion interface is called the Kirkendall effect. A similar effect exists in gases[48,14] and the speed of the marker motion can be used to determine the diffusion coefficient. Although it was not realized at the time, the Kirkendall effect in gases is analogous to the rise or fall of the water level in Graham's diffusion experiments.

The passage of a sound wave through a gas mixture produces a local partial separation of the components, caused mostly by pressure diffusion. The remixing by diffusion is out of phase with the sound wave, and thus furnishes a mechanism for absorption of sound energy. In other words, the absorption of an ultrasonic wave in a gas mixture is stronger than in either pure component, and the excess absorption depends on the diffusion coefficient, which could in principle thus be determined[49]. This method has been suggested for use at very high temperatures, where other methods fail[50].

Finally, a method has been proposed recently for measurement of vapor-gas diffusion coefficients[51] which is a clever combination of Graham's experiment and Stefan's evaporation technique. A bead of volatile liquid is placed in a long glass tube with a porous membrane at one end, and the bead is driven down the tube by the accumulation of vapor between the bead and the membrane. The vapor and outside gas interdiffuse through the membrane, as in Graham's experiment, and the motion of the bead can be used to determine the diffusion coefficient.

I   Indirect Methods

Kinetic theory shows that the viscosity of a gas mixture depends on the composition, the viscosities and molecular weights of the pure components, and two mixture quantities. One of these is the diffusion coefficient, and the other is a ratio of collision integrals known as A_{12}^*, which depends only weakly on temperature and intermolecular forces. It is fairly easy to get an independent theoretical estimate of A_{12}^*, at least for simple gases, and then the diffusion coefficient can be calculated by simple algebra if the viscosities are known[52]. This method gives very satisfactory diffusion coefficients if the viscosities are measured with high accuracy[53].

Kinetic theory also reveals several relations between the diffusion coefficient and the thermal diffusion factor, which describes how a gas mixture separates under the influence of a temperature gradient. Measurement of the (rather strong) composition dependence of the thermal diffusion factor in effect determines the diffusion coefficient[54]. The accuracy is only fair, because of the uncertainties in the measurements. A better method relates the temperature dependence of the diffusion coefficient to that of the thermal diffusion factor. A single isothermal measurement of the diffusion coefficient can thus be combined with thermal diffusion measurements to produce diffusion coefficients over a wide temperature range[55]. The final accuracy is good, because rather large uncertainties in the measurements appear only as much smaller uncertainties in the calculated diffusion coefficients.

Finally, kinetic theory shows how to calculate diffusion coefficients from intermolecular forces. At ordinary temperatures the best direct sources of information on forces between unlike molecules involve the diffusion coefficients themselves, and the method is useless. However, the short-range forces can be measured by the scattering of fast molecular beams, and this furnishes the basis for calculation of diffusion coefficients at very high temperatures[56], where direct experiments are still unavailable. Agreement between calculation and direct experiment is excellent in the small temperature range where the two overlap[57].

3 PHENOMENOLOGICAL DESCRIPTION

Here one seeks to describe how gases mix and flow, and what the variables are that influence this behavior. Ultimately such a description would take the form of some differential equations containing coefficients to be found by experiment. The task of a kinetic-molecular theory is then to relate these coefficients to atomic and molecular properties.

A   Modes of Gas Motion

Graham, in his studies of diffusion, effusion, and transpiration, had the genius to delineate clearly these three main mechanisms of gas transport, which usually occur in combination:

1) *Diffusion*, in which the different species of a mixture move under the influence of composition gradients. This is a continuum phenomenon in that molecule–molecule collisions dominate over molecule–wall collisions.

2) *Transpiration*, or laminar viscous flow, in which the gas acts as a continuum fluid driven by a pressure gradient. The pressure is high enough that molecule–molecule collisions dominate over molecule–wall collisions.

3) *Effusion*, now more commonly called free-molecule or Knudsen flow. Here the pressure is so low that collisions between molecules can be ignored compared to molecule–wall collisions.

The foregoing types of gas transport can also occur in combination. For instance, the combination of viscous flow and free-molecule flow leads to the phenomenon known as "viscous slip". Another combination is that between continuum diffusion and viscous flow, which gives rise to "diffusive slip", first discussed by Kramers and Kistenmaker[13]. Still another combination is that between free-molecule and continuum diffusion, which is important in the separation of gases by diffusion through porous barriers. This is the basis of "atmolysis", devised by Graham[4], and was studied intensively during World War II in connection with isotope separation[58]. Finally, all three mechanisms are present in a flowing, diffusing gas mixture in the transition regime between free-molecule and continuum behavior. The coupling between flow and diffusion is much more common than is generally realized, since almost all experimental arrangements involving diffusion also cause flow to occur.

B   Combined Transport

The problem now is to describe situations in which more than one mode of gas transport is operating. This is far from simple, and our current formulations are probably not exact, although more experimental work of high accuracy will probably be needed to show up deviations. Nevertheless, a quite satisfactory result can be obtained from remarkably simple arguments based on the foregoing classification of transport mechanisms. The astonishing thing is that one of the key points that makes the formulation simple is the idea of diffusion at *uniform* pressure—Graham's diffusion law again. If we start by thinking about diffusion in terms of equal countercurrent fluxes, as in the closed-tube apparatus, the simplicity disappears and only confusion seems to emerge.

The foregoing classification makes an important distinction between diffusive transport and viscous transport. The importance of the distinction lies in the fact that the two mechanisms contribute independently to the total transport, which is thus simply a sum of diffusive and viscous flow terms with no extra terms due to coupling between the two mechanisms. That is, we define pure diffusive transport to be at uniform pressure, where the viscous transport must be zero. To find the total flux when the pressure is nonuniform, we simply add the viscous flux due to the pressure gradient. Because of this simple additivity, diffusion at uniform pressure can be regarded as fundamental.[19]

The coupling between free-molecule and continuum diffusion is more subtle, and requires some explicit use of kinetic-theory arguments. The simplest argument is based on a discussion of momentum transfer in diffusion; this also leads to a generalization for diffusion in multicomponent mixtures, which is usually very laborious to obtain. A momentum-transfer theory was first devised by Maxwell[28] and independently by Stefan[59,60]. It was neglected and probably forgotten for many years, and then apparently rediscovered by Frankel[61] and by Present and deBethune[58,62] and further elaborated by Furry[63] and by Williams[64]. An excellent illustration of a momentum-transfer argument is the explanation given by Hoogschagen[16,17] for his uniform-pressure diffusion results, which were just Graham's diffusion law rediscovered. The explanation follows an earlier discussion of diffusive slip by Kramers and Kistemaker[13]. We define an average diffusion velocity, $\bar{V}_i$, for species i as

$$n_i \bar{V}_i = J_i, \tag{1}$$

where n_i is the number density and J_i is the flux. We then argue that the net force on the porous septum (or wall of a capillary tube) must be zero if there is no pressure gradient in the gas, and hence the total momentum transferred to the septum by all the molecular collisions is zero. The momentum transferred by the ith species per unit time is equal to the mean momentum transferred per molecular impact (proportional to $m_i \bar{V}_i$), multiplied by the number of molecular impacts per unit time (proportional to $n_i \bar{v}_i$, where $\bar{v}_i$ is the mean thermal speed of the molecules). The sum over both species must be zero,

$$(m_1 \bar{V}_1)(n_1 \bar{v}_1) + (m_2 \bar{V}_2)(n_2 \bar{v}_2) = 0, \tag{2}$$

which can be rearranged to

$$-\frac{J_1}{J_2} = -\frac{n_1 \bar{V}_1}{n_2 \bar{V}_2} = \frac{m_2 \bar{v}_2}{m_1 v_1} = \left(\frac{m_2}{m_1}\right)^{1/2}, \tag{3}$$

the last step following from the fact that $\bar{v}_i$ varies inversely as the square root of m_i. This is Graham's law of diffusion, and from the derivation its range of validity does not depend on any relation between the mean free path and the internal geometry of the porous medium. This is in contrast to the effusion law, which holds only in the free-molecule region.

However, this derivation of Graham's diffusion law is only approximate, since Eq. (2) involves replacing the average of a product by the product of two averages. I know of no fully rigorous derivation of the diffusion law, a remarkable situation in view of its manifestly simple form. The best derivation at present is based on the "dusty-gas" model, in which the porous septum is considered as a third, stationary, component of the gas mixture, and then the machinery of the Chapman-Enskog kinetic theory is applied[65].

At any rate, the above line of argument suggests that diffusive flows combine by additivity of momentum transfer. Together with the earlier remark that diffusive and viscous flows combine by additivity of fluxes, this suffices for a complete phenomenological description[19]. There is even a simple electrical analogue: diffusive flows combine like resistors in series in that voltage drops (momentum transfers) add, whereas diffusive and viscous flows combine like resistors in parallel in that currents (fluxes) add. Without going through all the algebra, we can write down the result in the following form:

$$\mathbf{J}_1 = -D_1 \nabla n_1 + x_1 \delta_1 \mathbf{J} - x_1 \gamma_1 (n B_0 / \eta) \nabla p, \tag{4}$$

where

$$\frac{1}{D_1} = \frac{1}{D_{1K}} + \frac{1}{D_{12}},$$

$$\delta_1 = \frac{D_1}{D_{12}} = \frac{D_{1K}}{D_{1K} + D_{12}},$$

$$\gamma_1 = \frac{D_1}{D_{1K}} = \frac{D_{12}}{D_{1K} + D_{12}} = 1 - \delta_1,$$

$$\mathbf{J} = \mathbf{J}_1 + \mathbf{J}_2.$$

Here D_{1K} is the Knudsen (free-molecule) diffusion coefficient, D_{12} is the continuum diffusion coefficient, B_0 is a geometric viscous-flow constant, and η is the viscosity. A similar equation holds for $\mathbf{J}_2$, with corresponding definitions of D_2, δ_2, and γ_2.

The extension to multicomponent mixtures by the same arguments is straightforward and yields[19]

$$-\nabla n_1 = \frac{1}{D_{1K}}\left[\mathbf{J}_1 + x_1\left(\frac{nB_0}{\eta}\right)\nabla p\right] + \frac{1}{D_{12}}(x_2\mathbf{J}_1 - x_1\mathbf{J}_2)$$

$$+ \frac{1}{D_{13}}(x_3\mathbf{J}_1 - x_1\mathbf{J}_3) + \cdots, \tag{5a}$$

$$-\nabla n_2 = \frac{1}{D_{2K}}\left[\mathbf{J}_2 + x_2\left(\frac{nB_0}{\eta}\right)\nabla p\right] + \frac{1}{D_{21}}(x_1\mathbf{J}_2 - x_2\mathbf{J}_1)$$

$$+ \frac{1}{D_{23}}(x_3\mathbf{J}_2 - x_2\mathbf{J}_3) + \cdots, \tag{5b}$$

a total of ν equations for a mixture of ν components, but only $\nu - 1$ of the equations are independent. For binary mixtures, these equations are equivalent to Eq. (4). No experimental tests of Eqs. (5) seem to have ever been carried out.

C Special Cases

Several interesting special cases follow from Eq. (4), including both of Graham's laws.

1 Effusion

Graham's law of effusion[11] is the low-pressure limit of Eq. (4). Since D_{1K} is independent of pressure and D_{12} is inversely proportional to pressure, we find the limiting form to be

$$\mathbf{J}_1 = -D_{1K}\,\nabla n_1. \tag{6}$$

Combined with the corresponding equation for $\mathbf{J}_2$ and the fact that $D_{iK} \propto \bar{v}_i \propto m_i^{-1/2}$, this yields

$$\frac{\mathbf{J}_1}{\mathbf{J}_2} = \frac{D_{1K}}{D_{2K}} = \left(\frac{m_2}{m_1}\right)^{1/2}. \tag{7}$$

2 Uniform-pressure diffusion

This is just Graham's (1833) diffusion experiment. We set $\nabla p = 0$ and $\nabla n_1 = -\nabla n_2$ in Eq. (4), and combine it with the corresponding equation for species 2 to yield

$$\frac{\mathbf{J}_1}{D_1} + \frac{\mathbf{J}_2}{D_2} = \mathbf{J}\left(x_1\frac{\delta_1}{D_1} + x_2\frac{\delta_2}{D_1}\right), \qquad (8)$$

which reduces to

$$-\frac{\mathbf{J}_1}{\mathbf{J}_2} = \frac{D_{1K}}{D_{2K}} = \left(\frac{m_2}{m_1}\right)^{1/2}. \qquad (9)$$

This looks like the effusion law, but follows without any special conditions on the pressure. Notice that the effusion law follows from only the first term on the right of Eq. (4), but the diffusion law involves the first two terms.

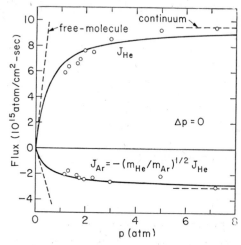

Figure 1 Uniform-pressure diffusion (Graham) for He and Ar in a porous graphite at 25 °C

It is interesting to see how the flux varies in uniform-pressure diffusion as the total pressure is varied. Some results are shown in Fig. 1 for the inter-diffusion of He and Ar through a low-permeability porous graphite specimen. The curves are from Eq. (4) and the points are experimental[66]. The results conform both with Graham's law of diffusion and with the predicted pressure dependence. Only one adjustable constant (D_{12}) occurs, everything else having been determined independently. Notice that the diffusion is still in the transition regime even at pressures of several atmospheres.

3 Transition diffusion

Here we ask how the flux varies with pressure in a closed-tube type of experiment, where $\mathbf{J}_1 = -\mathbf{J}_2$. Some results are shown in Fig. 2 for He—Ar diffusion in the same graphite specimen as in Fig. 1. The curves are from Eq. (4) and the points are experimental[67]. No adjustable constants are involved, the value of D_{12} having been obtained from the independent measurements at uniform pressure shown in Fig. 2.

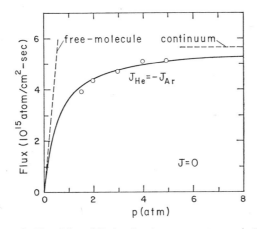

Figure 2   Equal flux diffusion for the same system as in Fig. 1

4 Diffusion pressure-effect (diffusive slip)

We have already mentioned that a pressure difference must exist during diffusion in a closed system. This pressure difference can be easily calculated from Eq. (4), and is shown in Fig. 3 together with the He—Ar measurements of Evans *et al.*[67] No adjustable constants are involved.

4   KINETIC-MOLECULAR THEORY

The Chapman-Enskog kinetic theory of gases, as based on the Boltzmann equation, is so well known that little more need be said about it here, thanks to the famous monographs by Chapman and Cowling[68] and by Hirschfelder,

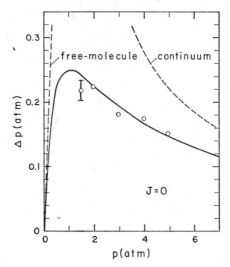

Figure 3   Pressure difference in diffusion (diffusive slip) for the same system as in Figs. 1 and 2

Curtiss, and Bird[69]. This theory deals only with continuum diffusion, where its primary use is to give a precise connection between the diffusion coefficient and intermolecular forces. Thus it can be used to obtain information on intermolecular forces from diffusion coefficients, or, in the reverse procedure, to predict diffusion coefficients from prior estimates of the intermolecular forces.

It has also been mentioned that kinetic theory furnishes several non-obvious relations among different transport coefficients, so that diffusion coefficients can be found from viscosity or thermal diffusion measurements, for example.

Rather than try to give any detailed account of the classical Chapman-Enskog theory, we will mention some closely related matters and some recent extensions.

A   Elementary Theories

Traditional mean-free-path theory gives a reasonable account of viscosity and thermal conductivity, but fails badly for diffusion, where it erroneously predicts a large composition dependence for the binary diffusion coefficient. This theory for many years co-existed with the rigorous Chapman-Enskog theory, seemingly inconsistent with it, even though there was an intuitive feeling that a connection had to exist. The connection was found by Monchick[70], who showed that an iterative solution of the Chapman-Enskog integral equation corresponded to a free-path or free-flight theory in which the trajectory of a molecule was followed back through successively more collisions. The extension to mixtures[71] showed that higher iterations did indeed get rid of the erroneous composition dependence of the diffusion coefficient. Unfortunately, the rate of convergence was painfully slow.

It is nevertheless esthetically pleasing to finally have all the transport theories neatly connected together.

B Pressure Dependence

All kinetic theories, even in their simplest forms, rigorously and correctly predict that the continuum diffusion coefficient is inversely proportional to density or pressure. This is a consequence of the binary collision assumption. In recent years there has been a great deal of work on the extension from dilute to moderately dense gases. A comprehensive list of references is given in the review article of Ernst *et al.*[72], and another recent review has been given by Cohen[73]. The results have been surprising, for divergences have appeared.

By analogy with the virial expansion for the equation of state, it was expected that the density dependence of the transport coefficients could be expressed as a power series in the number density,

$$nD_{12} = nD_{12}^0/(1 + \alpha_1 n + \alpha_2 n^2 + \cdots), \tag{10}$$

where nD_{12}^0 gives the binary collision limit, and $\alpha_1, \alpha_2, \ldots$ are "transport virial coefficients". To the dismay of theoreticians, all the coefficients α_i turned out to be divergent above a certain term. In three dimensions the first divergent coefficient is α_2, and in two dimensions it is α_1. The divergences have been shown to be logarithmic, and to be caused by long-range correlation resulting from recollisions between pairs of molecules. The first divergent term should thus be of the form

$$\alpha_i n^i \rightarrow \alpha_i' n^i (1 + \alpha_i'' \ln n). \tag{11}$$

Simple mean-free-path arguments suggest that a double series expansion in terms of n and $\ln n$ is needed[73].

At present it is not known whether the logarithmic terms are of sufficient numerical magnitude to be needed in fitting experimental data, or even whether there are still further contributions with another density dependence.

C   Quantum Effects

Quantum effects in collisions become significant when the deBroglie wave-length, $\lambda = h/\mu v$, approaches the molecular size or range-of-force para-meter, σ. Thus λ/σ is a measure of quantum effects, and collisions behave classically for $\lambda/\sigma \ll 1$. In kinetic theory we expect the same sort of criterion for classical behavior to hold, with $\mu v^2 \propto kT$; this is commonly expressed in terms of the deBoer parameter Λ^*,

$$\Lambda^* = h/[\sigma \,(2\mu\varepsilon)^{1/2}], \tag{12}$$

which is simply λ/σ for a colliding pair of molecules with reduced mass μ and kinetic energy equal to the depth, ε, of the intermolecular potential well. The larger the value of Λ^*, the more important are the quantum effects at a given value of the reduced temperature, $T^* = kT/\varepsilon$.

Quantum deviations for diffusion are illustrated in Figs. 4 and 5, where the diffusion coefficients for H_2—D_2 and ^3He—^4He are compared with theoretical calculations[74,75]. No adjustable constants are involved, the potential para-meters having been determined from second virial coefficient data. Quantum effects are seen to be about the same size as the (rather large) experimental uncertainties for H_2—D_2, but are much larger for ^3He—^4He.

D   Polyatomic Gases

The original Chapman-Enskog theory applies strictly only to molecules hav-ing spherically symmetric intermolecular forces and undergoing elastic colli-sions. To extend the results to polyatomic and polar gases, it is necessary to take into account nonspherical effects and inelastic collisions. This requires a reformulation of the whole problem from the beginning, but the methods of Chapman and Enskog can still be followed in a generalized form. Strictly speaking, new terms involving molecular angular velocities as well as trans-lational velocities should appear[76,77]. These extra "spin polarization" terms appear to have only a slight effect on the numerical value of the diffusion coefficient, as judged by calculations for the model of loaded spheres[78,79].

The formal kinetic theory of polyatomic gases (omitting spin polarization) has been worked out by Monchick *et al*[80,81,82] and by Alievskii and Zhda-nov[83]. The external appearance of the expression for the diffusion coefficient remains the same as in the original Chapman-Enskog theory, but the explicit expressions for the collision integrals are much more complicated. No numer-

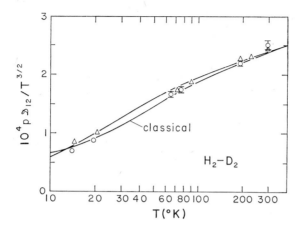

Figure 4 Quantum effects in H₂–D₂ diffusion; here $\Lambda^* = 1.5$ and
$\varepsilon/k = 36°$K

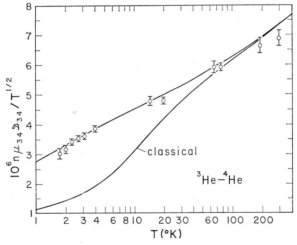

Figure 5 Quantum effects in ³He–⁴He diffusion; here $\Lambda^* = 2.9$ and
$\varepsilon/k = 10°$K

ical evaluations of these have yet been made for realistic models, but simple arguments suggest that inelastic effects on diffusion coefficients are small. This is in accord with the empirical fact that diffusion coefficients of polyatomic molecules can be correlated rather well on the basis of the original Chapman-Enskog theory.

The formal theory also shows that the relation between the diffusion coefficient and the mixture viscosity has the same mathematical form whether collisions are elastic or inelastic; only the explicit expression for the collision integral ratio A_{12}^* is changed. This validates the procedure of calculating diffusion coefficients for polyatomic gases from measured mixture viscosities according to the Chapman-Enskog formulas[84].

E   Multicomponent Diffusion

In the phenomenological Eqs. (5) for multicomponent diffusion, it was tacitly assumed that the D_{ij} are the same as the ones for the corresponding binary mixtures. That is, the D_{ij} are presumed to depend only on components i and j, and not on any of the other components of the mixture. This seems plausible for dilute gases, in which at most binary collisions are important, and is consistent with the fact that the binary D_{ij} are almost independent of composition, but is nevertheless an assumption. Detailed kinetic theory calculations are required for proof. These give very complicated expressions for the deviations of the multicomponent D_{ij} from the corresponding binary D_{ij}[81,83,85]. The only explicit calculations of the multicomponent deviations have been for the special case of the diffusion of a trace species through a stagnant binary mixture (as in Blanc's law for ion mobilities); these showed that deviations are usually small, of the order of experimental uncertainties, but that systems with large deviations are possible[79].

The multicomponent diffusion equations are very difficult to solve. Multicomponent diffusion is inherently more complex than binary diffusion, and even ternary diffusion exhibits new qualitative effects. The reason is that the diffusion of one pair of components can, so to speak, "drag along" another component. The motion of a third "solvent" gas even in the absence of a composition gradient was called *osmotic diffusion* by Hellund[86], who first studied it theoretically. Other related phenomena have been described by Toor[87]: *diffusion barrier*, in which the flux of a component is zero even though its gradient is not; and *reverse diffusion*, in which a component diffuses against its gradient. All three phenomena have been observed experimentally by Duncan and Toor[88].

A peculiar consequence of these multicomponent effects is that density inversions can develop in a diffusing system and cause it to become gravitationally unstable. Convection can then set in. The theoretical possibility of gravitational instability in ternary liquid diffusion was first pointed out by Wendt[89]. Instabilities in ternary gaseous diffusion were independently and unexpectedly discovered experimentally[90,91], and then sought and found in liquids[92]. A simple physical explanation of the instabilities can be given, even though the detailed mathematical theory is difficult. We have already mentioned that a pressure gradient must develop in a closed-tube diffusion apparatus. If a third gas is added as an initially uniform heavy solvent, the pressure difference can force the solvent upwards and make a layer of gas denser than the layer below. These instability phenomena are remarkably complex[91], and have only begun to be studied.

F   Dusty-gas Model

A clever model for gas flow and diffusion in porous media was first suggested by Maxwell[28], in which the porous medium is visualized as a collection of large particles fixed in space. This was forgotten, as with so many things in gaseous diffusion, and re-invented many years later by Deriagin (also spelled Derjaguin) and Bakanov[93], who proposed treating the particles of the porous medium as giant molecules by means of Chapman-Enskog theory. They calculated gas flow near the free-molecule region. The model was shortly thereafter independently re-invented yet once more[94], and applied to a variety of diffusion and flow problems[65]. By formally varying the mole fraction of the "dust" particles, the whole pressure range from the free-molecule to the continuum region could be covered. The results are the same as the phenomenological ones already discussed on the basis of momentum transfer, but have a better theoretical perdigree and supply more detail. This is the closest approach at present to a rigorous kinetic theory in the transition region. However, the work on the kinetic theory of dense gases suggests that the dusty-gas model should also show logarithmic density terms. All that can be said at present is that the limited experimental data do not seem to indicate that logarithmic terms are appreciable, even though they should theoretically exist in the transition region.

5 CONCLUSIONS

Predictions of future progress are notoriously dangerous, but it seems unsporting not to make a few here.

There will probably be continued progress on accurate determination of diffusion coefficients, especially over wide temperature ranges. This is unspectacular but necessary work.

Some progress will be made on treating nonspherical and inelastic collision effects, but it is painful to think that it may have to be done by brute force on fast computers.

Much needs to be done regarding diffusion at higher pressures, both theoretically and experimentally. Interesting things will probably be found in the critical region. Graham never worked in this field—he had an instinct for avoiding trouble.

The transition region is also beset with deep problems. Our present phenomenology works quite well in most cases, but there are uncomfortable loose ends around, such as logarithmic divergences. There are also occasional experimental indications of deviations from the $m^{-1/2}$ dependence of Graham's diffusion law. Then there are occasional spectacular failures like the diffusion pressure-effect in Ar—CO_2, which starts normally at very low pressures but reverses sign in the transition region and ends up backwards in the continuum region[15]. This is suspected to be a nonspherical-inelastic collision effect.

In short, the subject of diffusion in gases seems to be alive and interesting. It probably holds the record as the field in which more work has been forgotten and later rediscovered than in any other.

ACKNOWLEDGMENTS

Thanks are due T. R. Marrero for much help. This work was supported in part by the U.S. Army Research Office, Durham.

REFERENCES

1. T. Graham (1829). *Quart. J. Sci.* **2**, 74. Reprinted in *Chemical and Physical Researches* Edinburgh Univ. Press, Edinburgh (1876), pp. 28–35.
2. A. Fick, *Ann. Physik* **94**, 59; *Phil. Mag.* **10**, 30 (1855).
3. J. C. Maxwell (1867). *Phil. Trans. Roy. Soc.* **157**, 49. Reprinted in *Scientific Papers*. Dover Publications, New York (1962), Vol. 2, p. 26–78.
4. T. Graham (1863). *Phil. Trans. Roy. Soc.* **153**, 385. Reprinted in "Chemical and Physical Researches". Edinburgh Univ. Press. Edinburgh (1876), pp. 210–234.

5. Ts.M.Klibanova, V.V.Pomerantsev, and D.A.Frank-Kamenetskii, *Zh. Tekh. Fiz.* **12**, 14 (1942).

6. J.Loschmidt, *Sitzber. Akad. Wiss. Wien* **61**, 367 (1870a).

7. J.Loschmidt, *Sitzber. Akad. Wiss. Wien* **62**, 468 (1870b).

8. E.A.Mason, *Am. J. Phys.* **35**, 434 (1967).

9. E.A.Mason, and B.Kronstadt, *J. Chem. Educ.* **44**, 740 (1967).

10. T.Graham (1833). *Phil. Mag.* **2**, 175, 269, 351. Reprinted in *Chemical and Physical Researches*. Edinburgh Univ. Press, Edinburgh (1876), pp. 44–70.

11. T.Graham (1846). *Phil. Trans. Roy. Soc.* **136**, 573. Reprinted in *Chemical and Physical Researches*. Edinburgh Univ. Press, Edinburgh (1876), pp. 88–161.

12. R.B.Evans III, L.D.Love, and E.A.Mason, *J. Chem. Educ.* **46**, 423 (1969).

13. H.A.Kramers, and J.Kistemaker, *Physica*, **10**, 699 (1943).

14. K.P.McCarty, and E.A.Mason, *Phys. Fluids*, **3**, 908 (1960).

15. L.Waldmann, and K.H.Schmitt, *Z. Naturforsch.* **16a**, 1343 (1961).

16. J.Hoogschagen, *J. Chem. Phys.* **21**, 2096 (1953).

17. J.Hoogschagen, *Ind. Eng. Chem.* **47**, 906 (1955).

18. T.Graham (1849). *Phil. Trans. Roy. Soc.* **139**, 349. Reprinted in *Chemical and Physical Researches*. Edinburgh Univ. Press, Edinburgh (1876), pp. 162–210.

19. E.A.Mason, and R.B.Evans III, *J. Chem. Educ.* **46**, 358 (1969).

20. A. von Obermayer, *Sitzber. Akad. Wiss. Wien* **96**, 546, and previous papers (1887).

21. A.Lonius, *Ann. Physik*, **29**, 664 (1909).

22. L.E.Boardman and N.E.Wild, *Proc. Roy. Soc.* **A162**, 511 (1937).

23. H.F.Coward, and E.H.M.Georgeson. *J. Chem.. Soc.*, 1085 (1937).

24. I. Amdur, and J.W. Jr., Beatty, *J. Chem. Phys.* **42**, 3361, and previous papers (1965).

25. C.A.Boyd, N.Stein, V.Steigrimsson, and W.F.Rumpel, *J. Chem. Phys.* **19**, 548 (1951).

26. B.A.Ivakin, P.E.Suetin, and V.P.Plesovskikh, *Soviet Phys.-Tech. Phys.* **12**, 1403; *Zh. Tekh. Fiz.* **37**, 1913 (1967), and previous papers (1968).

27. J.Stefan, *Sitzber. Akad.Wiss. Wien*, **68**, 385 (1873).

28. J.C.Maxwell (1860). *Phil. Mag.* **20**, 21. Reprinted in *Scientific Papers*. Dover Publications, New York (1962), Vol. 1, pp. 392–409.

29. E.P.Ney, and F.C.Armistead, *Phys. Rev.* **71**, 14 (1947).

30. R.J.J. van Heijningen, J.P.Harpe, and J.J.M.Beenakker, *Physica* **38**, 1 (1968).

31. R.E.Walker, and A.A.Westenberg, *J. Chem. Phys.* **29**, 1139, and subsequent papers (1958).

32. T.A.Pakura, and J.R.Ferron, *Ind. Eng. Chem., Fundamentals* **5**, 553, and previous papers (1966).

33. J.C.Giddings, and K.L.Mallik, *Ind. Eng. Chem.* **59** (4), 18, and previous papers (1967).

34. S.P.Wasik, and K.E.McCulloh, *J. Res. Natl. Bur. Stds. (U.S.)* **73A**, 207 (1969).

35. E.Wicke, and R.Kallenbach, *Kolloid Z.* **97**, 135 (1941).

36. P.J.Bendt, *Phys. Rev.* **110**, 85 (1958).

37. S.Krongelb, and M.W.P.Strandberg, *J. Chem. Phys.* **31**, 1196 (1959).

38. R.A.Young, *J. Chem. Phys.* **34**, 1295, and previous work (1961).

39. R.E.Walker, *J. Chem. Phys.* **34**, 2196 (1961).

40. R.S.Yolles, and H.Wise, *J. Chem. Phys.* **48**, 5109, and previous papers (1968).

41. B.Khouw, J.E.Morgan, and H.I.Schiff, *J. Chem. Phys.* **50**, 66, and previous papers (1969).

42. R. Browning, and J. W. Fox, *Proc. Roy. Soc.* **A 278**, 274 (1964).

43. K. S. Yun, S. Weissman, and E. A. Mason, *Phys. Fluids* **5**, 672 (1962).

44. P. Harteck, and H. W. Schmidt, *Z. Phys. Chem.* **B 21**, 447 (1933).

45. L. Waldmann, *Z. Phys.* **124**, 2, and previous papers (1947).

46. E. A. Mason, L. Miller, and T. H. Spurling, *J. Chem. Phys.* **47**, 1669 (1967b).

47. E. A. Mason, M. Islam, and S. Weissman, *Phys. Fluids* **7**, 1011, and previous papers (1964).

48. L. Miller, and P. C. Carman, *Nature* **186**, 549 (1960).

49. R. Holmes, and W. Tempest, *Proc. Phys. Soc.* **75**, 898 (1960).

50. E. H. Carnevale, L. C. Lynnworth, and G. S. Larson, *J. Chem. Phys.* **46**, 3040 (1967).

51. T. Katan, *J. Chem. Phys.* **50**, 233, (1969).

52. S. Weissman, and E. A. Mason, *J. Chem. Phys.* **37**, 1289, and subsequent papers (1962).

53. J. Kestin, and J. Yata, *J. Chem. Phys.* **49**, 4780, and previous papers (1968).

54. E. A. Mason, and F. J. Smith, *J. Chem. Phys.* **44**, 3100 (1966).

55. B. K. Annis, A. E. Humphreys, and E. A. Mason, *Phys. Fluids*, **11**, 2122 (1968).

56. I. Amdur, and E. A. Mason, *Phys. Fluids* **1**, 370 (1958).

57. R. E. Walker, and A. A. Westenberg, *J. Chem. Phys.* **31**, 519 (1959).

58. R. D. Present, and A. J. de Bethune, *Phys. Rev.* **75**, 1050 (1949).

59. J. Stefan, *Sitzber. Akad. Wiss. Wien* **63**, 63 (1871).

60. J. Stefan, *Sitzber. Akad. Wiss. Wien* **65**, 323 (1872).

61. S. P. Frankel, *Phys. Rev.* **57**, 661 (1940).

62. R. D. Present (1958). *Kinetic Theory of Gases.* McGraw-Hill, New York.

63. W. H. Furry, *Am. J. Phys.* **16**, 63 (1948).

64. F. A. Williams, *Am. J. Phys.* **26**, 467 (1958).

65. E. A. Mason, A. P. Malinauskas, and R. B. Evans III, *J. Chem. Phys.* **46**, 3199, and previous papers (1967a).

66. R. B. Evans III, G. M. Watson, and J. Truitt, *J. Appl. Phys.* **33**, 2682 (1962).

67. R. B. Evans III, G. M. Watson, and J. Truitt, *J. Appl. Phys.* **34**, 2020 (1963).

68. S. Chapman, and T. G. Cowling, *The Mathematica Theory of Non-Uniform Gases*, 2nd ed., Cambridge Univ. Press, London and New York (1952).

69. J. O. Hirschfelder, C. F. Curtiss, and R. B. Bird (1954). *Molecular Theory of Gases and Liquids.* Wiley, New York.

70. L. Monchick, *Phys. Fluids* **5**, 1393 (1962).

71. L. Monchick and E. A. Mason, *Phys. Fluids* **10**, 1377 (1967).

72. M. H. Ernst, L. K. Haines, and J. R. Dorfman, *Rev. Mod. Phys.* **41**, 296 (1969).

73. E. G. D. Cohen (1969). In *Transport Phenomena in Fluids* (H. J. M. Hanley, ed.), pp. 157–207. Dekker, New York.

74. L. Monchick, E. A. Mason, R. J. Munn, and F. J. Smith, *Phys. Rev.* **139**, A 1076 (1965).

75. D. E. Diller, E. A. Mason, *J. Chem. Phys.* **44**, 2604 (1966).

76. Yu. Kagan, and A. M. Afanas'ev, *Soviet Phys.—JETP* **14**, 1096, *Zh. Eksperim. Teor. Fiz.* **41**, 1536 (1961) (1962).

77. L. Waldmann, *Z. Naturforsch.* **18a**, 1033 (1963).

78. S. I. Sandler, and J. S. Dahler, *J. Chem. Phys.* **47**, 2621 (1967).

79. S. I. Dandler, and E. A. Mason, *J. Chem. Phys.* **47**, 4653 (1967).

80. L. Monchick, K. S. Yun, and E. A. Mason, *J. Chem. Phys.* **39**, 654 (1963).

81. L. Monchick, R. J. Munn, and E. A. Mason, *J. Chem. Phys.* **45**, 3051 (1966).

82. L. Monchick, S. I. Sandler, and E. A. Mason, *J. Chem. Phys.* **49**, 1178 (1968).
83. M. Ya. Alievskii, and V. M. Zhdanov, *Soviet Phys.—JETP* **28**, 116, *Zh. Eksperim. Teor. Fiz.* **55**, 221 (1968) (1969).
84. S. Weissman, *J. Chem. Phys.* **40**, 3397 (1964).
85. V. Zhdanov, Yu. Kagan, and A. Sazykin, *Soviet Phys.—JETP* **15**, 596, *Zh. Eksperim. Teor. Fiz.* **42**, 857 (1962).
86. E. J. Hellund, *Phys. Rev.* **57**, 737 (1940).
87. H. L. Toor, *Am. Inst. Chem. Engrs. J.* **3**, 198 (1957).
88. J. B. Duncan, and H. L. Toor, *Am. Inst. Chem. Engrs. J.* **8**, 38 (1962).
89. R. P. Wendt, *J. Phys. Chem.* **66**, 1740 (1962).
90. L. Miller, and E. A. Mason, *Phys. Fluids* **9**, 711 (1966).
91. L. Miller, T. H. Spurling, and E. A. Mason, *Phys. Fluids* **10**, 1809 (1967).
92. L. Miller, *J. South African Chem. Inst.* **19**, 125 (1966).
93. B. V. Deriagin, and S. P. Bakanov, *Soviet Phys.-Tech. Phys.* **2**, 1904, *Zh. Tekh. Fiz.* **27**, 2056 (1957).
94. R. B. Evans III, G. M. Watson, and E. A. Mason, *J. Chem. Phys.* **35**, 2076 (1961).

Simple stochastic theory as an "explanation" of thermal diffusion in gases

P. G. WRIGHT

Dept. of Chemistry, The University, Dundee

ABSTRACT

The phenomenon of thermal diffusion, discovered in liquid systems during Graham's lifetime, has been known in gases for over fifty years. It provides a well-known challenge to expository accounts of diffusive phenomena—of whether it is possible to give a convincing elementary argument indicating that in gaseous systems thermal diffusion will indeed occur.

A further suggestion on this point is here advanced. More than forty years ago, Einstein's and Smoluchowski's stochastic treatment of Brownian motion was extented by Chapman to systems wherein there is a gradient of temperature, and from this elaborated stochastic theory it is possible to infer that it would be remarkable in the extreme if thermal diffusion in gases did not occur. The argument is, moreover, at a reasonably "elementary" mathematical level, as judged by the standard of degree courses in physical sciences.

The argument rests solely on general relations for diffusive migration, and special assumptions relating to particular models and mechanisms do not enter. The treatment indicates that thermal diffusion might well occur either in the "normal" or in the reversed sense, reversal being associated with a particularly rapid dependence of the random molecular displacements on temperature.

1 INTRODUCTION

The phenomenon of thermal diffusion, the establishment of a gradient of composition in a mixture subjected to a gradient of temperature, was discovered in liquid systems by Ludwig during Graham's lifetime. The occurrence of thermal diffusion in gaseous mixtures has now been known for over fifty years, but the phenomenon is one whose existence is still not infrequently found astonishing.

Perhaps it ought not. The diffuse flux of molecules is a vector, and at a given point it will take a value which can depend only on conditions very near that point. In a mixture at uniform pressure, with accelerations and external forces absent, a diffusive flux can depend only on the composition and temperature at neighbouring points. Therefore (if derivatives of composition and temperature higher than the first may be neglected), the only vector available to be equated to the flux of molecules of species i is of the form

$$\sum_j H_{ij} \operatorname{grad} x_i + H_{Ti} \operatorname{grad} T$$

(where x_j is the mole fraction of species j). Hence, the flux must be equal to an expression of this form; and H_{Ti} can only be zero if some special property of the material constrains it to vanish. In principle, no term with the correct invariant properties should be suppressed unless some mechanical reason, or some consideration of orders of magnitude, justifies its exclusion.

Arguments along such lines have been advanced by Truesdell[1] *(inter alia)*, and rest on the same principle as the reasoning used by Brillouin[2] in deciding what was the most general possible dependence of the velocity-distribution function (of a pure gas) on first and second spatial derivatives of density, temperature, and velocity of macroscopic flow. (Such ideas have a long history, extending back to the classical formulation of the hydrodynamics of a viscous fluid: see for example the treatment given by Stokes[3].)

It nevertheless seems useful to examine any reasonable suggestion for a not too complicated but moderately cogent argument indicating that the phenomenon of thermal diffusion will (or, better, must) occur.

A "lack of an adequate elementary physical explanation of thermal diffusion in gases", in particular of a truly *adequate* treatment in terms of mean free paths, has long been lamented. Pursuing this line of attack, Monchick and Mason[4] have elaborated a "free-flight theory" which they advance as "the first rigorous elementary explanation of thermal diffusion".

There is also a quite different approach which, though mathematically reasonably simple, leads to equations which require that thermal diffusion can fail to occur only if certain terms cancel in a way that is not *a priori* obviously to be expected. This approach is a stochastic one, basically that developed by Einstein and by Smoluchowski in treating Brownian motion. The necessary extension to systems in which the temperature is not uniform was obtained by Chapman[5], who thereby predicted the occurrence of thermal diffusion for particles suspended in a fluid, and discussed thermal diffusion in gases along similar lines.

Equivalent equations were deduced by Le Claire[6] for diffusion in solids. It has been pointed out[7,8] that these equations may be used in discussing thermal diffusion in solids, but the potentiality of the equations for giving a reasonably simple account of thermal diffusion in gases has perhaps received undeserved neglect.

What is obtainable is an algebraic, rather than a physical, explanation (or half-explanation). Its merit lies in being algebraically not too grossly more involved than the treatment by Fürth[9] in terms of mean free paths, and the rather similar treatment given by Gillespie[10], while avoiding the arbitrariness of assumptions made in those accounts. General relations for diffusive migration are all that is required, and special assumptions do not enter. The result is not indeed a firm requirement that thermal diffusion must occur, but a conclusion that it would be remarkable if it did not.

2 THE STOCHASTIC RELATIONS APPLIED TO A GASEOUS MIXTURE WHOSE TEMPERATURE IS NON-UNIFORM

Consider for simplicity a system in which all gradients are parallel to the z axis ("vertical"). Let J_i be the resultant number of molecules of species i crossing a given horizontal plane per unit area per unit time, in an upward direction. Then[5,6,8], at least in a steady state

$$J_i = n_i \frac{\bar{\xi_i}}{\tau} - n_i \frac{\partial}{\partial z} \frac{\overline{\xi_i^2}}{2\tau} - \frac{\overline{\xi_i^2}}{2\tau} \frac{\partial n_i}{\partial z}$$

where n_i is the number of molecules of species i per unit volume, ξ_i is the vertical component of the displacement undergone by a molecule of species i in an interval of time τ (further specified in the Appendix), and the averages are to be taken for molecules which at the beginning of the interval were situated at the given plane.

Now an equation is sought in which J_i is expressed in terms of the gradients of composition and temperature. In the final term above, then, $\partial n_i/\partial z$ is to be expressed in terms of $\partial x_i/\partial z$ and $\partial T/\partial z$. Now

$$\frac{\partial n_i}{\partial z} = \frac{\partial}{\partial z}(nx_i) = n \frac{\partial x_i}{\partial z} + x_i \frac{\partial n}{\partial z} = n\left(\frac{\partial x_i}{\partial z} + x_i \frac{\partial \log_e n}{\partial z}\right)$$

(where n is the total number of molecules per unit volume) and for an ideal gaseous system at uniform pressure, since $n \propto T^{-1}$, this is equal to

$$n \left(\frac{\partial x_i}{\partial z} - x_i \frac{\partial \log_e T}{\partial z} \right) = n \left(\frac{\partial x_i}{\partial z} - \frac{x_i}{T} \frac{\partial T}{\partial z} \right)$$

The flux J_i for an ideal gaseous system at uniform pressure is therefore given by

$$J_i = n \left\{ x_i \frac{\bar{\zeta}_i}{\tau} - x_i \frac{\partial}{\partial z} \frac{\overline{\zeta_i^2}}{2\tau} - \frac{\overline{\zeta_i^2}}{2\tau} \frac{\partial x_i}{\partial z} + \frac{x_i}{T} \frac{\overline{\zeta_i^2}}{2\tau} \frac{\partial T}{\partial z} \right\}$$

The last two terms involve gradients of composition and temperature explicitly. The first two terms can involve both implicitly. In particular, $(\partial/\partial z)(\overline{\zeta_i^2}/2\tau)$ will be expected to be non-zero if there is a gradient of temperature: in the hotter parts of the gas, random molecular displacements will be expected to be greater than in the cooler parts.

Any needless arbitrariness is avoided by assuming the first two terms to have the form (for a binary mixture)

$$\frac{\bar{\zeta}_i}{\tau} = C_i \frac{\partial x_i}{\partial z} + B_{Ti} \frac{\partial T}{\partial z}$$

$$\frac{\partial}{\partial z} \frac{\overline{\zeta_i^2}}{2\tau} = \theta_i \frac{\partial x_i}{\partial z} + \eta_{Ti} \frac{\partial T}{\partial z}$$

where η_{Ti} will certainly be positive, but the signs of the other coefficients, C_i, B_{Ti}, θ_i, are uncertain. (Only terms linear in the gradients are retained.)

In terms of these coefficients, then (but for convenience putting $B_{Ti} = C_{Ti}/T$ and $\eta_{Ti} = \theta_{Ti}/T$)

$$J_i = n \left\{ -\left(\frac{\overline{\zeta_i^2}}{2\tau} - x_i C_i + x_i \theta_i \right) \frac{\partial x_i}{\partial z} - \left(\theta_{Ti} - \frac{\overline{\zeta_i^2}}{2\tau} - C_{Ti} \right) \frac{x_i}{T} \frac{\partial T}{\partial z} \right\}$$

For a steady state with the flux J_i vanishing, this equation implies that if there is a gradient of temperature there will also be a gradient of composition— unless the terms in the second parenthesis should happen to cancel to zero. That is, it implies that thermal diffusion will occur, unless a certain coefficient vanishes.

Chapman's stochastic expression, then, indicates that it would be absence, rather than occurrence, of thermal diffusion that should occasion surprise.

3   THE STOCHASTIC TREATMENT AND GENERAL FEATURES
OF THERMAL DIFFUSION IN GASES

It is evident that for an ideal binary gaseous mixture the steady state in a gradient of temperature is given by the equation

$$\left(\frac{\overline{\zeta_1^2}}{2\tau} - x_1 C_1 + x_1 \theta_1\right)\frac{\partial x_1}{\partial z} = -\left(\theta_{T1} - \frac{\overline{\zeta_1^2}}{2\tau} - C_{T1}\right)\frac{x_1}{T}\frac{\partial T}{\partial z}.$$

The corresponding equation for species 2 conveys no further information (other than a connection among the various coefficients), but by taking the two together a more symmetrical form

$$\left\{x_2 \frac{\overline{\zeta_1^2}}{2\tau} + x_1 \frac{\overline{\zeta_2^2}}{2\tau} + x_1 x_2 \left[(\theta_1 + \theta_2) - (C_1 + C_2)\right]\right\}\frac{\partial x_1}{\partial z}$$

$$= -\frac{x_1 x_2}{T}\left\{(\theta_{T1} - \theta_{T2}) - \frac{\overline{\zeta_1^2}}{2\tau} + \frac{\overline{\zeta_2^2}}{2\tau} - (C_{T1} - C_{T2})\right\}\frac{\partial T}{\partial z}$$

can be obtained.

Following Chapman[5], the r.h.s. of this equation can be expressed in a perhaps more informative way.

The coefficients θ_i and θ_{Ti} are altogether different in nature from those, C_i and C_{Ti}, which originate in the "drift" term $\overline{\zeta_i}/\tau$. In $(\partial/\partial z)(\overline{\zeta_i^2}/2\tau)$, it is legitimate to use the value of $\overline{\zeta_i^2}$ corresponding to a uniform gas: any corrections due to non-uniformity introduce only terms of the second order which may be neglected.

Hence, for a binary mixture at uniform pressure

$$\frac{\partial}{\partial z}\left(\frac{\overline{\zeta_i^2}}{2\tau}\right) = \left(\frac{\partial}{\partial x_i}\right)_{T,p}\left(\frac{\overline{\zeta_i^2}}{2\tau}\right)\frac{\partial x_i}{\partial z} + \left(\frac{\partial}{\partial T}\right)_{p,x_i}\left(\frac{\overline{\zeta_i^2}}{2\tau}\right)\frac{\partial T}{\partial z}$$

and therefore

$$\theta_i = \left(\frac{\partial}{\partial x_i}\right)_{T,p}\left(\frac{\overline{\zeta_i^2}}{2\tau}\right), \quad \theta_{Ti} = T\left(\frac{\partial}{\partial T}\right)_{p,x_i}\left(\frac{\overline{\zeta_i^2}}{2\tau}\right).$$

The r.h.s. specified above is thus equal to

$$-\frac{x_1 x_2}{T}\left\{-\frac{\overline{\zeta_1^2}}{2\tau}\left[1 - \left(\frac{\partial \log \overline{\zeta_1^2}/2\tau}{\partial \log T}\right)_{p,x_1}\right]\right.$$

$$\left. + \frac{\overline{\zeta_2^2}}{2\tau}\left[1 - \left(\frac{\partial \log \overline{\zeta_2^2}/2\tau}{\partial \log T}\right)_{p,x_1}\right] - C_{T1} + C_{T2}\right\}\frac{\partial T}{\partial z}.$$

3   Sherwood I (1426)

Now, if molecules of species 1 are heavier or larger than those of species 2, $\bar{\zeta_1^2}/2\tau$ will be expected to be smaller than $\bar{\zeta_2^2}/2\tau$. Then, the sign of the r.h.s. is likely to be dictated by that of

$$C_{T2} + \frac{\bar{\zeta_2^2}}{2\tau}\left[1 - \left(\frac{\partial \log \bar{\zeta_2^2}/2\tau}{\partial \log T}\right)_{p,x_1}\right]$$

(the set of terms relating to displacements of molecules of species 2). If this were determined by the "obvious" term $\bar{\zeta_2^2}/2\tau$, then the r.h.s. would be of the *opposite* sign to $\partial T/\partial z$. That is, in the *cooler* regions there would be an enhanced proportion of the heavier or larger species. This is indeed the more usual situation.

The possibility is also correctly indicated that thermal diffusion might proceed in the reverse sense. If

$$\left(\frac{\partial \log \bar{\zeta_2^2}/2\tau}{\partial \log T}\right)_{p,x_1} > 1 + \frac{C_{T2}}{\bar{\zeta_2^2}/2\tau}$$

then the above situation would be expected to be reversed. That is, there would be a tendency of the heavier (or larger) species 1 to migrate to hotter regions.

The stochastic treatment, then, correctly indicates that thermal diffusion might well occur either in the "normal" or in the "reversed" sense, and suggests that reversal might be associated with a too rapid dependence of the random molecular displacements on the temperature at the position from which the motion is deemed to start. Migration of the lighter molecules away from a hotter region would then not be counterbalanced by influx from cooler regions, nor would their arrival in a cooler region from hotter regions be counterbalanced by their migration in the reverse sense.

These points in the stochastic treatment correspond roughly to the requirement of the Chapman-Enskog kinetic theory that the sense in which thermal diffusion occurs is governed by the sign of

$$2 - \left(\frac{\partial \log D_{12}}{\partial \log T}\right)_p.$$

4 COMMENTS

Each of the three terms in J_i

$$n_i \frac{\bar{\zeta_i}}{\tau} - n_i \frac{\partial}{\partial z} \frac{\bar{\zeta_i^2}}{2\tau} - \frac{\bar{\zeta_i^2}}{2\tau} \frac{\partial n_i}{\partial z}$$

makes a contribution to thermal diffusion. If only the final term were present, as in Einstein's treatment of Brownian motion, thermal diffusion would of necessity occur. It is the other two terms which make the phenomenon one which, according to circumstance, may occur in one sense, or the reverse, or may even be absent. As such, these terms, respectively representing "drift" $(n_i \bar{\zeta}_i / \tau)$ and dependence of the random walk on local temperature $(-n_i [\partial/\partial z] [\bar{\zeta}_i^2 / 2\tau])$, whose importance in the context of ordinary diffusion has been emphasized with especial force by Le Claire[6], are crucial in thermal diffusion.

Because of the very general nature of the reasoning, many important results seem not to be deducible from a stochastic treatment. In particular, no analogue arises of the valuable relation[11,12] between thermal diffusion and the contributions to the thermal conductivity made by the translational motions of the individual molecular species. Such *translational* contributions require an analysis extending down to times comparable with the interval between consecutive molecular encounters, whereas the shortest times considered in the stochastic theory are much larger than this.

On the other hand, the (less detailed) stochastic treatment is very easily extended to imperfect gases.

ACKNOWLEDGEMENTS

I am grateful to Dr. K.E.Grew, Dr. A.E.Humphreys, and especially Prof. T.G.Cowling, F.R.S., for helpful comments on, and criticism of, an ealier version of this MS.

None of them is necessarily to be taken as agreeing with any expression of opinion here advanced.

APPENDIX: DEDUCTION OF THE BASIC RELATION

A system is considered in which all gradients are parallel to the z axis. The coordinates x, y then become irrelevant to the discussion of transport phenomena (save for effects arising at the boundary). [Vector ideas are hereby brought in implicitly, without necessitating the use of vector notation.]

It is assumed that there exists a time τ_0 which, while short, is long enough (compared with the interval between consecutive molecular encounters) for the displacement undergone by a molecule in an interval of duration τ_0 to have no appreciable correlation with its velocity at the beginning of that interval, nor with its initial angular momentum and internal motion, nor with the initial proximity (or otherwise) of other molecules.

Consider the probability that a molecule of a particular species, if at time t it lies at a height between z and $z + \delta z$, will at time $t + \tau$ ($\tau \geq \tau_0$) have undergone a displacement whose vertical component lies between ζ and $\zeta + \delta\zeta$. This probability may be written

$$\phi\,(z, t, \zeta, \tau)\,\delta\zeta.$$

(Any dependence on such variables as temperature, concentrations, and gradients thereof, is implicitly allowed for in the dependence on z and t. Dependence on the temperature is expected to be significant: larger displacements are reasonably likely to become more probable as the temperature is increased. Dependence on t is eliminated by considering a steady state.)

Then, in the interval from time t to time $t + \tau$:

The number, per unit area, of molecules of this particular species crossing the plane $z = 0$ downwards from above will be the number, per unit area of this plane, for which $\zeta < 0$ and $0 < z < |\zeta|$ (i.e. for which the displacement is downward and the initial position was sufficiently close to the plane); that is

$$\int_{-\infty}^{0} \int_{0}^{-\zeta} n\,(z, t)\,\phi\,(z, t, \zeta, \tau)\,dz\,d\zeta$$

where $n\,(z, t)$ is the number of molecules of the particular species per unit volume, at height z at time t.

The number, per unit area, of molecules of this species crossing upwards from below is similarly

$$\int_{0}^{\infty} \int_{-\zeta}^{0} n\,(z, t)\,\phi\,(z, t, \zeta, \tau)\,dz\,d\zeta.$$

Therefore, the resultant number, per unit area, of molecules of the particular species crossing the plane $z = 0$ in an upward direction, in the interval of time t to $t + \tau$, is

$$-\int_{-\infty}^{0} \int_{0}^{-\zeta} n\phi\,dz\,d\zeta + \int_{0}^{\infty} \int_{-\zeta}^{0} n\phi\,dz\,d\zeta = \int_{-\infty}^{\infty} \int_{-\varepsilon}^{0} n\phi\,dz\,d\zeta.$$

Now, because τ is small, very few molecules cross the plane $z = 0$ if initially they were far from it. Thus, for molecules which do not start close to the plane, all contributions to the integral are negligible. The integrand may therefore be replaced by a Taylor's expansion about $z = 0$, truncated after two terms. This gives

$$\int_{-\infty}^{\infty} \int_{-\zeta}^{0} \left\{ n\,(0,\,t)\,\phi\,(0,\,t,\,\zeta,\,\tau) + z\left[\frac{\partial}{\partial z}\,n\phi\right]_{\text{at }z=0} \right\} dz\,d\zeta$$

$$= \int_{-\infty}^{\infty} \left\{ +\zeta\,n\,(0,\,t)\,\phi\,(0,\,t,\,\zeta,\,\tau) - \frac{\zeta^2}{2}\left[\frac{\partial}{\partial z}\,n\phi\right]_{\text{at }z=0} \right\} d\zeta$$

$$= n\,(0,\,t) \int_{-\infty}^{\infty} \zeta\phi\,(0,\,t,\,\zeta,\,\tau)\,d\zeta - \frac{1}{2}\int_{-\infty}^{\infty} \zeta^2 \left[\frac{\partial}{\partial z}\,n\phi\right]_{\text{at }z=0} d\zeta.$$

The first integral is equal to $\bar{\zeta}$, the mean vertical displacement of molecules initially situated at $z = 0$. The second [if absolutely convergent] may be evaluated by interchanging the order of integration and differentiation, which gives

$$\left[\frac{\partial}{\partial z}\left(n\int_{-\infty}^{\infty} \zeta^2\phi\,d\zeta\right)\right]_{\text{at }z=0}.$$

Thus, the resultant number of molecules of the species concerned which in the interval of time from t to $t + \tau$ cross the plane $z = 0$ upwards, per unit area, is

$$n\,(0,\,t)\,\bar{\zeta} - \frac{1}{2}\left[\frac{\partial}{\partial z}\,n\,\overline{\zeta_2}\right]_{\text{at }z=0} \qquad \text{[Chapman's eq. (42)]}$$

or

$$n\,(0,\,t)\,\bar{\zeta} - \frac{1}{2}\,n\,(0,\,t)\,\frac{\partial\overline{\zeta^2}}{\partial z} - \frac{1}{2}\,\overline{\zeta^2}\,\frac{\partial n}{\partial z}. \qquad \text{[Le Claire's eq. (14)].}$$

For each species in the mixture, then, the resultant number of molecules crossing a given horizontal plane upwards per unit area per unit time is given by an equation

$$J = n\,\frac{\bar{\zeta}}{\tau} - \frac{1}{2}\,n\,\frac{\partial}{\partial z}\,\frac{\overline{\zeta^2}}{\tau} - \frac{1}{2}\,\frac{\overline{\zeta^2}}{\tau}\,\frac{\partial n}{\partial z}.$$

REFERENCES

1. Truesdell, C., *Arch. Rat. Mech. Anal.* **1**, 245 (1952).
2. Brillouin, M., *Ann. Chim.* **20**, 440 (1900).
3. Stokes, G.G., *Trans. Camb. Phil. Soc.* **8**, 287 (1845) [Papers, vol. 1, p. 75].
4. Monchick, L., and Mason, E.A., *Physics of Fluids* **10**, 1377 (1967).
5. Chapman, S., *Proc. Roy. Soc. A* **119**, 34, 55 (1928).

6. Le Claire, A.D., *Phil. Mag.* **3**, 921 (1958).

7. Lidiard, A.B., *Disc. Faraday Soc.* **28**, 209 (1959).

8. Allnatt, A.R., and Rice, S.A., *J. Chem. Phys.* **33**, 573 (1960).

9. Fürth, R., *Proc. Roy. Soc. A* **179**, 461 (1942).

10. Gillespie, L.J., *J. Chem. Phys.* **7**, 530 (1939).

11. Zhdanov, V., Kagan, Yu., and Sazykin, A., *Soviet Physics J.E.T.P.* **15**, 596 (1962) [*Zh. Eksp. Teor. Fiz.* **42**, 857].

12. Monchick, L., Munn, R.J., and Mason, E.A., *J. Chem. Phys.* **45**, 3051 (1966).

1.3

Diffusion in dense fluids

J. H. DYMOND

Department of Chemistry, The University, Glasgow

ABSTRACT

The self-diffusion coefficients of gases at densities greater than critical are quantitatively predicted for temperatures down to near the normal boiling temperature on the basis of the van der Waals theory of diffusion which is, in fact, equivalent to the hard-sphere theory. This shows the importance for real systems of correlated events which have been observed in computer experiments on hard sphere diffusion. The contribution of correlated events to the mutual diffusion coefficient of binary dense gas mixtures is found to depend significantly on the relative molecular masses. The results are in qualitative agreement with preliminary computer results on hard-sphere mixtures.

1 INTRODUCTION

Existing theories for the diffusion coefficient in dense fluids differ in their description of the trajectories of the molecules. However, as a result of molecular dynamics studies, it is possible to check the validity of the various models. For example, it appears that the activation model, which provides a simple explanation of the linear dependence of the logarithm of the diffusion coefficient on reciprocal temperature, does not have a sound physical basis[1]. A theory which has been extensively discussed[2], which gives a better description of molecular motion in dense fluids is based on the Brownian motion approximation. This postulates that the molecules undergo many deflections (soft collisions) as a result of a nett attraction by neighbouring molecules between successive hard-core collisions. However, molecular dynamics calculations of the number of soft and hard collisions for molecules

39

with a realistic interaction potential showed[3] that even under the conditions corresponding to the normal liquid state they were comparable in number. This result indicates that for a more realistic picture of molecular motion in dense fluids the van der Waals model, recently applied to transport properties[4], should be used.

The van der Waals concept of a fluid is of an assembly of molecules for which the interaction potential is made up of a hard core plus a weak long-range attractive force. For real systems the potential does have a steep repulsive part and the range of the attractive region can be considered large with respect to the interparticle spacing at densities greater than the critical density. Furthermore, the attractive energy can be considered weak relative to the kinetic energy whenever the temperature is greater than the attractive well-depth or, roughly, the critical temperature. It has been found that for temperatures and densities above critical the van der Waals model serves well for equilibrium properties[5] providing that the core size is allowed to diminish as the temperature increases, a consequence of the somewhat soft repulsive energy of real molecules. An extremely important consequence of this model for diffusion is that the molecules move in straight lines between core collisions, because the attractive energy forms a uniform potential surface. The van der Waals theory of diffusion is thus equivalent to the hard-sphere theory. At high temperatures and densities, this model corresponds closely with reality. For temperatures and densities less than critical, the van der Waals assumption of uniformity of the attractive potential energy surface begins to break down. Once roughness of this surface causes the molecular trajectories to be deflected this free-flight approximation between hard-core collisions is no longer valid. However, the differences between the computed hard-sphere values and the experimental results under these conditions will give a measure of the importance of the soft collisions.

For the calculation of the hard-sphere diffusion coefficients, the molecular chaos approximation is removed by using the results of recent computer studies[6]. In this way[7] it is possible to delineate the region of applicability of the van der Waals model, and also to substantiate that the unexpected long persistence of velocity currents found in hard-sphere systems at intermediate densities, which leads to enhanced diffusion coefficients, also occurs in real systems. At still higher densities correlated events lead to an opposite deviation from the kinetic theory.

The experimental tests of the hard-sphere deviations from kinetic theory are unfortunately severely limited at the present time. Only for methane[8] are sufficiently accurate diffusion data available at temperatures and densities much greater than critical. Measurements on methane have also been made along the coexistence curve, and a comparison of the computed hard-sphere diffusion coefficients with the experimental data along the liquid branch of the curve provides an estimate of the importance of soft collisions at these low temperatures. Diffusion coefficient data for binary dense gas mixtures have recently been obtained by the chromatographic peak broadening method[9], and in the final section the van der Waals model is applied to these mixtures to determine the mass ratio dependence of the hard-sphere deviations from kinetic theory.

In this following section a brief discussion is given of the empirical methods which have been used to extend the low density transport theory to high densities by means of van der Waals ideas. This is done primarily to point out that agreement at low density should not be forced. Under these conditions, molecules aggregate in clusters and the concept of a uniform potential field is no longer valid.

2 EXTENSIONS TO LOW DENSITY THEORY

For the calculation of the diffusion coefficient for a system of dense hard spheres, the approximate Enskog theory is used[10]. This theory just scales in time the solution of the Boltzmann equation valid at low densities. The collision rate is higher in a dense hard-sphere system because the distance two spheres have to travel in order to collide is significantly decreased by the diameter of the spheres, while at low densities the diameter is so small compared with the average distance travelled between collisions that the spheres can be considered as point particles. This argument by the virial theorem leads to the result that the ratio of the high density diffusion coefficient to the low density value is inversely proportional to the ratio of the compressibility factor minus 1 to the low density value B/V, where B is the second virial coefficient;

$$\frac{D}{D_0} = \frac{B/V}{pV/NkT - 1}.\tag{1}$$

It is evident from this derivation that correlations between successive hard-sphere collisions have been neglected. In the Enskog theory, a sphere is considered as always colliding with other spheres approaching in a random direction and a velocity chosen at random from the Maxwell-Boltzmann distribution appropriate to the given temperature. This molecular chaos approximation is exact only in the limit of low density, but recent molecular dynamics studies on hard-sphere systems[6] have made it possible to remove this approximation and quantitatvely check the validity of the van der Waals theory.

There are several possible methods of application of equation (1) to real systems but certain of them, which are just extensions to low density theory, fail at high densities because van der Waals ideas are incorrectly applied. For a real system the low density diffusion coefficient is given by the solution of the Boltzmann equation with an appropriate potential of interaction, that is

$$D_0 = (3/8\pi\varrho)\,(\pi mkT)^{1/2}(1/\sigma^2\Omega^{(1,1)^*})$$

where ϱ is the fluid density, σ the molecular diameter and $\Omega^{(1,1)^*}$ a collision integral involving the dynamics of a molecular encounter.

One method of applying Enskog's theory to real systems is to define an effective core size σ_{eff} from the dilute gas data such that $\sigma_{eff}^2 = \sigma^2\Omega^{(1,1)^*}$ will give agreement with the experimental results. When σ_{eff} is used to calculate diffusion coefficients at higher densities from equation (1) the agreement with experiment is poor. This is shown in Fig. 1 in the case of methane using data[8] at 298 °K, about one and a half times the critical temperature. For high densities where agreement with the van der Waals model might be expected, large deviations are found. The reason is that the core sizes were not chosen to be consistent with the van der Waals model for other properties such as equilibrium properties.

An alternative method of applying Enskog's high density hard-sphere theory to real systems, based on the van der Waals model, is that suggested by Enskog himself[10]. Consideration of the internal pressure lead to the following generalisation of equation 1

$$\frac{D}{D_0} = \frac{[B + T\,(\partial B/\partial T)]/V}{(\partial p/\partial T)_v\,(V/Nk) - 1} \tag{2}$$

The form of the numerator is dictated by the requirement that the expression reduce to unity at low density. Furthermore, the expression reduces to equation (1) for hard spheres. Application of this equation to methane (using Lennard-Jones parameters $\varepsilon/k = 148$ °K, $\sigma = 3.82$Å) leads to agreement with experiment up to densities slightly above critical, but there are large deviations at high densities as shown in Fig. 1. Nevertheless the agreement

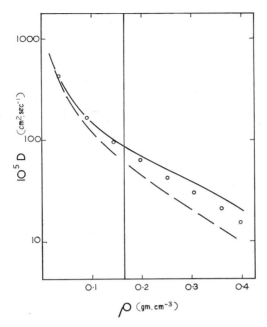

Figure 1 The self-diffusion coefficient of methane at 298.2 °K as a function of density. The experimental data are given by the solid curve. The theoretical curves correspond to using an effective diameter (dashed line) and to using internal pressure (circles). The vertical line indicates the critical density

over the wide low density region is remarkable. It turns out that the factor on the right-hand side of equation 2 is within 10% of unity in the density region where agreement with experiment is found. This observation is equivalent to that made earlier[11] that the uncorrected low-density theory holds over this wide density range. The failure of the theory at high densities is again due to an improper application of van der Waals ideas.

3   VAN DER WAALS THEORY OF DIFFUSION

The logically consistent method of applying van der Waals theory to the calculation of diffusion coefficients is to use the theory for equilibrium properties to determine hard core diameters to utilize in equation (1). The core sizes can be evaluated in either of two ways, depending on the external pressure. For dense gases, for which pVT data are available, the procedure is to plot pV/NkT against $1/T$ at a given density. The core size is then determined at a given temperature from the high temperature intercept of the slope by comparison with the known hard-sphere values of pV/NkT^{12}. One way to judge whether the temperature and density are sufficiently high for van der Waals theory to be applicable is to observe whether density independent values of the core size are obtained. In fact, for methane the core diameter is found to be practically independent of density at densities above the critical, but it is noticeably temperature dependent, decreasing with increase in tempera-ture, as seen from Fig. 2. This is just a consequence of the somewhat soft repulsive energy of real molecules.

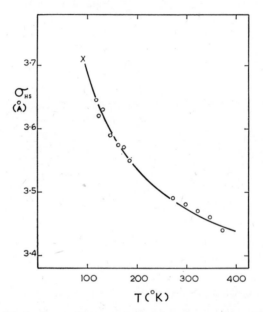

Figure 2 The hard core diameter as a function of temperature. The highest
point is obtained from the Lindemann melting model

Another method of calculating the core diameter within the framework of the van der Waals theory is based on the idea that when the pressure is low, as for normal liquids, the hard-sphere repulsive pressure has to be balanced almost exactly by the attractive term in the van der Waals equation:

$$\frac{a}{NkTV} = \left(\frac{pV}{NkT}\right) HS.$$

The constant a can be determined[13] from the observation that the diffraction data for a number of liquid metals can be fitted by assigning a constant value for the packing fraction at the melting point[14]. Although the melting temperature may be rather low for the applicability of the van der Waals equation, it is interesting to observe from Fig. 2 that, assuming liquid methane shows the same packing fraction as liquid metals, the hard sphere diameter determined in this way is consistent with the values from pVT data.

The core sizes used for the calculation of the diffusion coefficients of methane were taken from the smooth plot of Fig. 2. The resulting values for the diffusion coefficient are compared with the experimental data in Fig. 3, for temperatures both above and below the critical temperature. All points

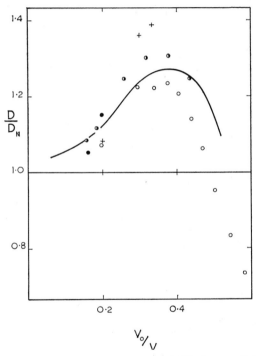

Figure 3 The ratio of the experimental self-diffusion coefficient to the kinetic theory coefficient as a function of density, where the volume is measured relative to the close-packed volume V_0. The experimental points are for methane at 298.2 °K ◑[8] and along the liquid branch of the co-existence curve ○[11]; krypton ●[15] and carbon dioxide +[16]. The solid curve represents the same ratio of diffusion coefficients for hard spheres determined on the computer and extrapolated to an infinite particle system

are subject to a possible variation of up to 5% as a result of uncertainties in the core sizes used together with experimental uncertainties. Also shown are the calculated diffusion coefficients for krypton at 35°C, obtained using a core size of 3.40 Å[4]. They differ from the experimental data[15] to the same extent as the predicted coefficients for methane in the corresponding density range. Calculated values for the self-diffusion coefficient of carbon dioxide at 35°C are almost 40% less than the experimental results of O'Hern and Martin[16] at densities about twice critical. Although there is some uncertainty in the experimental data (the values of O'Hern and Martin are in disagreement with the earlier results, obtained by Drickamer and co-workers[17,18] which, however, show high scatter) the general agreement with the methane points is good.

For all these fluids, the ratio of experimental to calculated (Enskog) diffusion coefficient follows closely the same ratio given by hard spheres, where the experimental hard-sphere values were obtained on the computer[6]. This striking correspondence between the hard-sphere curve and the points obtained for real systems provides a clear illustration of the applicability of van der Waals model. Furthermore, it demonstrates the importance of hard-sphere correlated events to the self-diffusion in real fluids. These correlated events are of two kinds. At high densities the decreased diffusion coefficient can be ascribed to back-scattering caused by the high probability of reversal of velocity of a molecule upon collision with its nearest neighbours. In the subsequent collision its velocity is again reversed with greater than even probability as the particle collides with spheres on the other side of the shell of neighbours. It is then positively correlated with its initial velocity, but only weakly so, even at the highest density where the effect is expected to be greatest. The first anticorrelating collision thus gives the predominating correction to the Enskog theory. At intermediate densities, there is a significant increase in diffusion arising from an unexpected persistence of velocity currents. This has been shown[19] to arise as a result of vortex motion of relatively "hot" molecules. After about eight or nine collision times the velocity of such molecules is more positively correlated with their initial velocity than it would be if the molecular chaos approximation was valid under such conditions.

The difference between the methane points corresponding to high density data at a temperature about one and a half times critical and the results along the liquid branch of the coexistence curve are small which suggests that the van der Waals theory may be applied at temperatures down to 30% below critical for diffusion. However, any conclusion regarding the quantitative measure of the contributions of soft collisions must await precise diffusion data at different temperatures over the whole density region, if possible for a simple fluid such as argon.

4 DIFFUSION IN DENSE FLUID MIXTURES

On the basis of the van der Waals model, the mutual diffusion coefficient D_{12} for a binary dense fluid mixture can be calculated by a method analogous to that used for self-diffusion coefficients for dense fluids. For hard spheres of diameter σ_1 and σ_2, the low density mutual diffusion coefficient is given by the expression[20]

$$D_{0(12)} = (3/8n\sigma_{12}^2) \, [kT \, (m_1 + m_2)/2\pi m_1 m_2]^{1/2}$$

where n is the sum of the particle number densities and σ_{12} can be written as half the sum of σ_1 plus σ_2.

An extension to this low density kinetic theory expression has been given by Thorne[21], but this is just up to the first order in density. For high density diffusion, the Enskog theory relates D_{12} to the low density coefficient by the unlike pair contact distribution function $g_{12}(\sigma)$. This assumes that collisions between like molecules do not affect D_{12}. In the absence of extensive $g_{12}(\sigma)$ values from molecular dynamics studies, for a wide range of size and number ratio, the approximate expression derived by Lebowitz[22] on the basis of the Percus-Yevick theory is used. In terms of number densities n_1 and n_2 this may be written

$$g_{12}(\sigma) = \left\{ 1 - \frac{\pi}{6} [n_1\sigma_1^3 + n_2\sigma_2^3] + \frac{\pi}{2} \left(\frac{\sigma_1\sigma_2}{\sigma_1 + \sigma_2} \right) (n_1\sigma_1^2 + n_2\sigma_2^2) \right\} \times$$

$$\times \left\{ 1 - \frac{\pi}{6} (n_1\sigma_1^3 + n_2\sigma_2^3) \right\}^{-2} \tag{3}$$

When one component is present only in trace concentration, the expression reduces to

$$g_{12}(\sigma) = \left\{ 1 - \frac{\pi}{6} n_1\sigma_1^3 \left(\frac{\sigma_1 - 2\sigma_2}{\sigma_1 + \sigma_2} \right) \right\} \left(1 - \frac{\pi}{6} n_1\sigma_1^3 \right)^{-2}. \tag{4}$$

There are two corrections to be applied to the resulting expression for D_{12}. Firstly, there is the correction arising from use of the approximate Percus-Yevick theory and secondly there is the removal of the molecular chaos approximation. The magnitude of the first correction can be estimated by comparing the values of pV/NkT predicted on the basis of the Percus-Yevick theory[22] using the virial equation of state with the 'experimental' molecular dynamics results obtained for an equimolar mixture with radius ratio 3 : 1[23]. The results indicate that D_{12} calculated from the low density expression together with equation (3) or (4) should be decreased by about 2%. To determine the importance of correlated events in the diffusion of binary dense gas mixtures, diffusion coefficients were calculated for the systems recently investigated by means of the chromatographic peak broadening method[9]. The values of the core sizes used, obtained from equilibrium data, are summarised in Table 1. For discussion of the results, the mixtures are conveniently divided

Table 1   Core sizes in angstroms from pV/NkT data

	T (°K)	
	298.2	308.2
Helium	2.20	
Argon	3.14	3.13
Hydrogen	2.48	
Nitrogen	3.31	3.30
Krypton		3.40
Methane	3.48	
Carbon Tetrafluoride	4.11	
Carbon Dioxide		3.61

into two groups, those for which the trace gas has a higher molecular weight than the carrier gas, and those for which the carrier gas molecule is the heavier. Diffusion coefficients calculated for the group one mixtures at carrier gas densities greater than critical are smaller than experimental values, the discrepancy increasing with increasing density over the whole density region for which experimental data are available. In Fig. 4 it is shown that the ratio of the experimental diffusion coefficient to the calculated (Enskog) value for different trace gases in helium exhibits the same density dependence, to a good approximation, as the results obtained from diffusion experiments using Kr[85] as radioactive tracer[15]. Comparison with Fig. 3 indicates that for these mixtures there is an enhanced correlation above that observed for single component systems at these intermediate densities. This can be explained qualitatively by the heavier molecules in the mixture pushing their

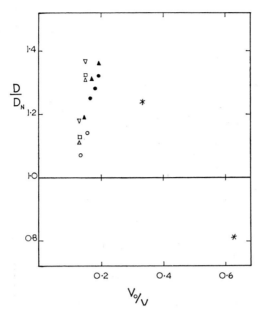

Figure 4 The ratio of the experimental mutual diffusion coefficient to the kinetic theory coefficient as a function of density for binary mixtures where the trace gas has the higher molecular weight. Volumes are measured relative to the close-packed volume V_0. The experimental points are for Ar–He ◯, N_2–He △, CH_4–He ☐ and CH_4–He ▽ [9]; Kr–Ar ●, and Kr–N_2 ▲ [15]. Computed hard-sphere values for equal sized particles, mass ratio 4 : 1, determined for a 108 particle system are represented by stars

way through the lighter particles by virtue of their larger thermal momentum.

An opposite deviation from hard-sphere kinetic theory is evident with the group two mixtures as seen from Fig. 5. Values for the ratio of experimental to calculated (Enskog) diffusion coefficient at any given density appear to vary, depending on the system considered, though this is possibly just a reflection of greater uncertainty in these experimental data. However, the definite conclusion can be drawn from these results that for light molecles diffusing in a dense gas composed of heavy particles, the diffusion coefficient is less than expected (Enskog) and this difference increases as the density increases. A possible partial explanation for this is that the light molecules become trapped inside a loose cage formed by the neighbouring molecules and possess insufficient momentum to break out easily. It is unlikely to be due to a size effect alone since these light molecules, which have the smaller core sizes, would be expected to diffuse more quickly. However, it is possible that this size effect is not entirely insignificant. The importance of this effect with respect to the mass effect can be determined from experimental data when accurate measurements of D_{12} are made for systems such as hydrogen–helium, neon–hydrogen or methane–argon (or nitrogen). It also would be very useful to have diffusion data for mixtures at high densities, for example for trace gases in argon or nitrogen at pressures above 1000 Atmospheres.

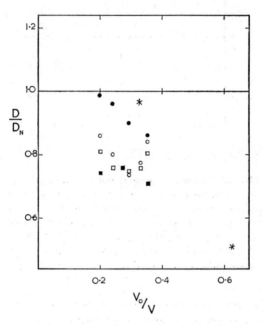

Figure 5 The ratio of the experimental mutual diffusion coefficients to the kinetic theory coefficients as a function of density for binary mixtures where the trace gas has the lower molecular weight. The experimental points are for He–N$_2$ $\bigcirc$, H$_2$–N$_2$ $\square$, He–Ar $\bullet$ and H$_2$–Ar $\blacksquare$[9]. Computed hard-sphere values for equal sized particles mass ratio 1 : 4, determined for a 108 particle system are represented by stars

However, a full understanding of the dependence of the hard-sphere deviations from kinetic theory on size ratios and mass ratios must await the results of molecular dynamics calculations on binary systems. Preliminary runs have been made at volumes corresponding to 1.6 and 3 times the close-packed volume for mixtures of equal-sized particles where the mass ratio is 4 : 1

and 1 : 4. The results, shown in Figs. 4 and 5, qualitatively confirm the deviations from kinetic theory deduced from experimental diffusion results at intermediate densities.

ACKNOWLEDGEMENTS

We wish to thank Dr. B.J.Alder for helpful discussion and Imperial Chemical Industries Limited for the award of a Fellowship.

REFERENCES

1. B.J.Alder and T.Einwohner, *J. Chem. Phys.* **34**, 3399 (1965).
2. A.R.Allnatt and S.A.Rice, *J. Chem. Phys.* **34**, 2156 (1956); S.A.Rice and A.R.Allnatt, *ibid* **34**, 2144 (1956); H.T.Davis, S.A.Rice and J.V.Sengers, *ibid* **35**, 2210 (1961).
3. B.J.Alder and T.Einwohner, *J. Chem. Phys.* **49**, 1458 (1968).
4. J.H.Dymond and B.J.Adler, *J. Chem. Phys.* **45**, 2061 (1966).
5. E.B.Smith and B.J.Alder, *J. Chem. Phys.* **30**, 1190 (1959).
6. B.J.Alder and T.E.Wainwright, *Phys. Rev. Letters* **18**, 988 (1967).
7. J.H.Dymond and B.J.Alder, *J. Chem. Phys.* **48**, 343 (1968).
8. P.H.Oosting, Doctoral Dissertation, Amsterdam (1968).
9. Z.Balenovic, M.N.Myers and J.C.Giddings, *J. Chem. Phys.* (in press).
10. S.Chapman and T.G.Cowling, *The Mathematical Theory of Non-Uniform Gases* (Cambridge University Press, New York, 1939), Chap. 16.
11. N.J.Trappeniers and P.H.Oosting, *Phys. Letters* **23**, 445 (1966).
12. F.H.Ree and W.E.Hoover, *J. Chem. Phys.* **40**, 939 (1964).
13. P.Ascarelli and A.Paskin, *Phys. Rev.* **165**, 222 (1968).
14. N.W.Ashcroft and J.Lekner, *Phys. Rev.* **145**, 83 (1966).
15. L.Durbin and R.Kobayashi, *J. Chem. Phys.* **37**, 1643 (1962).
16. H.A.O'Hern and J.J.Martin, *Ind. Eng. Chem.* **47**, 2081 (1955).
17. W.L.Robb and H.G.Drickamer, *J. Chem. Phys.* **19**, 1504 (1951).
18. K.D.Timmerhaus and H.G.Drickamer, *J. Chem. Phys.* **20**, 981 (1952).
19. B.J.Alder (private communication).
20. See ref. 10, Chap. 14.
21. See ref. 10, p. 292 *et seq.*
22. J.L.Lebowitz, *Phys. Rev.* **133**, 895 (1964).
23. B.J.Alder, *J. Chem. Phys.* **40**, 2724 (1964).

Self-diffusion in gaseous CO_2 and H_2O and the inter-diffusion coefficient of CO_2/H_2O mixtures

F. L. SWINTON

Department of Pure and Applied Chemistry,
University of Strathclyde

ABSTRACT

An isotopic tracer technique has been used to measure the self-diffusion coefficients of gaseous CO_2 and H_2O and also the inter-diffusion coefficient of gaseous CO_2/H_2O mixtures. The experimental temperature range was 300 to 530 K and low gas pressures were used in the range 0 to 0.2 bar. The two-bulb apparatus and the original method of obtaining diffusion coefficients from the experimental tracer concentration/time plots are described briefly. The results for the self-diffusion coefficients of CO_2 are of high precision and are compared with the results of other workers.

It is found that, over this temperature range, a simple Lennard-Jones 12-6 potential represents the $D_{CO_2}^{self}$ data better than more sophisticated intermolecular potentials. The results for $D_{H_2O}^{self}$ and D_{H_2O/CO_2}^{inter} are, unfortunately, of lower precision but have been analysed using the values for the collision integrals for the Stockmayer potential computed by Mason and Monchick. Satisfactory agreement is obtained for both sets of experimental data when certain values for the potential parameters for H_2O recommended by Mason and Monchick from an analysis of viscosity data are used.

The fact that no measurements of the self-diffusion coefficient of water vapour appear to have been made is somewhat surprising when the other major transport properties, namely the viscosity and thermal conductivity, of this most important substance have been studied by many workers for many decades. The present work was primarily an attempt to rectify this situation by constructing an apparatus capable of measuring the self-diffusion coefficient of water vapour at low pressures over the temperature range 360 to 530 K.

53

The apparatus was first tested by measuring the self-diffusion coefficient of carbon dioxide in the same temperature range.

EXPERIMENTAL

The apparatus was of the two-bulb, Ney and Armistead[1] type, and will be fully described elsewhere[2], only a brief description of its design and operation will be given here.

The two bulbs were identical in size and each was an ionization chamber so that the concentration of radioactive species in each section could be monitored continuously. The all-metal two-bulb diffusion cell is shown diagrammatically in Fig. 1. The ceramic insulators B, B', of the central elec-

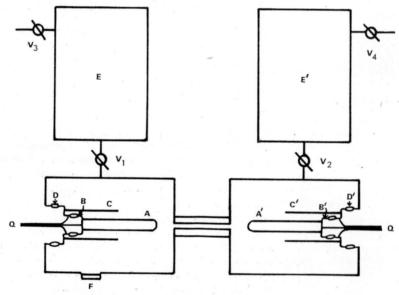

Figure 1 The diffusion cell. A, A' electrodes, B, B' ceramic insulators, C, C' guard rings, D, D' metal–glass–metal seals, E, E' glass expansion chambers, F pressure transducer, Q co-axial cable. V_1 to V_4 all-metal valves

trodes A, A' were shielded by guard rings C, C'. The guard-rings, in turn, were insulated from the main body of the ionization chambers by the large metal–glass–metal seals D, D'. PTFE-insulated co-axial screened cable led from the central electrodes to electrometers so that the ionization currents could be measured. Ekco type N616B electrometers were used which had a facility for 'backing-off' the voltage registered so that the electrometers could be used at maximum sensitivity at all times. A polarizing voltage of -100 volt was applied to the chambers and ionization currents were typically in the range 10^{-10} to 10^{-12} amp. The internal volume of each chamber was approximately 200 cm^3 and the two chambers were connected by a carefully machined and polished capillary of 40 mm length with an internal diameter of 2.5 mm. The exact dimensions of the chambers were determined to ± 0.01 per cent.

The gas pressure in the diffusion cell was measured using a Bell and Howell pressure transducer and a Bolton Paul type C61 transducer meter. Pressures in the range 0.05 to 0.2 bar could be measured to ± 0.1 per cent. The two ionization chambers were connected to the two glass expansion chambers E, E' via the all-metal Hoke valves V_1, V_2 and vapour could be admitted to the expansion chambers through two other Hoke valves, V_3 and V_4. The ionization chambers, expansion chambers and the four valves were all enclosed in an air thermostat which controlled the temperature to better than 0.05 K in the range 300 to 540 K.

EXPERIMENTAL PROCEDURE AND ANALYSIS OF RESULTS

At the beginning of a diffusion experiment the diffusion cell and the expansion chambers were evacuated and valves V_1 to V_4 closed. Active vapour was next admitted to E through V_3 and inactive vapour, at approximately the same pressure, was admitted to E' through V_4. Time was allowed for both vapour samples to reach the thermostat temperature and then valves V_1 and V_2 were opened simultaneously and, after 30 seconds, closed again. Some premixing occurred through the capillary due to initial pressure difference but this was immaterial. Values of the ionization current were measured at gradually increasing intervals of from 1 min to 10 min over a period of from 2 to 5 hours. For the initial experiments the ionization currents in both chambers were monitored, the current in one chamber decreasing and the current in the other chamber increasing during the course of an experiment. It was found that the results obtained with decreasing current were always of higher precision and so, in later experiments, only these measurements were made.

The theory of the two-bulb diffusion apparatus has been discussed by Ney and Armistead[1] and Visner[3]. The basic equation is:

$$\ln (i_\infty - i_t) = -D\alpha t + \beta \qquad (1)$$

where i_t and i_∞ are the ionization currents at times $t = t$ and $t = \infty$, D is the diffusion coefficient, α is an apparatus constant which can be determined precisely from the known dimensions and geometry of the apparatus and β is another constant, so that $\ln (i_t - i_\infty)$ should be a linear function of t. When calculating α, the apparatus constant, the usual end corrections were made when estimating the effective length of the capillary tube and, following Brown and Murphy[4], a further small correction term was included to take account of the fact that the actual concentration difference across the ends of the diffusion capillary was slightly less than the difference between the mean concentrations in the two ionization chambers.

i_∞ was not determined experimentally but was used as an adjustable parameter.

The experimental (i_t, t) data were plotted in accordance with equation (1) using a reasonable trial value of i_∞ and a computer program was used that adjusted i_∞ until the overall standard deviation of the experimental points from linearity was minimised. The standard deviation: i_∞ function usually exhibited an extremely sharp minimum so that i_∞ and hence D could be determined with high precision.

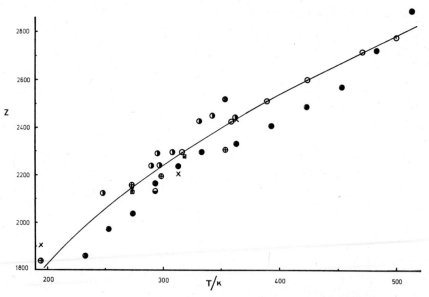

Figure 2 The self-diffusion coefficient of carbon dioxide. The function
$$Z = D_{CO_2}^{self} \, T^{-3/2}/10^8 \text{ cm}^2\text{sec}^{-1}\text{K}^{-3/2} \text{ against } T/K$$

——— Lennard-Jones 12-6 potential, ⊙ present results, × reference 5, ⊕ reference 6, ▢ reference 7, ◑ reference 8, ● reference 9, ◓ reference 10

EXPERIMENTAL RESULTS

Measurements on the self-diffusion of carbon dioxide were made using C_{14}-labelled tracer and the results are given in Table 1 and illustrated in Fig. 2 where the product $DT^{-3/2}$ is shown as a function of temperature. All other previous measurements[5-10] of $D_{CO_2}^{self}$ are shown in Fig. 2 except the

Table 1 The self-diffusion coefficient of carbon dioxide

T/K	$D_{self}^{CO_2}/cm^2sec^{-1}$	$DT^{-3/2}/cm^2sec^{-1} K^{-3/2} 10^{-8}$
315.8	0.1289	
	0.1285	
	0.1285	
	0.1290	
	Av = 0.1287	2294
358.8	0.1648	
	0.1645	
	0.1643	
	0.1651	
	Av = 0.1647	2423
389.2	0.1921	
	0.1923	
	0.1932	
	0.1927	
	Av = 0.1926	2508
423.4	0.2261	
	0.2255	
	0.2269	
	Av = 0.2262	2598
470.3	0.2758	
	0.2771	
	0.2781	
	0.2764	
	Av = 0.2769	2715
500.1	0.3093	
	0.3134	
	0.3087	
	Av = 0.3105	2776

Table 2 Self-diffusion coefficient of water vapour

T/K	$D_{self}^{H_2O}/cm^2 sec^{-1}$	$DT^{-3/2}/cm^2 sec^{-1} K^{-3/2} 10^{-8}$
363.2	0.267	
	0.275	
	0.269	
	0.273	
	0.272	
	Av = 0.271	3920
379.2	0.289	
	0.285	
	0.300	
	0.279	
	0.300	
	Av = 0.291	3940
433.2	0.383	
	0.385	
	0.400	
	0.402	
	0.394	
	Av = 0.393	4360
466.2	0.445	
	0.474	
	0.457	
	0.461	
	Av = 0.459	4560
517.2	0.574	
	0.564	
	0.575	
	0.581	
	0.540	
	Av = 0.567	4820

data of Pakura and Ferron[11] which were obtained at high temperatures in the range 1200 to 1700 K. It is seen that the present measurements are of high precision and lie on a smooth curve. The line drawn through the present data is the theoretical one obtained by using a Lennard-Jones 12-6 potential with force constants $\sigma = 3.595$ Å, $\varepsilon/k = 270$ K. It is surprising that, even for this limited range of data, a simple two-parameter potential gives a better fit than the various three-parameter potentials that were tested.

The self-diffusion coefficient of water vapour was measured using tritiated water as the tracer species. Preliminary experiments showed that HTO could be used successfully in the ionization chambers. The ionization current proved to be directly proportional to the partial pressure of HTO present, at least over a range of pressure of from 0.01 to 0.3 bar. The results of the twenty-four successful experiments on this system are tabulated in Table 2 and it is seen that, inexplicably, the precision is much lower than the preci. sion of the results for carbon dioxide, ranging from $\pm 1.5 \%$ at 363 K to $\pm 4 \%$ at 513 K. In most of the experiments at the higher temperatures a slight rise in the total pressure was noted as the experiment neared completion. This pressure increase was either caused by outgassing of the ionization chambers or was caused by a slow reaction of the water vapour with the various me- tals composing the diffusion cell.

The experimental results are illustrated in Fig. 3 where it is seen that the experimental precision is such that the points are best represented by a

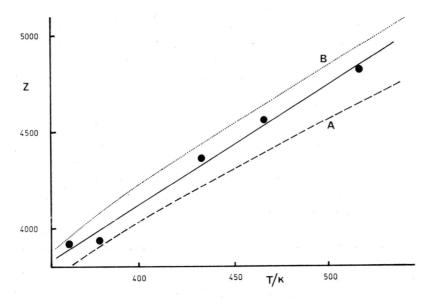

Figure 3 The self-diffusion coefficient of water vapour. The function
$$Z = DT^{-3/2}/10^{-8} \text{ cm}^2\text{sec}^{-1}\text{K}^{-3/2} \text{ against } T/K$$
− − − Theoretical equation. Set A, ·········· Theoretical equation. Set B,
● present results

straight line although, on theoretical grounds, it is to be expected that such a graph should be slightly curved. There appear to be no other experimental measurements of this function in the literature.

The theory of the transport properties of polar molecules has been developed by Mason and Monchick[12,13] who have calculated and tabulated the collision integrals for an assembly of molecules interacting according to a Stockmayer (12-6-3) potential. The transport (and thermodynamic) properties can be calculated once three parameters, ε/k, σ and δ_{max} have been evaluated for the particular substance. ε/k and σ are the depth of the potential well and the collision diameter respectively while δ_{max} is related to the permanent dipole moment of the molecules. Mason and Monchick[13] have analysed the experimental viscosity of steam which is available over a wide temperature range. The analysis yielded several sets of ε/k, σ and δ_{max} values, each set of parameters representing the experimental viscosities within the (fairly large) experimental error. Mason and Monchick were unable to distinguish between the different sets and were unable to recommend a 'best' set of parameters. The two theoretical curves, A and B drawn in Fig. 3 represent $D_{self}^{H_2O}$ calculated using the two sets of parameters, $\varepsilon/k = 775$, $\sigma = 2.52$ and $\delta_{max} = 1.0$ (curve A) and $\varepsilon/k = 260$, $\sigma = 2.80$ and $\delta_{max} = 2.5$ (curve B). The agreement between the experimental points and both curves is satisfactory bearing in mind the fact that the Mason and Monchick theory, based as it is on the Stockmayer potential might only be expected to apply to weakly polar molecules.

Twenty-one successful experiments were carried out to measure the H_2O/CO_2 inter-diffusion coefficient covering approximately the same temperature range as the measurements on water vapour. The experimental technique used was to fill the apparatus with inactive CO_2 at the required pressure and to start the experiment by injecting a small amount of highly active water vapour into one of the diffusion chambers. Inter-diffusion coefficients are known to be slightly concentration dependent and the coefficient measured in the present case was D_{H_2O/CO_2} $(C_{H_{20}} \rightarrow 0)$. It would be interesting to measure D_{H_{20}/CO_2} $(C_{H_{20}} \rightarrow 1)$ and so determine the maximum variation of D_{12} with composition.

The experimental measurements are given in Table 3 and shown in Fig. 4. The experimental measurements of previous workers are shown in Fig. 4.[14-16] These earlier measurements were all obtained over only a small temperature range and were measured using the Stefan technique. A possible source of experimental error when the Stefan method is used for the CO_2/H_2O system

Table 3 H_2O/CO_2 Inter-diffusion coefficient. $C_{H_{20}O}$

T/K	D_{12}/cm^2sec^{-1}	$D_{12}T^{-3/2}/cm^2sec^{-1}$ $K^{-3/2}$ 10^{-8}
357.2	0.238	
	0.237	
	0.242	
	0.238	
	0.241	
	Av = 0.239	3540
390.2	0.283	
	0.289	
	0.286	
	0.287	
	Av = 0.286	3710
412.2	0.312	
	0.313	
	0.318	
	0.313	
	Av = 0.314	3750
481.2	0.432	
	0.434	
	0.422	
	0.428	
	Av = 0.429	4060
521.2	0.492	
	0.500	
	0.494	
	0.487	
	Av = 0.493	4140

is the relatively high solubility of CO_2 in liquid water. Another disadvantage of the Stefan method is that D_{12} is measured at some undetermined overall concentration. The present results are seen to be in reasonable agreement with the older data.

The theoretical curve in Fig. 4 has been calculated using the interaction parameters (set A for H_2O) used previously for the like interactions and the usual combining rules $\varepsilon_{12} = (\varepsilon_{11} \cdot \varepsilon_{22})^{1/2}$ and $\sigma_{12} = (\sigma_{11} + \sigma_{22})/2$. In the

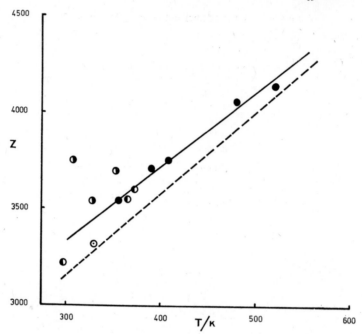

Figure 4 The CO_2/H_2O inter-diffusion coefficient. The function
$$Z = DT^{-3/2}/10^{-8} \text{ cm}^2\text{sec}^{-1}\text{K}^{-3/2} \text{ against } T/K$$
● — present results, – – – – theoretical equation, ◑ reference 15, ◐ reference 16, ⊙ reference 14

present case, because CO_2 is non-polar, the third mixed parameter, δ_{12}, is zero. The agreement between theory and experiment is reasonable.

It is obvious that measurements with a precision and accuracy at least ten times greater than the present results and extended where possible to lower reduced temperatures are required if any real test is to be made of the combining rules for mixtures containing polar molecules.

The present work was carried out under a National Engineering Laboratory Research Contract and the author expresses thanks to the Director for permission to publish this paper. The author also thanks I. Arnott and J. Carrothers for invaluable experimental assistance.

REFERENCES

1. E. P. Ney and F. C. Armistead, *Phys. Rev.*, **71**, 14 (1947).
2. I. Arnott and F. L. Swinton (to be published).
3. M. Brown and E. G. Murphy, *Trans. Faraday Soc.*, **61**, 2442 (1965).

4. I. Visner, *Gaseous Self-diffusion and Flow in Capillaries at Low Pressure* (Carbide and Carbon Chemicals Co., K-688), (1951).

5. A. Amdur, J. W. Irvine, E. A. Mason and J. Ross, *J. Chem. Phys.*, **20**, 436 (1950).

6. E. B. Winn, *Phys. Rev.*, **80**, 1024 (1950).

7. E. R. S. Winter, *Trans. Faraday Soc.*, **47**, 342 (1951).

8. R. P. Wendt, J. N. Mundy, S. Weissman and E. A. Mason, *Phys. Fluids*, **6**, 572 (1963).

9. K. Schäfer and P. Reinhard, *Z. Naturforschg.*, **18a**, 187 (1963).

10. L. Miller and P. C. Carman, *Trans. Faraday Soc.*, **60**, 33 (1964).

11. T. A. Pakura and J. R. Ferron, *J. Chem. Phys.*, **43**, 2917 (1965).

12. L. Monchick and E. A. Mason, *J. Chem. Phys.*, **35**, 1676 (1961).

13. E. A. Mason and L. Monchick, *J. Chem. Phys.*, **36**, 2746 (1962).

14. W. L. Crider, *J. Amer. Chem. Soc.*, **78**, 924 (1956).

15. M. Trautz and W. Muller, *Ann. Physik*, **22**, 233 (1935).

16. F. A. Schwertz and J. E. Brow, *J. Chem. Phys.*, **19**, 640 (1951).

Diffusion in liquids

Diffusion in liquids

H. J. V. TYRRELL

Chelsea College of Technology, London

This is not an easy field of study in any sense. It took nearly eighty years from the time when Thomas Graham worked on diffusion in liquids before precise data on diffusion coefficients began to be collected for binary liquid systems. There is, too, a deceptive simplicity about the phenomenological description of the process as given originally by Fick[1,2] which has led to confusion about definitions which is not wholly absent even from the modern literature. Even when armed with our modern accurate, diffusion coefficients, correctly defined in terms of the experimental reference plane, how far can we relate them to fundamental molecular quantities? The answer is, of course, "not very far", though this relative ignorance about the relationship of macroscopic properties of liquids to molecular properties is certainly not confined to diffusion. We can explain very well[3] the diffusion coefficients found for very dilute solutions of 1 : 1 electrolytes in water, and this can be done with sufficient precision for diffusion coefficient measurements to provide a useful way of measuring solute activities in dilute electrolyte solutions[4]. In more concentrated solutions of electrolytes it is necessary to introduce "hydration numbers"[5], always a potent source of controversy, in order to rationalise the experimental data. The most fundamental approach would be to calculate the transport coefficients, including diffusion coefficients, by statistical-mechanical methods. It is always necessary at the present time to use approximations if theory is to be compared with experiment. The most useful methods to date currently available for this purpose are probably those due to Bearman and Kirkwood[6], and to Rice and Allnatt[7]; in the long run, methods using time-dependent perturbation functions[8] may be more effective. In terms of frequency of citation, the Stokes-Einstein equation has been the most favoured way of 'explaining' diffusion data, both in electrolytes and in non-electrolytes. The calculation of "Stokes radii, r", from the equation:

$$D = kT/6\pi\eta r \tag{1}$$

has been a favourite occupation, and the interpretation of changes in Stokes radii with changes in experimental conditions has occupied much attention. The inadequate theoretical base of equation (1), and the fact that it had been recognised at an early date[9] that, in molecular systems, the numerical factor 6 in equation (1) might more correctly be replaced by a factor close to 4 has often been forgotten. A very different theory was based on Eyring's theory of absolute reaction rates[10], and assumed that diffusion occurred in a series of activated jumps. The formulation of this theory was marked by insufficient clarity in the definition of the diffusion coefficient (reference 3, pp. 136 ff.), and there are now reasons for believing that diffusion occurs by the cooperative motion of many molecules rather than by the activation of a single one.

It is not my purpose to discuss experimental methods except to say that adequate methods are now available for most systems. I propose rather to discuss briefly the problems of the definition of diffusion coefficients, and then to deal with some advances made by the application of more modern concepts of the phenomenology of diffusion. These arise from non-equilibrium thermodynamics, and from mechanical analogies which lead to the definition of frictional coefficients. It is these which are calculable in principle from statistical theory, and their use has led to a deeper insight into diffusion processes than can be obtained from the mere inspection of diffusion coefficients.

THE DEFINITION OF DIFFUSION COEFFICIENTS

Fick, having failed to deduce a diffusion law from a consideration of the nature of inter-molecular forces, argued by analogy from Fourier's Law of heat conduction, and wrote the flow density (J) in moles per unit area per unit time for a solute in a binary solution as a linear function of the molarity gradient (dc/dx) for the solute:

$$J = -D\,(dc/dx). \tag{2}$$

Simple arguments then lead to the equation now usually called Fick's Second Law of Diffusion:

$$dc/dt = d/dx\,(D\,dc/dx). \tag{3}$$

Most methods of measuring diffusion coefficients depend upon equation (3), integrated subject to the appropriate boundary conditions, and with the assumption that D is independent of concentration. Fick believed this assumption to be correct, and felt D to be completely comparable with the coefficient of thermal conduction derived from Fourier's Law. In reality the situation is more complex. Not only are diffusion coefficients often strongly concentration-dependent, but equation (2) is incomplete without a definition of the plane with respect to which the flow-densities are measured. Furthermore it refers only to one component while diffusion phenomena are primarily regarded as characteristic of multi-component mixtures, and the choice of molarity gradient as the driving force is an arbitrary one. This arbitrary choice incidentally determines the dimensions of D as area per unit time.

These problems can be resolved with the help of non-equilibrium thermodynamics. It is now well-known that the rate of entropy production, dS_i/dt, arising from irreversible processes within a system can be written in the form of a sum of products of generalised flows J_k with the generalised forces X_k giving rise to the flows, (equation (4)):

$$T\,dS_i/dt = \sum_k J_k X_k. \tag{4}$$

T is the absolute temperature of the system. The appropriate force for diffusion of component j in a one dimensional system is $-(\partial \mu_j/\partial x)_{T,P}$, the gradient of chemical potential. It is customary to assume that these forces are related linearly to the flows by equations of the form:

$$J_k = \sum_j L_{kj} X_j \quad (k = 0, 1, 2, \ldots). \tag{5}$$

L_{kj} are the phenomenological coefficients, and have the property that $L_{kj} = L_{jk}$ (Onsager relationship). For isothermal diffusion in a two-component system of solvent (suffix 0) and solute (suffix 1):

$$\left. \begin{aligned} J_0 &= L_{00}X_0 + L_{01}X_1 \\ J_1 &= L_{10}X_0 + L_{11}X_1. \end{aligned} \right\} \tag{6}$$

These equations (6) have four phenomenological coefficients. However, they are not independent because:

a) The Onsager relation applies.

b) J_0 and J_1 must be related in a manner which depends on the definition of the reference plane.

c) The gradients of chemical potential X_0 and X_1 are related through the Gibbs-Duhem equation.

For a two-component system there is only one independent phenomenological coefficient, while for an n component system there are $n(n-1)/2$ such coefficients. Thus for a two-component system such as that envisaged by Fick, it would be desirable to define a single coefficient to characterise the diffusion process. To avoid confusion this coefficient should retain the dimension area per unit time.

By analogy with equation (1) we can write a Fick Law equation for each component:

$$\left.\begin{aligned} J_0 &= -D_0 \, dc_0/dx \\ J_1 &= -D_1 \, dc_1/dx. \end{aligned}\right\} \tag{7}$$

If the flows are measured with respect to a plane across which no net transfer of volume occurs, and a superscript v is used on J, D, to denote this, then, if v_0, v_1 are the partial molar volumes,

$$v_0 J_0^v + v_1 J_1^v = 0. \tag{8}$$

From equations (7), (8),

$$v_0 D_0^v \, dc_0/dx + v_1 D_1^v \, dc_1/dx = 0. \tag{9}$$

By definition, and by using the Gibbs-Duhem equation,

$$v_0 \, dc_0 + v_1 \, dc_1 = 0. \tag{10}$$

Hence,

$$D_0^v = D_1^v. \tag{11}$$

Thus, definitions of diffusion coefficients according to equation (7), coupled with the definition of a volume-fixed reference plane, equation (8), lead to the definition of a single *mutual diffusion coefficient* which can be written D_{01}^v such that

$$\left.\begin{aligned} J_0^v &= -D_{01}^v \, dc_0/dx \\ J_1^v &= -D_{01}^v \, dc_1/dx. \end{aligned}\right\} \tag{12}$$

Other diffusion coefficients can be defined in terms of other reference planes (cf. reference (3), Chapter 3). A more important matter is the relationship between such formally-defined coefficients to experimental diffusion coefficients which are most commonly obtained from an integrated form of equation (3). No single answer can be given to this. Experimentally, flows in a diffusing system are measured with respect to a cell-fixed frame of reference. We can write[11] for a component k, the flow density J_k^c in the cell-fixed frame as:

$$J_k^c = J_k^v + c_i \vec{u}_{vc} \tag{13}$$

where $\vec{u}_{vc}$ is the velocity of the volume-fixed frame with respect to the cell-fixed frame. In free diffusion $\vec{u}_{vc}$ is zero if either (i) the molar volumes of all components are independent of concentration, or (ii) the concentration gradient across the diffusion boundary is small. Even if (i) does not apply, the normal techniques for measuring differential diffusion coefficients in liquid systems require that the second condition is fulfilled. Such experimental coefficients can therefore be equated directly with volume-fixed diffusion coefficients. Coefficients with respect to any other well-defined frame of reference can then be calculated fairly easily[11].

CALCULATION OF PHENOMENOLOGICAL COEFFICIENTS

For a binary system, D_{01}^v, which can be identified with the experimental mutual diffusion coefficient D, is related to the single phenomenological coefficient, L, characteristic of such a system. Since the flow density J is related to the diffusion coefficient in terms of a molarity gradient, and to L in terms of a chemical potential gradient, the relationship between these two gradients is important. If y_k, f_k are activity coefficients defined respectively on the molarity (c) and mole fraction (N) scales, the following identities apply:

$$(\partial \mu_k / \partial x)_{T,P} = c_k^{-1} RT \left(1 + \partial \ln y_k / \partial \ln c_k \right)_{T,P} \tag{14}$$

$$= N_k^{-1} RT \left(1 + \partial \ln f_k / \partial \ln N_k \right)_{T,P} \tag{15}$$

$$J = -L \left(\partial \mu / \partial x\right)_{T,P} = -D \left(\partial c / \partial x\right) \tag{16}$$

$$D = c^{-1} RT \left(1 + \partial \ln y / \partial \ln c \right)_{T,P} L. \tag{17}$$

If D is an experimental coefficient identifiable with a volume-fixed coefficient, the simplest form of activity coefficient term to use in calculating is therefore

$(1 + \partial \ln y/\partial \ln c)$. Both statistical theory[6] and experiment suggest that for binary mixtures of similar species, viscosity η, the quantity

$$D\eta\, (1 + \partial \ln y_1/\partial \ln c_1)^{-1}_{T,P} = c_1^{-1} RTL \tag{18}$$

should be almost independent of concentration. On the other hand, Glasstone, Laidler, and Eyring[10], whose diffusion coefficients are not well-defined, concluded on the basis of absolute reaction rate theory that,

$$D\eta\, (1 + \partial \ln f_1/\partial \ln N_1)^{-1}_{T,P} \tag{19}$$

should be a linear function of mole fraction. This apparent contradiction can be resolved from the thermodynamic identity (cf. equations (14), (15))

$$(1 + \partial \ln f_1/\partial \ln N_1)^{-1}_{T,P} = \frac{\bar{v}}{v_0}\,(1 + \partial \ln y_1/\partial \ln c_1)^{-1}_{T,P} \tag{20}$$

where $\bar{v}$ is the mean molar volume. It follows from equations (18)–(20) that,

$$D\, (1 + \partial \ln f_1/\partial \ln N_1) = (c_1 v_0)^{-1} RTL\bar{v}. \tag{21}$$

If $(c_1^{-1} RTL)$ and the molar volumes of both components are independent of concentration, Glasstone, Laidler, and Eyring's conclusion follows from equation (12). Other examples of difficulties caused by imprecision in the formal definition of diffusion coefficients, and a lack of clarity in their relationship with experimental coefficients, can be found in the extensive literature on the significance of "intrinsic diffusion coefficients" in liquid diffusion[6,12,13].

The calculation of the single phenomenological coefficient for a binary mixture of non-electrolytes from experimental mutual diffusion coefficients is trivial and of little value. However, in electrolyte solutions, and in multi-component systems, this is not true, and the work of Miller in particular[14] has shown their value for the understanding of such systems. A very simple example of the way in which the thermodynamic theory can provide a clearer understanding of a problem is shown in the case of the well-known Nernst equation for the diffusion coefficient of an electrolyte at infinite dilution. If the valencies and limiting mobilities of the ions are represented by z_+, z_-, and u_+, u_-, respectively, and F is the Faraday then the Nernst equation states that

$$D = \frac{|z_+| + |z_-|}{|z_+|\,|z_-|}\,\frac{RT}{F}\,\frac{u_+ u_-}{u_+ + u_-}. \tag{22}$$

The thermodynamic treatment (reference 3, p. 56) shows that this is only valid when the cross-coefficients L_{jk} ($j \neq k$), corresponding to ion–ion interactions are neglected. If these are to be included, a more complex equation results.

Frictional coefficients

An alternative, still basically phenomenological, approach is that based on the idea that the driving force for diffusion can be linearly related to the relative velocity of each pair of components through a *frictional coefficient*. This is a characteristic concept in mechanics; such a coefficient has the dimension MT^{-1}. There are several possible formulations of this concept as applied to diffusion. Possibly the earliest was due to Onsager[14] who defined a matrix of *frictional coefficients* ζ_{jk} essentially as the inverse matrix of the phenomenological coefficients L_{jk} defined by equation (5). A flow density J_k is the product of the molar concentration c_k of component k and its velocity $\vec{u}_k$ in some reference frame. Onsager[14] therefore defined these frictional coefficients according to equation (23):

$$\operatorname{grad} \mu_j = -\sum_k \zeta_{jk} c_k (\vec{u}_j - \vec{u}_k). \tag{23}$$

The coefficients ζ_{jk} have the dimensions $L^3 M^{-1} T^{-1}$, i.e. dimensionally they are not frictional coefficients at all. The advantages of this formulation have been stressed particularly by Laity[15], and ratios of these frictional coefficients can be calculated for binary systems in terms of integrals involving intermolecular potentials and radial distribution functions[6,7]. Other formulations are shown in equations (24) and (25):

$$\operatorname{grad} \mu_j = -\sum_k r_{jk} N_k (\vec{u}_j - \vec{u}_k) \tag{24}$$

$$\operatorname{grad} \mu_j = -\sum_k \varphi_{jk} (\vec{u}_j - \vec{u}_k). \tag{25}$$

The coefficients r_{jk}, due to Klemm[16], and φ_{jk}, introduced by Lamm[17], are, dimensionally, frictional coefficients per mole; they bear much the same relation to one another as do the activity coefficient and the activity. All these frictional coefficients are independent of the reference plane, and, for an n-component system, $n(n-1)/2$ such coefficients are required.

The relation between mutual and self-diffusion coefficients for a binary system

These frictional coefficients defined by equations (23)–(25) are particularly useful in the discussion of the relation between the three diffusion coefficients which can be measured for a two component system. The mutual diffusion coefficient D_{01} has already been discussed. If, to a mixture of the two components, a small amount of one of these components 'labelled' in some way but otherwise identical, is added, the system becomes a three-component one for the purposes of the phenomenological description. The labelling can be achieved by the use of isotopic species, or by distinguishing the direction of the nuclear spins, as in the Carr-Purcell "spin-echo" experiment[18]. The rate at which a concentration gradient of the labelled species is dissipated in an otherwise uniform solution is measured by a diffusion coefficient which is not D_{01}; it is variously referred to as the tracer diffusion coefficient, the self-diffusion coefficient, or as Mills[19] has suggested, the intradiffusion coefficient. Here it will be termed the self-diffusion coefficient and denoted D_{jj} simply because the distinctions drawn by Mills between these terms, while valid, are unimportant in the present context. For an unlabelled binary system, the definitions (23)–(25) can be used to relate frictional coefficients to the mutual diffusion coefficient D_{01}. Using Onsager's definition (equation (23)), it is simple to show that:

$$D_{01} = \frac{\bar{v}RT}{\zeta_{01}} (\partial \ln a_1 / \partial \ln N_1)_{T,P} \qquad (26)$$

ζ_{01} can therefore be found from the experimental mutual diffusion coefficient provided that the thermodynamic factor is known. If labelled solute is added (denoted by suffix 1*), and the system treated as a ternary one, the mutual diffusion coefficient for components 1 and 1* can be regarded as the self-diffusion coefficient of component 1, D_{11}. Then

$$D_{11} = \frac{RT\bar{v}}{N_0\zeta_{01} + N_1\zeta_{1*1}} \qquad (27)$$

where $\bar{v}$ is the mean molar volume, and ζ_{1*1} is the frictional coefficient between labelled and unlabelled component 1; this is identified with ζ_{11}. Similarly for the other component,

$$D_{00} = \frac{RT\bar{v}}{N_0\zeta_{00} + N_1\zeta_{01}}. \qquad (28)$$

The two self-diffusion coefficients, while independent of the thermodynamic properties of the solution, involve two frictional coefficients while D_{01} involves only ζ_{01}. In principle, all three frictional coefficients can be found from the three experimental diffusion coefficients and the activity coefficients of the solution as a function of concentration. In practice, rather accurate experimental data are required if reliable values of frictional coefficients are to be obtained[20,21].

From equations (26)–(28), it is easy to show[20,22] that:

$$(N_0 D_{11} + N_1 D_{00}) \left(\frac{\partial \ln a_1}{\partial \ln N_1} \right)_{T,P} = D_{01} \left(2 - \frac{N_1 \zeta_{11} D_{11} + N_0 \zeta_{00} D_{00}}{\bar{v} RT} \right) \quad (29)$$

$$= D_{01} \left[\frac{N_0 \zeta_{01}}{N_0 \zeta_{01} + N_1 \zeta_{11}} + \frac{N_1 \zeta_{01}}{N_1 \zeta_{01} + N_0 \zeta_{00}} \right]. \quad (30)$$

This equation takes a particularly simple form if

$$\zeta_{01}^2 = \zeta_{00} \zeta_{11}. \quad (31)$$

When this is true,

$$D_{01} = (N_0 D_{11} + N_1 D_{00}) \left(\frac{\partial \ln a_1}{\partial \ln N_1} \right)_{T,P}. \quad (32)$$

Equation (32) is similar to a relation which can be derived from kinetic theory for dilute gas mixtures[23],

$$D_{01} = (N_0 D_{11} + N_1 D_{00}) \quad (33)$$

and identical with an equation, long-current, often somewhat incorrectly referred to as the "Hartley-Crank equation", and derived using unstated or physically obscure assumptions[12,20,22,24]. Any deviations from it for real systems will arise because the geometric mean rule, expressed by equation (31), is not obeyed. Such deviations could be calculated if the ratios $\zeta_{01}/\zeta_{00}, \zeta_{01}/\zeta_{11}$ could be found from statistical mechanics. Approximate calculations of this kind are only possible at present for simple molecular models. On the assumption that no volume change occurs on mixing, and that radial distribution functions are independent of concentration, the Kirkwood-Bearman[6] theory of transport processes leads to equations (34) and (35).

$$D_{00}/D_{11} = v_1/v_0 \quad (34)$$

$$D_{01} = D_{11} \left(\frac{\partial \ln a_1}{\partial \ln c_1} \right)_{T,P}. \quad (35)$$

Combination of these gives equation (32), and equations (34) and (35) must be consistent with the geometric mean rule (31) though not required by it. The self-diffusion coefficient of argon in pure liquid argon, and of krypton in liquid mixtures with argon have been measured[25], their ratio being 1.16. Bearman[26] has calculated the ratio of the molar volume of krypton to that of argon derived from the densities of the pure liquid as 1.10. The ratio of the diameter of the krypton molecule to that of the argon molecule lies between 1.03 and 1.05 depending on the method used for measuring the diameters. Evidently, equation (35) represents the data for this simple system better than any based on the Stokes-Einstein equation (equation (1)).

The validity of the geometric mean rule itself has been studied theoretically by Loflin and McLaughlin[22]. Frictional coefficient ratios were calculated from another statistical theory of transport processes, the Rice-Allnatt theory, for a two-component system where the intermolecular potential $\Phi_{jk}(R_{jk})$ for species j, k, takes the Lennard-Jones 12:6 form[27]:

$$\Phi_{jk}(R_{jk}) = 4\varepsilon_{jk}\left[\left(\frac{\sigma_{jk}}{R_{jk}}\right)^{12} - \left(\frac{\sigma_{jk}}{R_{jk}}\right)^{6}\right]. \tag{36}$$

It was found (m_{01} is the reduced mass) that:

$$\frac{\zeta_{00}}{\zeta_{01}} = \frac{\sigma_{00}\varepsilon_{00}}{\sigma_{01}\varepsilon_{01}} \cdot \frac{m_0}{m_{01}} \frac{D_{00}}{D_{00} + D_{11}}. \tag{37}$$

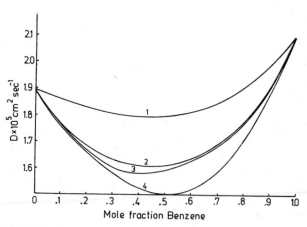

Figure 1 (from T. Loflin and E. McLaughlin, *J. Phys. Chem.*, **73**, 186 (1969)). Comparison of calculated and experimental values of the mutual diffusion coefficient of the benzene-*cyclo*-hexane system at 25°: 1, experimental; 2, Rice-Allnatt theory; 3, equation 30; 4, Vignes correlation (A. Vignes, *Ind. Eng. Chem., Fundam.*, **5**, 189 (1966))

With equation (37), it is possible to calculate the expression in brackets on the right-hand side of equation (30) in terms of the constants of the Lennard-Jones equation, and the self-diffusion coefficients. Hence, in principle, the mutual diffusion coefficient D_{01} could be found from the self-diffusion coefficients. This worked well for the *n*-octane-*n*-dodecane system at 25°. However, for this system, equation (32) applies equally well, and it must therefore obey the geometric mean rule closely. Equation (32) is not obeyed by the system benzene-*cyclo*hexane, but, in this case, the combination of equations (30) and (37) to calculate D_{01} gave no significant improvement (Fig. 1).

The product $\zeta_{00}\zeta_{11}$ can be calculated from eqation (37) and the similar one for ζ_{11}/ζ_{01}. If the usual combining rules for the Lennard-Jones parameters are used, namely:

$$\varepsilon_{01}^2 = \varepsilon_{00}\varepsilon_{11} \tag{38}$$

$$\sigma_{01} = \tfrac{1}{2}(\sigma_{00} + \sigma_{11}) \tag{39}$$

it follows that:

$$\zeta_{00}\zeta_{11} = 4\zeta_{01}^2 \, \frac{\sigma_{00}\sigma_{11}}{(\sigma_{00} + \sigma_{11})^2} \, (m_0 + m_1)^2 \, \frac{D_{00}D_{11}}{(D_{00} + D_{11})^2}. \tag{40}$$

Equation (40) reduces to the geometric mean rule for systems where the molecules have similar diameter and mass; these properties are associated with systems for which $D_{00} \sim D_{11}$. For other more complex binary systems, deviations from the geometric mean rule for frictional coefficients should provide a stringent test for theories of transport coefficients in view of the relative failure of the application of the Rice-Allnatt theory to the benzene-*cyclo*hexane system.

FRICTIONAL COEFFICIENTS AND VISCOSITY

Statistical theories of transport processes also yield expressions for the coefficient of viscosity. Detailed calculations are no more successful than are those of diffusion coefficients, and it would be helpful to include the viscosity coefficient, itself a kind of frictional coefficient, within the general phenomenological scheme. This has been done by Dullien[28] using the molar frictional coefficients φ_{jk} of equation (25). The following equations apply:

$$D_{01} = \frac{RTN_1}{\varphi_{01}}\left(\frac{\partial \ln a_1}{\partial \ln N_1}\right)_{T,P} = \frac{RTN_0}{\varphi_{10}}\left(\frac{\partial \ln a_1}{\partial \ln N_1}\right)_{T,P} \tag{41}$$

$$D_{00} = RT/(\varphi_{00} + \varphi_{01}) \tag{42}$$

$$D_{11} = RT/(\varphi_{11} + \varphi_{10}). \tag{43}$$

The coefficient φ_{jj} is defined as $(\varphi_{j*j} + \varphi_{jj*})$ where the asterisk denotes the labelled species, and it can be shown to be the frictional coefficient per two moles of species j[28]. An average molar frictional coefficient η^* can be defined as:

$$\eta^* = N_0 \left(\varphi_{01} + \tfrac{1}{2}\varphi_{00}\right) + N_1 \left(\varphi_{10} + \tfrac{1}{2}\varphi_{11}\right). \tag{44}$$

A value for η^* in terms of experimental quantities can then be found from equations (42), (43) and (44):

$$\eta^* = RT \left(N_0/D_{00} + N_1/D_{11}\right)/2. \tag{45}$$

It is noteworthy that this does not contain D_{01}.

The viscosity of a solution is also a measure of the internal friction of a liquid in laminar flow. Suppose that, in such a liquid, the average distance of momentum transfer perpendicular to the flow direction is δ, and that the velocity gradient in this direction is $d\vec{u}/dx$. Then the relative velocity of the planes across which momentum is transferred is $(\delta \, d\vec{u}/dx)$. It is possible to define, by analogy with equation (25), a frictional coefficient ξ for viscosity such that,

$$\text{Driving force for viscous flow} = \xi\delta \, d\vec{u}/dx. \tag{46}$$

Consider a cube of solution with base, area A, lying in one shear plane, and height δ. The volume $(A\delta)$ of this cube encloses $\Sigma \, n_j$ moles of all species. The average molar volume, $\bar{v}$, is therefore:

$$\bar{v} = A\delta/\Sigma n_j. \tag{47}$$

The conventional definition of the coefficient of viscosity η (Newton's Law) relates the applied shear stress, or the driving force for viscous flow per unit area, to the velocity gradient $d\vec{u}/dx$:

$$\text{Applied shear stress} = \eta \, d\vec{u}/dx. \tag{48}$$

Comparison of (46), (47) and (48) shows that:

$$\eta = \xi\delta/A = (\xi/\Sigma \, n_j) \, \delta^2/\bar{v}. \tag{49}$$

The quantity $(\xi/\Sigma n_j)$ appearing in equation (49) is an average frictional coefficient per mole. If it is identified with η^* as defined by (44),

$$\eta/\eta^* = \delta^2/\bar{v}. \tag{50}$$

For a pure component with self-diffusion coefficient D_{00}^0, it follows from equation (45) that,

$$\eta^* = RT/2D_{00} \tag{51}$$

and from equations (50) and (51) that,

$$2D_{00}^0\eta_0 = \delta^2 RT/v^0 \tag{52}$$

where η_0 and v^0 are respectively the viscosity and molar volume of the pure component. Equation (52) is consistent with equation (34) if δ^2/η is the same for the two liquids being compared.

Equation (52) should apply to self-diffusion in a gas. The elementary kinetic theory of gases shows that the average distance of momentum transport δ is $\frac{2}{3}$ of the mean free path λ. Using the simple kinetic theory equations for the self-diffusion coefficient, the viscosity, and the mean free path, and applying them to equation (52) shows[28] that:

$$\delta = \tfrac{2}{3}\lambda.$$

The assumption that $\xi/\Sigma n_j$ can be identified with η^* is therefore justified in this case. For a close-packed fluid, i.e. a liquid, δ must be related to the molecular radius r. Dullien[28] has for simplicity assumed that $r = 0.5\,(v^0/N)^{1/3}$. For a series of pure liquids, it is possible to calculate δ from equation (52) and to find the ratio δ/r. The results of such calculations are shown in Table 1.

Table 1   Values of the ratio δ/r as calculated by Dullien[28]. Data refer to room temperature except where otherwise stated

Liquid	η_0 (centipoise)	D_{00} (cm²sec⁻¹ × 10⁵)	δ (Å)	δ/r
Ethyl iodide	0.593	2.05	2.94	1.15
Butyl iodide	0.867	1.35	3.30	1.15
Nitromethane	0.670	2.72	2.84	1.27
Ethanol	1.090	1.02	2.30	1.00
Methanol	0.562	2.30	2.07	1.02
Water	0.897	0.52	1.84	1.19
Carbon tetrachloride	0.975	1.18	3.03	1.11
Bromethane	3.58	3.96	2.91	1.16
Benzene	0.605	2.18	3.08	1.16
Brombenzene	0.650	1.12	3.34	1.20
Argon (84.2°K)	2.80	2.07	2.15	1.24
Mercury	15.80	1.79	1.85	1.27

Since the liquids studied are isotropic δ should refer to the distance between molecular centres averaged over all possible distances, or $4r/\pi$ for a simple cubic lattice. Hence, if the liquid packing can be regarded as roughly equivalent to that in a simple cubic lattice, δ/r should be about 1.27. Table 1 shows that the experimental values are close to this, especially for the liquids with spherical molecules. The average value is about 1.2, and this result could form the basis of an empirical equation relating viscosity, self-diffusion coefficient, and molar volume. Similar calculations were carried out by Dullien for solutions, and he found that the values obtained lay between those for the pure components. This shows that frictional coefficients and viscosities must vary with concentration in a similar manner, a conclusion supported by the results of Schonert's calculations on the benzene-*cyclo*hexane system[21].

The best known of the equations relating the diffusion coefficient to viscosity is equation (1), the Stokes-Einstein equation. This was derived for a sphere moving in a continuum and is a limiting form corresponding to the case where there is no 'slip' in the hydrodynamic sense at the surface of the sphere. For small molecules in a molecular environment, Sutherland[9] suggested that 'slip' might occur easily and that the following equation might be obeyed more accurately:

$$D = kT/4\pi\eta r. \tag{53}$$

There is a good deal of evidence to suggest that this is so (ref. 3, p. 159, (29)). For a pure liquid with $r = 0.5 \, (v^0/N)^{1/3}$, $\delta/r = 1.2$, it can be shown[28] from equation (52), that,

$$D_{00} = \frac{kT}{3.8\pi\eta_0 r}. \tag{54}$$

Thus Dullien's concept of an average frictional coefficient leads to an equation almost identical with equation (53). The experimental data in Table 1 shows that δ/r varies slightly from one system to another, and it would be expected that δ/r would increase as the internal pressure of the liquid decreases. Although the form of equations (53) and (54) are almost identical, Dullien's treatment does not imply a constant numerical factor in equation (54). The value of this factor depends on δ/r, which in turn depends on the degree of expansion of the liquid*.

Frictional Coefficients as Molar Properties

In thermodynamic studies of non-ideal systems the comparison of the plot of some molar property against composition, against a similar plot for an ideal system can be extremely informative. The first problem in applying this

---

* For an extension of this treatment to multicomponent systems, see J. G. Albright, *J. Phys. Chem.*, **73**, 1280 (1969).

technique to diffusion processes is to establish a norm of "ideal behaviour". For this purpose, Klemm's formulation of frictional coefficients is convenient (equation 24). The three diffusion coefficients are related to the frictional coefficients r_{jk} as follows:

$$D_{01} = \frac{RT}{r_{01}} (\partial \ln a_1 / \partial \ln N_1)_{T,P} \tag{55}$$

$$D_{00} = \frac{RT}{N_0 r_{00} + N_1 r_{01}} \tag{56}$$

$$D_{11} = \frac{RT}{N_1 r_{11} + N_0 r_{10}}. \tag{57}$$

The coefficients r_{10}, r_{01} are, of course, equal.

From equation (57),

$$\text{Lim.} (D_{11}) = RT/r_{11}^* \tag{58}$$
$$\scriptstyle N_1 \to 1$$

$$\text{Lim.} (D_{11}) = RT/r_{10}^* \tag{59}$$
$$\scriptstyle N_0 \to 1$$

where the asterisk denotes frictional coefficients r_{jk} characteristic of the pure component k. Define an "equivalent random solution" with self-diffusion coefficients D_{kk}^0 represented by:

$$D_{kk}^0 = RT/(N_k r_{kk}^* + N_j r_{kj}^*) \quad (j, k = 0, 1; j \neq k) \tag{60}$$

and a mutual diffusion coefficient D_{01}^0 such that:

$$D_{01}^0 = N_1 D_{00}^0 + N_0 D_{11}^0 \tag{61}$$

$$= RT \left[\frac{N_1}{N_0 r_{00}^* + N_1 r_{01}^*} + \frac{N_0}{N_1 r_{11}^* + N_0 r_{10}^*} \right]. \tag{62}$$

All the three coefficients defined by equations (60)–(62) can be calculated from experimental data. Dullien[28] defined Lamm frictional coefficients φ_{01}^0, φ_{11}^0, φ_{00}^0 characteristic of the "equivalent random solution" as:

$$\varphi_{01}^0 = RTN_1/D_{01}^0 \tag{63}$$

$$(\varphi_{00}^0 + \varphi_{01}^0) = RT/D_{00}^0 \tag{64}$$

$$(\varphi_{11}^0 + \varphi_{10}^0) = RT/D_{11}^0. \tag{65}$$

Comparison of equations (62) and (63) shows that, in general, φ_{01}^0 is not a simple function of mole fraction, as is clearly shown from Dullien's calculations on benzene-alcohol mixtures (Fig. 3). It only becomes so if the geometric mean rule is obeyed, a restriction which is not made in Dullien's treatment. From equations (60), (64),

$$\varphi_{00}^0 + \varphi_{01}^0 = N_0 r_{00}^* + N_1 r_{01}^*. \tag{66}$$

Since φ_{01}^0 is generally a complicated function of composition, φ_{00}^0 will not normally be linear in mole fraction, and it will become equal to r_{00}^* at $N_0 = 1$.

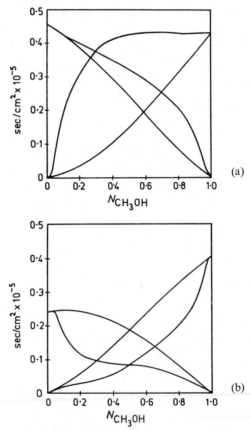

Figure 2   Lamm frictional coefficients for the benzene–methanol system as functions of composition (from F. A. L. Dullien, *Trans. Faraday Soc.*, **59**, 856 (1963)). (a) Upper curves φ_{jj}/RT, lower curves φ_{jj}^0/RT. (b) Upper curves φ_{ij}/RT, lower ucrv s φ_{ij}/RT. $\varphi_{11} = \varphi_{11}^0 = \varphi_{01} = \varphi_{01}^0 = 0$ when $N=0$ (MeOH = component 1)

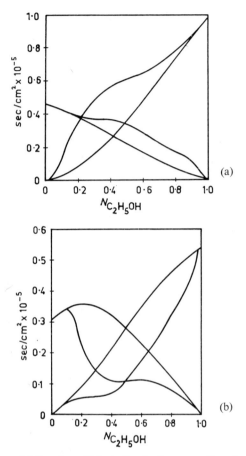

Figure 3 Lamm frictional coefficients for the benzene–ethanol system as functions of composition (from F.A.L.Dullien, *Trans. Faraday Soc.*, **59**, 856 (1963)). (a) Upper curves φ_{jj}/RT, lower curves φ_{jj}^0/RT. (b) Upper curves φ_{ij}^0/RT, lower curves φ_{ij}/RT. $\varphi_{11} = \varphi_{11}^0 = \varphi_{01} = \varphi_{01}^0 = 0$ when $N_1 = 0$ (EtOH = component 1)

Dullien has calculated (in the present notation) φ_{00}^0 and φ_{11}^0, and φ_{10}^0 and φ_{01}^0 as functions of mole fraction. These plots, together with the equivalent experimental values of the Lamm frictional coefficients, also calculated by Dullien, are shown in Figs. 2 and 3 for benzene–methanol and benzene–ethanol mixtures respectively. As methanol is added to benzene φ_{11}, increases very much more rapidly than φ_{11}^0, and φ_{10} decreases more rapidly than φ_{10}^0. There are therefore more alcohol–alcohol neighbours and fewer benzene–alcohol neighbours than would be expected from the "equivalent random solution". In these dilute solutions of methanol in benzene, φ_{00} is almost identical with φ_{00}^0, i.e. the number of benzene–benzene neighbours essentially that expected for the model. The almost constant value of φ_{11} (methanol = component 1) in the concentrated methanol solutions suggests that any more alcohol added at this stage goes into existing large clusters of alcohol molecules. The deviation of φ_{00} from φ_{00}^0 is in the same sense but occurs to a less extent; the clustering of the benzene molecules is forced to occur because of the positive tendency of the alcohol molecules to associate together. The ethanol–benzene system behaves similarly but the deviations from the model are smaller, i.e. ethanol clusters are less readily formed in benzene solution.

The most interesting aspect of this treatment is that the conclusions reached are very similar to those obtained from quite different experimental data. The thermodynamic activity coefficients of the benzene–methanol system show that, at low temperatures, these mixtures are very close to showing an upper consolute temperature[30], while this is not true for benzene–ethanol mixtures. In addition, the shapes of the excess heat of mixing curves can be interpreted most simply[31] by assuming that the addition to benzene to the alcohol breaks down the hydrogen-bonded structure more readily in ethanol than in methanol. Comparison of the thermal diffusion coefficients found for these two systems[30] can also be interpreted in this way. In dilute solutions of alcohol in benzene the separation of the components in a thermal gradient depends mainly on the shift in the monomer-oligomer equilibrium, and the alcohol migrates to down the temperature gradient to the region where the oligomers are most stable. In a solution of benzene in excess alcohol, the direction of separation is governed almost entirely by the heat of transfer Q^* of the benzene in the associated solvent. This has a sign such that the benzene now migrates down the temperature gradient. Thus at some composition the sign of separation is changed (Fig. 4). This change, characteristic of the transition from benzene dissolved in an associated alcohol solvent, to mainly monomeric alcohol dissolved in benzene, occurs at a mole fraction of benzene of 0.66 for solutions of methanol, and of 0.32 for solutions of ethanol. Evidently benzene is more efficient at breaking down the associated structure of ethanol than that of methanol, which is just the conclusion reached on the other evidence including the analysis of the diffusion data. In view of this consistency, further study of this method of analysing the diffusion coefficients of binary systems seems to be desirable, particularly with a view to simplifying the reference model. If, in addition to the restrictions represented by equations (60) and (61), the geometric mean rule is adopted, the reference coefficient φ_{jk}^{0} become linear functions of mole fraction. This would certainly simplify the calculations without apparently detracting from the usefulness of the method.

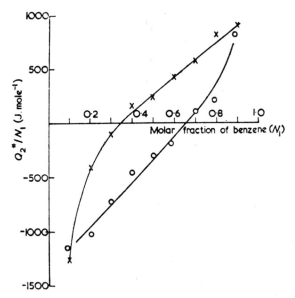

Figure 4 Variation of the ratio of heat of transfer of the solute to the molar fraction of solvent (Q_2^*/N_1) as a fraction of concentration for mixtures of benzene with primary alcohols at 25° (X, ethanol; O, methanol) (from G. Farsang and H. J. V. Tyrrell, *J. Chem. Soc. (A)*, 1839 (1969))

REFERENCES

1. A. Fick, *Pogg. Annalen*, **94**, 59 (1855).
2. H. J. V. Tyrrell, *J. Chem. Educ.*, **41**, 397 (1964).
3. e.g. H. J. V. Tyrrell, *Diffusion and Heat Flow in Liquids*. Butterworth Scientific Publications, London, 1961, p. 170.
4. H. S. Harned and J. A. Shropshire, *J. Amer. Chem. Soc.*, **80**, 2618, 2967 (1958).
5. e.g. R. A. Robinson and R. H. Stokes, *Electrolyte Solutions*. Butterworth Scientific Publications, London, 1959. Chapter 11.
6. R. J. Bearman, *J. Phys. Chem.*, **65**, 1961 (1961), and earlier papers cited therein.
7. S. A. Rice and A. R. Allnatt, *J. Chem. Phys.*, **34**, 409 (1961).
8. cf. R. Zwanzig, *Ann. Rev. Phys. Chem.*, **16**, 67 (1965).
9. W. Sutherland, *Phil. Mag.*, **9**, 781 (1905).

10. S. Glasstone, K. J. Laidler, and H. Eyring, *The Theory of Rate Processes*. McGraw-Hill. New York, 1941.

11. J. G. Kirkwood, R. L. Baldwin, P. J. Dunlop, L. J. Gosting and G. Kegeles, *J. Chem. Phys.*, **33**, 1505 (1960).

12. e.g. G. S. Hartley and J. Crank, *Trans. Faraday Soc.*, **43**, 801 (1949); S. Prager, *J. Chem. Phys.*, **21**, 1344 (1953); P. C. Carman and L. H. Stein, *Trans. Faraday Soc.*, **52**, 619 (1956).

13. R. Mills, *J. Phys. Chem.*, **67**, 600 (1963).

14. L. Onsager, *Ann. N.Y. Acad. Sci.*, **46**, 241 (1945).

15. R. W. Laity, *J. Phys. Chem.*, **63**, 80 (1959).

16. A. Klemm, *Z. Naturf.*, **8a**, 397 (1953).

17. O. Lamm, *Acta Chem. Scand.*, **6**, 1331 (1952).

18. H. Y. Carr and E. M. Purcell, *Phys. Rev.*, **94**, 630 (1954); D. C. Douglass and D. W. McCall, *J. Phys. Chem.*, **62**, 1102 (1958).

19. J. G. Albright and R. Mills, *J. Phys. Chem.*, **69**, 3120 (1965).

20. H. J. V. Tyrell, *J. Chem. Soc.*, 1599 (1963).

21. H. Schonert, *J. Phys. Chem.*, **73**, 752 (1959).

22. T. Loflin and E. McLaughlin, *J. Phys. Chem.*, **73**, 186 (1969).

23. J. E. Mayer and M. G. Mayer, *Statistical Mechanics*, Wiley. New York, 1940, p. 30.

24. R. M. Barrer, *J. Phys. Chem.*, **61**, 178 (1957).

25. G. Cini-Castagnoli and F. P. Ricci, *Nuovo Cimento*, **15**, 795 (1960).

26. R. J. Bearman, *J. Phys. Chem.*, **66**, 379 (1962).

27. e.g. J. A. Barker, *Lattice Theories of the Liquid State*. Pergamon. London, 1963. Chapter 4.

28. F. A. L. Dullien, *Trans. Faraday Soc.*, **59**, 856 (1963).

29. A. D. Osborne, H. J. V. Tyrrell and M. Zaman, *Trans. Faraday Soc.*, **60**, 395 (1964).

30. G. Farsang and H. J. V. Tyrrell, *J. Chem. Soc. (A)*, 1839 (1969).

31. J. S. Rowlinson, *Liquids and Liquid Mixtures*. Butterworths Scientific Publications. London, 1959. p. 178 ff.

Self-diffusion in liquid gallium metal

S. LARSSON and A. LODDING

Physics Department, Chalmers University of Technology
Gothenburg, Sweden

ABSTRACT

A new capillary method for diffusion measurements in liquids permits several successive readings on a sample without interruption of anneal. Errors from solidification and sectioning are avoided, and the reproducibility of the measured diffusion coefficients is kept within $\pm 2.5\%$.

The technique has been applied to the diffusion of ^{72}Ga in liquid gallium between 20° (supercooled) and 283 °C. Below ca 70 °C the Arrhenius plot yields an 'activation energy' of about 1.1 kcal/mole, in good agreement with an earlier investigation of Petit and Nachtrieb. However, at higher temperatures the slope rises and at 280 °C the 'activation energy' is about 2.1 kcal/mole. This confirms predictions from isotope thermotransport experiments. The high temperature region is best represented as linear with absolute temperature, by $10^5 D = 2.3 \times 10^{-2}$ (T-250). Below 70 °C the corresponding approximation is $10^5 D = 1.0 \times 10^{-2}$ (T-130). A fair agreement is found with the Ascarelli and Paskin theory based on the hard-sphere approximation, on the Percus-Yevick equation of state and on the empirically derived structural similarity of all simple liquids at their melting points. However, for Ga the agreement applies only if the 'scaling' temperature lies some 50° below the m.p.

Another theoretical approach, based on small density fluctuations, fits reasonably well in the high temperature region; however, especially at lower temperatures, this treatment may have to be combined with a consideration of 'free volume'.

INTRODUCTION

The self-diffusion of liquid gallium has been measured once before, by Petit and Nachtrieb[1]. This study, conducted over a rather narrow range of temperature (30–100 °C) yielded the result $D = 1.1 \times 10^{-4} \exp(-1120/RT)$ (cm^2/sec).

Several reasons motivated a new investigation of gallium self-diffusion:

a   The interpretation[2] of recent measurements at this laboratory of isotope thermotransport in liquid metals[3] led to the conclusion that the ratio of
the mean displacement to the diameter of the diffusing species should be
about 0.1. While this was clearly implied by the results in Li, K, Rb and In,
gallium suggested a much smaller ratio. To reach agreement with the other
metals, the mean activation energy of Ga self-diffusion in the range 60–500 °C
would have to be, not 1.12 kcal/mole, but about twice that figure. We therefore suspected that the temperature dependence of D might increase greatly
above the temperatures of the Petit-Nachtrieb investigation.

b   All liquid metals show reasonable agreement between the temperature
behaviour of self-diffusion and that of viscosity, via a Stoke-Einstein type
relation[4]. Again gallium is an exception: according to experiments T/η exhibits an activation energy close to 2 kcal/mole.

c   The understanding of liquid diffusion requires accurate measurements
within wide ranges of temperature. From this point of view gallium is ideal.
The new technique developed at this laboratory permits precise determinations of liquid diffusivity up to about 580 K. This corresponds to
$(T_{max} - T_{min}) (T_{min}^{-1} - T_{max}^{-1}) \simeq 0.53$ for gallium, which can be supercooled.
This compares favourably with the widest range studies hitherto, those in
Hg[5], In[6], Sn[7], K[8] and Li[9], for which the corresponding figures are 0.47, 0.85,
0.82, 0.24 and 0.21, respectively.

d   Liquid gallium can be easily handled in glass apparatus, is not toxic or
volatile, and presents minimal radioactive hazard. It is therefore an especially
suitable metal for testing a new experimental technique.

EXPERIMENTAL

The technique is in principle an adaptation for liquids of the classical thin
layer method used for solid diffusion studies. A small drop of radioactive
metal is introduced, by aid of a syringe, at the closed end of a precision bore
glass capillary, 0.8 mm i.d. and about 10 cm in length. The capillary is then
filled, by means of a vacuum procedure, with inactive metal. The diffusivity
is about 4 powers of 10 greater in the liquid than in the solid. The penetration
depth is proportional to $D^{1/2}$, and so 0.5 mm of active liquid at the end of
a 10 cm long inactive column corresponds to a surface layer of only about
5 micron thickness on the butt end of a solid cylinder of 1 mm depth. The

solution of the diffusion equation is therefore the same as in the solid analogy (see, e.g., ref. 10), and the activity along the capillary, x cm from the
closed end at time t, varies according to

$$I = \text{const. exp}\,(-x^2/4Dt).\tag{1}$$

The diffusion coefficient D can therefore be evaluated with good precision by
measuring the slope of the straight line in the log vs. x^2 plot at any time t.

The activity was recorded by means of a scintillation crystal and a single-
channel analyser. The crystal was placed (see Fig. 1) on one side of a narrow

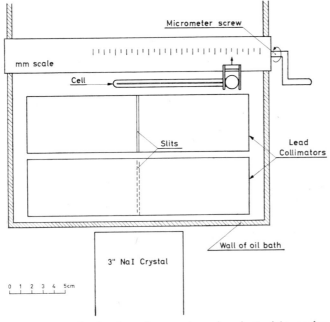

Figure 1   Schematic drawing of apparatus, viewed at right angles to
capillary

collimator slit in a lead block. The capillary was held horizontally in a carefully controlled temperature bath on the other side of the slit, and could be
moved by measurable steps perpendicular to the slit. By this arrangement
the variation of I with x could be recorded on one sample at several times t,
before the geometrical conditions of a "semi-infinite" diffusion column were
infringed. At each x the background activity was efficiently subtracted from
the total count, by lowering a lead screen between the capillary and detector

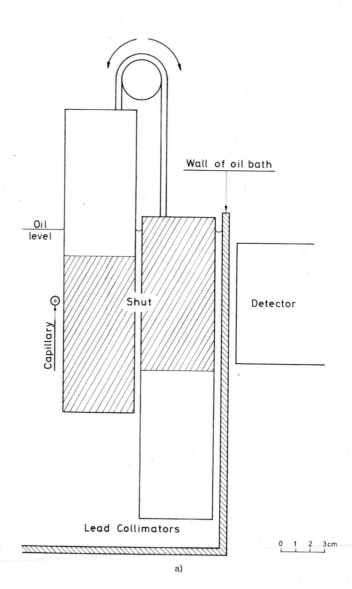

Wall of oil bath

Oil
level

Capillary

Shut

Detector

Lead Collimators

0   1   2   3cm

a)

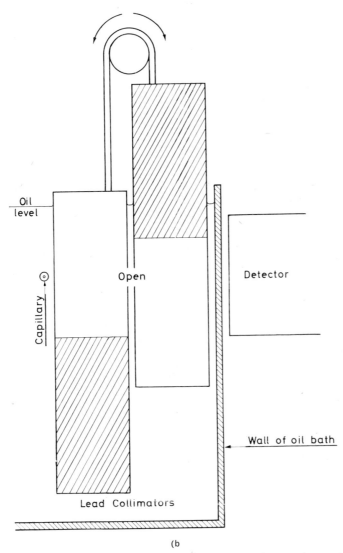

Figure 2 View along capillary, showing screening curtain

and making a second reading (see Fig. 2). Certain corrections to D had to be made to account for the finite width of the collimator slit. These corrections, which are discussed in a separate publication[11] were facilitated by making successive readings during one anneal. The great advantages of the method

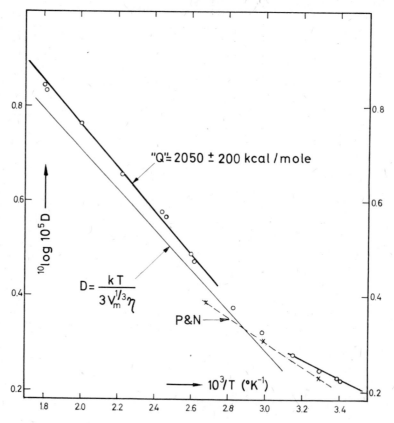

Figure 3 Self-diffusion coefficient of molten gallium, Arrhenius plot. Crosses denote points from an earlier investigation[1]. Also shown is diffusivity as computed from experimental viscosity

arise from the avoidance of heating, cooling and solidifying the metal column at any stage of the anneal; the convective effects of contraction and elongation of the column is known to influence the apparent diffusion coefficient[12]. The knowledge of the penetration profile at all times is also a great improvement over commonly used methods of liquid diffusion study[13,14]. While the

Anderson-Saddington technique, e.g. applied to liquid indium[6] gave a re-producibility of some 10%, the present method allows about 2.5% and is still being improved.

A detailed description of the apparatus appears elsewhere[11].

RESULTS

The self-diffusion coefficients of liquid gallium are plotted logarithmically vs. inverse temperature in Fig. 3 and linearly against T in Fig. 4. The results of the earlier investigation[1] are also shown. It is obvious that the agreement with the earlier data is very good. A certain discrepancy occurs at the highest of the three temperatures of the earlier study. This is probably due to the fact that while two or three separate anneals were made at the two lower temperatures, only one was made at the highest; the scatter in the earlier study

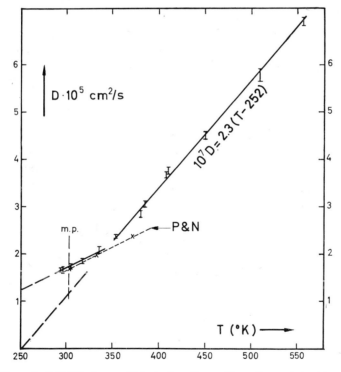

Figure 4   Self-diffusion coefficient of molten–gallium, linear plot vs. temperature. Crosses denote results of the investigation of Petit and Nachtrieb[1]

was of sufficient order of magnitude to align even the highest point with our present results.

The conspicuous feature of our data is the rapid increase in slope above ca 70 °C, while above 130° or 140 °C the slope appears to approach a constant value. In the Arrhenius plot an "activation energy" of about 2.1 kcal/mole is attained, while in the linear diagram the high temperature results are well represented by a straight line, $D = 2.3 \times 10^{-7}$ (T-252) (cm^2/sec). The corresponding approximation for the low temperature portion is $D = 1.0 \times 10^{-7}$ (T-130).

The fact that the results, at least in the high temperature region, are well represented by a linear relationship, has also been noted for other metals[15]. Several attempts have been made to derive theoretically a linear dependence of T on D[15,16]. However, it should be remembered that extrapolation to the origin cannot be made. For all high temperature characteristics there is a sizable intercept on the T-axis, for Ga about 250 K, for other metals[17] a value just under the melting point.

An increase in "activation energy" with temperature has been also observed in the most precise of earlier investigations of liquid diffusivity, and is qualitatively predicted by all hitherto postulated theories. However, the abrupt increase observed by us in gallium, from 1 to 2 kcal/mole in less than 200°, is particularly striking.

That the curvature is especially strong at the lower temperatures is particularly well demonstrated by the present results. An analysis of the results for other metals[15,17] does suggest the same behaviour, but hitherto the accuracy has not been sufficient to allow definite conclusions to be reached.

In Fig. 3 a plot has also been made of the diffusivity as calculated from the experimental viscosity[18] via a modified Stokes-Einstein relation, $D\eta = $ const. T. It is seen that the present results give much better agreement than did the earlier results of Petit and Nachtrieb[2].

The predictions based on thermotransport studies[3] have been confirmed by the present high temperature data. Substituting the "activation energy" value of 2.2 kcal/mole into the theoretical evaluation[2,19] of liquid thermotransport, from the gallium studies at a mean temperature of about 300 °C one can derive the value of 0.096 ± 0.012 for the ratio of mean displacement length to particle diameter. This is of the same order as for other metals, and has a bearing on the following discussion of liquid diffusion.

DISCUSSION

Several theories of liquid diffusion, based on widely differing premises, have been developed during recent years. They have undoubtedly contributed to the understanding of liquid structure. Thus it has become recognised[20,21] that diffusive motion in liquids is not activated in the solid state sense, and that a high degree of cooperation exists between neighboring atoms. Most recent workers[15,21-23] have suggested that the mean displacements are probably considerably smaller than interatomic distances, and that the "jump" frequencies may approach those of thermal oscillations. These views have been supported by experimental evidence, such as activation volume measurements[24], thermotransport[3] and neutron scattering[25].

In spite of increasing insight into the basic problem no particular theory has hitherto been able to predict the diffusivity and its temperature dependence quantitatively for any group of liquids. Most theories contain arbitrarily adjustable, not readily measurable parameters.

Possibly the most successful recent theory has been presented by Ascarelli and Paskin[26]. It is based on a dense gas hard sphere approximation, on the Percus-Yevick equation of state, and on the empirical conclusion from diffraction studies that all simple liquids at their melting points show about the same packing fraction. The equation describing the self-diffusion coefficient becomes

$$D = 0.14 d_m \, (\pi kT/M)^{1/2} \, (\xi_m/\xi)^{1/2} /[(10 T_m \varrho/T \varrho_m) - 1] \qquad (2)$$

where d is the atomic diameter, ξ the packing fraction, M the atomic mass k Boltzmann's constant, and T the temperature. The index m denotes "at the melting point". ϱ is the density.

This equation has the virtue of avoiding arbitrary parameters, except the empirical and probably approximate $\xi_m = 0.45$, and a minor back scattering correction. The formula is also satisfactory in expressing the diffusion coefficients at the melting points in the form $D_m = \text{const.} \, (T d_m^2/M)^{1/2}$ with the constant rather closely fitting all available experimental data. The same relation and same agreement has been obtained earlier[27], simply by assuming that (a) in non-activated displacement a particle will move into an adjacent void with the mean thermal velocity and (b) the mean void size is the same for all metals at their melting points (again as suggested by diffraction data, expansion on melting and other evidence). One can then write

$$D_m = \frac{1}{6} \left(\frac{3kT_m}{M} \right)^{1/2} d_m \cdot \beta \qquad (3)$$

where the factor β is the same for all metals. This formulation[27] is at the melting point analogous toe quation 2. X-ray diffraction data[28] indicate that the free volume at the m.p. in liquid metals is about 10%, so that, if diffusive motion is cooperative, the mean displacement is about 0.1 d_m. This is supported by the interpretation of the thermotransport experiments[2]. Finally, when $\beta = 0.1$ is substituted into equation (3) it is found, see Table 1, that indeed all available experimental D_m data are well represented. Thus self-diffusion results imply the same thing as do thermotransport results.

Table 1   Theoretical and experimental self-diffusion coefficients at the melting points of ten liquid metals

Metal	α	$D_{0_{theor}}$ (cm^2/sec)	$D_{0_{exp}}$ (cm^2/sec)
Ag	2.3	2.6	2.7
K	2.3	3.9	3.8
Na	2.3	4.2	4.2
Li	2.4	6.9	6.6
Zn	2.6	2.7	2.6
Cd	2.8	2.0	2.1
Ga	3.1	1.7	1.7
Sn	3.5	1.9	2.1
In	3.7	1.8	1.8
Hg	4.0	0.9	1.0

One can note that, with the Percus-Yevick general equation for $\xi(T)$, equation (2) can be made into a D/D_m versus T/T_m relation applicable to all simple liquids. This characteristic is plotted in Fig. 5. As a description of the behaviour of D in gallium, it appears barely acceptable. How, the melting properties of Ga are rather anomalous; there is a contraction of 3.1% on melting; the heat of melting L is more than $2RT_m$; and the metal supercools easily. One may question the justification of accepting T_m as a reducing temperature for Ga in equation (2), as the melting point is a property of the solid rather than the liquid. It may be more correct to replace T_m by for example, the lowest supercooling temperature T_0, a fairly good fit is obtained if $T_0 = 250$ K.

However, the prediction[26] of the temperature dependence of D is less successful for other metals, as is seen from the comparison in Fig. 5 of the experimental diffusivities of K[8] and of Hg[5]. Structural differences between different classes of metal must probably be taken into account.

Our current approach to the liquid diffusion problem[17] continues the reasoning suggested in connection with equation (3) above. The premises are: (a) to remain liquid, a metal (or another simple substance) needs a free volume of at least 1/12 of the molar volume. This is suggested by many diverse experiments and implies that in close-packed order every atom has one neighbour site free. (b) The heat of fusion is expended to attain this geometry. (c) The share of thermal energy available for diffusion is determined

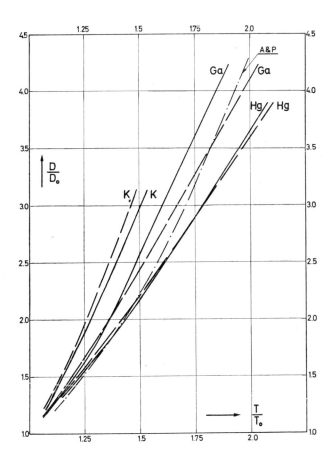

Figure 5 Comparison of experimental temperature dependence of *D* in Ga, K[8] and Hg[5] with theory. A & P denotes the predictions of the Ascarelli-Paskin[26] formulation, the other dashed lines are plotted from equation (5). The fully drawn lines are experimental results

by cooperative activation of groups of atoms. (d) Mean displacement is given by the minimum free volume obtained in melting, plus the whole increase above the freezing point of the amplitude of thermal agitation.

These conditions lead to

$$D = \frac{6\sqrt{2}}{\alpha} \frac{kT + L/3}{\sqrt{MK}} \exp\left(-\frac{6}{\alpha} \frac{T_0}{T}\right) \tag{4}$$

where K is the force constant of compressibility, L the heat of fusion and α the ratio of the atomic radius to the mean thermal amplitude at T_0, such that $4\alpha^2 (3kT_0 + L) = \frac{1}{2} Kd_m^2$. Table 1 lists approximate values of α for ten liquid metals.

It can be seen, that for temperatures above ca. $6T_0/\alpha$ equation (4) becomes nearly linear in T. Also, for the temperature dependence one obtains

$$D/D_0 = \left(\frac{T}{T_0} + \frac{L}{3kT_0}\right) \bigg/ \left(1 + \frac{L}{3kT_0}\right) \exp\left[\left(\frac{6}{\alpha}\right)\left(1 - \frac{T_0}{T}\right)\right]. \tag{5}$$

This relation is plotted in Fig. 5 for three metals, and compared with experiment. The agreement for K, Ga and Hg appears acceptable. For gallium again the agreement is further improved, if the melting point is replaced by 250 K for T_0. Indeed, if the data of any of the order metals in Table 1 are plotted in Fig. 5, one finds that the systematics follow α as required by theory. One might therefore conclude at this still semi-empirical stage of the theoretical treatment, that the assumptions employed were reasonable.

ACKNOWLEDGEMENTS

This work has been supported by a grant from the Swedish Natural Science Research Council. Fil. Kand. C. Roxbergh and ing. H. Olsson have rendered valuable assistance at several stages of the project.

REFERENCES

1. J. Petit and N. H. Nachtrieb, *J. Chem. Phys.* **24**, 1027 (1956).
2. A. Lodding, *Z. Naturf.* **21a**, 1348 (1966).
3. A. Lodding and A. Ott, *Z. Naturf.* **21a**, 1344 (1966).
4. H. A. Walls and W. R. Upthegrove, *Acta Met.* **12**, 461 (1964).
5. R. E. Meyer, *J. Phys. Chem.* **65**, 567 (1961).
6. A. Lodding, *Z. Naturf.* **11a**, 200 (1956).
7. C. H. Ma and R. A. Swalin, *J. Chem. Phys.* **36**, 3041 (1962).
8. J. Rohlin and A. Lodding, *Z. Naturf.* **17a**, 1081 (1962).

9. A. Ott and A. Lodding, *Z. Naturf.* **20a**, 1578 (1965).
10. P. G. Shewmon, *Diffusion in Solids*, McGraw-Hill, New York (1963).
11. S. Larsson and A. Lodding (to be published in *Z. Naturf.*).
12. A. Nordén-Ott and A. Lodding, *Z. Naturf.* **22a**, 215 (1967).
13. J. S. Andersson and K. Saddington, *J. Chem. Soc.*, *Suppl.* 381 (1949).
14. G. Careri, A. Paoletti and M. Vicentini, *Nuovo Cim.* **10**, 1088 (1958).
15. N. H. Nachtrieb, *Adv. Phys.* 309 (1967).
16. R. J. Reynik, *Appl. Phys. Letters* **9**, 239 (1966).
17. A. Lodding (to be published).
18. *Liquid Metals Handbook*, Navexos P 733 (Rev.), p.42 and 43 (1954).
19. A. Lodding, *Adv. Chem. Series*, **89**, 264 (1969).
20. M. H. Cohen and D. Turnbull, *J. Chem. Phys.* **31**, 1164 (1959).
21. R. A. Swalin, *Acta Met.* **7**, 736 (1959).
22. R. A. Swalin, *Z. Naturf.* **23a** (1968).
23. N. H. Nachtrieb and S. A. Rice, *Adv. Phys.*, 324 (1967).
24. N. H. Nachtrieb, *Liquid Metals and Solidification*, p.49, Amer. Soc. Met., Cleveland (1958).
25. P. A. Egelstaff, *Adv. Phys.* **11**, 203 (1962).
26. P. Ascarelli and A. Paskin, *Phys. Rev.* **165**, 223 (1968).
27. A. Ott and A. Lodding, *Z. Naturf.* **20a**, 1578 (1965).
28. K. Furukawa, *Sci. Rep. Tohoku Univ.*, *Ser. A.*, **12**, 368 (1960).

2.3

The study of diffusion processes in fluids using the pulsed magnetic field-gradient, pulsed nuclear magnetic resonance technique

K. J. PACKER, C. REES and D. J. TOMLINSON

School of Chemical Sciences, University of East Anglia,
Norwich, Norfolk

ABSTRACT

An introduction is given to the principles involved in the pulsed magnetic field-gradient spin echo method for studying diffusion in fluids. Some of the apparatus required for the experiment is discussed and illustrated and a comparison is made between spin echo and other methods for studying diffusion and between the steady gradient and pulsed gradient spin echo techniques. Examples are given of the measurements of self-diffusion coefficients in simple binary systems and of the study of restricted diffusion in a variety of materials.

INTRODUCTION

The effects on a spin-echo pulsed nuclear magnetic resonance (NMR) experiment of the translational diffusion of the spins in an inhomogeneous external magnetic field were first noted by Hahn[1] and then treated further by Carr and Purcell[2]. Since that time the self-diffusion coefficients* of many molecular species have been studied by this method[3-10]. Recently Stejskal

---

* With McCall and Douglass[8] we prefer to use the term self-diffusion to include the diffusion of a particular molecular species under conditions of uniform macroscopic concentration whether in a single or a multicomponent system.

et al.[11-17] have extended the technique to make it capable of measuring smaller diffusion coefficients and to make the study of restricted and anisotropic diffusion simpler.

This paper is intended to serve as an introduction to these newer techniques and to discuss and illustrate their use in studying diffusion in a variety of systems.

THE PULSED FIELD-GRADIENT (PFG) SPIN ECHO EXPERIMENT

The innovation evaluated by Stejskal and his co-workers was that of using time varying magnetic field-gradients to label the positions of the nuclear magnetic moments rather than a constant gradient. The particular form of time varying gradient investigated in detail was that of rectangular pulses.

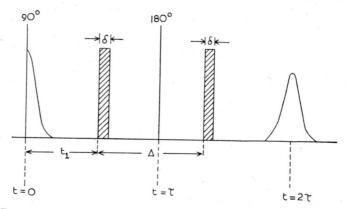

Figure 1 The sequence of events in the pulsed field-gradient spin echo method for the study of self-diffusion

The sequence of events in the PFG spin echo experiment is shown in Fig. 1. At time $t = 0$ a pulse of electromagnetic radiation of the appropriate frequency, amplitude and polarisation is applied to the nuclear spin in the sample, which is in a strong magnetic field H_0 about which the spins precess at a frequency $\omega_0 = -\gamma H_0$. The effective part of the electromagnetic radiation is the magnetic vector $H_1(\omega_0)$ which is rotating with ω_0 in the plane perpendicular to H_0. The effect of this pulse, shown in Fig. 2 in a frame of reference rotating at ω_0 about the $H_0(Z)$ direction, is to turn the resultant nuclear magnetisation, which is the vector sum of the individual nuclear moments, through an angle $\theta = \gamma |H_1| t_p$, where t_p is the duration of the pulse. The pulse at $t = 0$ is chosen so that $\theta = 90°$ which corresponds to equalising the populations of the two allowed energy states (for $I = \frac{1}{2}$ nuclei) and to bringing the spins into phase with each other along the y' axis (see Fig. 2: x' and y' are the x and y axes in the rotating frame). As time goes on

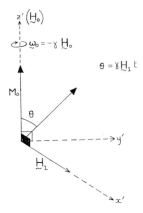

Figure 2 The effect of the $H_1(\omega_0)$ field on the equilibrium nuclear magnetisation shown in a frame of reference rotating about H_0 with frequency

$$\omega_0 \; (= -\gamma H_0)$$

the spins get out of phase in the $x'y'$ plane due to inhomogeneities in H_0 which produce a distribution of precession frequencies in the sample about the mean value ω_0. The NMR signal following the 90° pulse thus dies away. At time t_1 a pulse of linear, homogeneous magnetic field-gradient of duration δs and amplitude g gauss cm^{-1} is applied to the sample. The effect of this gradient pulse and subsequent H_1 and gradient pulses on the phase angles of the spins in the $x'y'$ plane and the effect of diffusion of the spins is shown schematically in Fig. 3*. The field gradient pulse at time t_1 gives to each spin a phase angle shift in the $x'y'$ plane which depends in a linear manner on the spin's coordinate along the gradient direction (see Fig. 3b), provided δ is small. This pulse serves to label the positions of the spins along the gradient direction at this time. At a time $\tau > t_1$ a second H_1 pulse is applied to the spins such that $\theta = 180°$. Fig. 3c shows that the effect of this pulse is to turn each component of the nuclear magnetisation in the $x'y'$ plane (of which only three are shown in Fig. 3) through 180° about the H_1 axis (x' in this case). At the time $t = t_1 + \Delta$ a second field-gradient pulse is applied,

---

* The effects of the inhomogeneities in H_0 are not shown for the sake of clarity.

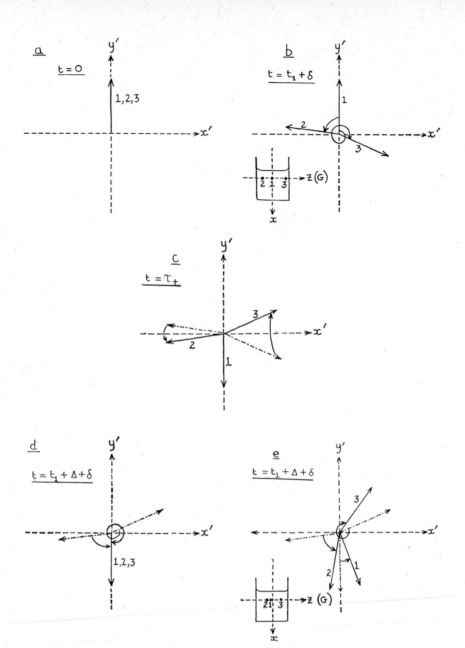

Figure 3 The effects of the $H_1(\omega_0)$ and magnetic field-gradient pulses on the nuclear magnetisation during the PFG experiment a) 90° H_1 pulse at $t = 0$; b) the first field-gradient pulse at $t = t_1$. The small diagram shows the position of the three representative spins in the gradient direction; c) 180° H_1 pulse at $t = \tau$; d) the second field-gradient pulse at $t = t_1 + \Delta$ when the spins have not diffused; e) as for d) except that the spins have diffused in the gradient direction. The new positions are shown in the small diagram

identical in all respects to the first. Fig. 3d indicates that if the spins have not diffused in the gradient direction in the time Δ then the second field-gradient pulse, together with the 180° $\mathbf{H}_1$ pulse at $t = \tau$, exactly reverses the phase shift produced by the first field-gradient pulse for every component of the nuclear magnetisation. In a real situation the components come into phase again to produce a spin echo at $t = 2\tau$. This echo will be attenuated relative to the amplitude of the NMR signal immediately following the 90° $\mathbf{H}_1$ pulse by the natural transverse relaxation processes of the nuclei, the effects of which are described by a relaxation time, T_2. We shall not concern ourselves with these except to note that they can limit the values of τ and Δ which may be used. Fig. 3e shows the effect of the second field-gradient pulse when the spins have diffused to new positions in the gradient direction. Each spin in this case receives a phase angle shift which is different from that it received at the time of the first gradient pulse and this can be seen to result in an incomplete refocussing of the components of the nuclear magnetisation at $t = 2\tau$. The echo amplitude is thus attenuated by the diffusion. For spins undergoing unrestricted diffusion Stejskal and Tanner[11] showed that

$$\ln \frac{E\,(2\tau)^*}{E\,(2\tau)} = \ln R = -\gamma^2 g^2\, \delta^2 D \left(\Delta - \frac{1}{3}\delta \right) \qquad (1)$$

where $E\,(2\tau)^*$ is the echo height at time 2τ in the presence of the gradient pulses and $E\,(2\tau)$ is the echo height obtained without the field-gradient pulses.

In this context unrestricted diffusion may be defined as that which satisfies the probability function:

$$P\,(x_0;\,x,\,t) = (4\pi Dt)^{-1/2} \exp - \left[\frac{(x - x_0)^2}{4Dt} \right] \qquad (2)$$

for all relevant values of t. $P\,(x_0;\,x,\,t)$ is the probability that a spin initially at position x_0 will have a position x after time t and D is the self-diffusion coefficient of the spins. In terms of the PFG spin-echo experiment the criterion for the spins to appear to undergo unrestricted diffusion is that their average (rms) displacement in the gradient direction in time Δ should be substantially less than the dimensions between any barriers to diffusion.

EXPERIMENTAL

Apart from a spin echo spectrometer the apparatus required for the PFG experiment consists of the coils to produce the homogeneous linear field-

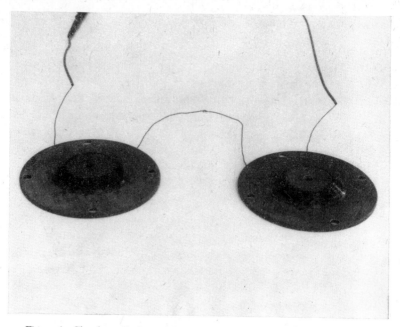

Figure 4 Circular coils for pulsing the gradient in H_0 in the direction of H_0

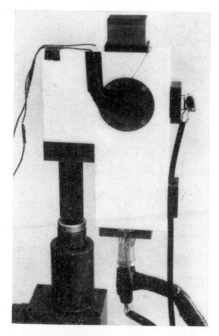

Figure 5 A sample probe used for the PFG spin echo experiment. The gradient coils in Fig. 4 can be seen mounted in the sides of the probe body

gradient and a current source capable of being pulsed. Tanner[12] has discussed the design of coils for this purpose and has described a transistor switch which can be used to pulse the current source. Figs. 4 and 5 show a pair of coils and the complete sample probe used for the PFG experiment in the authors' laboratory. The coils are to Tanner's design and are roughly conical in shape, this being the result of optimising the homogeneity of the gradient taking into account image currents in the magnet pole pieces. The probe body is made of PTFE and all screws and fittings are nylon or Tufnol (a resin-bonded fibre). This is to reduce the effects of eddy currents induced by the pulses in nearby conductors which would prolong the field-gradient pulses. The coils shown in Fig. 4 produce a gradient in H_0 in the direction of H_0 at 7.37 gauss cm^{-1}amp^{-1}. Gradients >200 gauss cm^{-1} have been

Figure 6   Rectangular coils for pulsing the gradient in H_0 in the direction perpendicular to H_0

used although this is not the maximum obtainable. Fig. 6 shows one of a pair of coils which we have designed and built to produce H_0 gradient pulses in the plane perpendicular to H_0. These are for use in experiments involving flow and studies on magnetically oriented liquid crystal samples.

A typical measurement of D for a sample showing unrestricted diffusion involves the measurement of the ratio R (see equation 2) as a function of one of the variables g, δ or Δ. Obtaining sufficient data for a precise determination of D in such a system might typically take 2 hours. In practice the rate at which the pulse sequence of Fig. 1 may be repeated is limited to $\sim(10T_1)^{-1}\,s^{-1}$ or less, where T_1 is the spin-lattice relaxation time of the nuclear magnetisation. This condition ensures that the spin system is at thermal equilibrium at the start of each pulse sequence.

DISCUSSION

A   General

By comparison with other methods of measuring self-diffusion coefficients in fluids (e.g. using radio-tracers) the spin echo techniques have the advantage of being quick, capable of unlimited repetition using the same sample, requiring only relatively small samples, being capable of covering wide ranges of temperature and pressure with little difficulty and of producing as near to ideal labelling of the molecules as is possible. Disadvantages are the limited sensitivity of the NMR experiment, the cost of the apparatus, the requirement that the sample should in general have only one component containing the magnetic nuclei and that these nuclei should usually have a spin quantum number $I = \frac{1}{2}$. This last requirement arises from the fact that most nuclei with $I > \frac{1}{2}$ have too short T_2 values to allow diffusion in a field gradient to produce an effect before the phase relationships between the nuclei are completely randomised by the natural T_2 processes.

Relative to the steady field-gradient (SFG) spin echo experiment the PFG technique has several advantages. The main ones lie in the range of diffusion coefficients accessible to measurement and in the study of restricted diffusion.

Self-diffusion coefficients as small as $10^{-9}\,cm^2\,s^{-1}$ have been reported using this method[15] whereas the SFG method is normally limited to values of D greater than $10^{-7}\,cm^2\,s^{-1}$. The main factor determining this limitation in the SFG experiment is the effect of the large gradients, necessary for the

Diffusion processes in fluids   109

measurement of small diffusion coefficients, on the duration of the spin echo. Large gradients lead to short duration echos which require a wide bandwith receiver for accurate amplification. Wide bandwiths lead to lower signal/noise ratios and hence more scatter in the measurements. An echo of short duration arises from a large range, $\Delta\omega$, of nuclear precession frequencies in the sample which thus necessitates an increase in $|\mathbf{H}_1|$ to maintain the condition $\gamma |\mathbf{H}_1| \gg \Delta\omega$. This condition simply ensures that all spins in the sample are effectively "at resonance" during the $\mathbf{H}_1$ pulses and are thus all subjected to the required values of θ. The PFG method permits the $\mathbf{H}_1$ pulses and the spin echo to occur at times of relatively homogeneous $\mathbf{H}_0$ and thus avoids these problems.

In the study of systems exhibiting restricted diffusion the PFG experiment has the advantage of defining quite clearly the time, Δ (if $\Delta \ll \frac{1}{3}\delta$), over which diffusion is being studied. Δ may be varied over the accessible range (determined by T_2) without varying any other experimental parameter, whereas in the SFG experiment the only time variable is τ and changing this changes the effect of the transverse relaxation processes (T_2) on the echo height as well as allowing the gradient to act for a longer period. Despite these factors the SFG experiment has been used to study restricted diffusion[18,19] and many of the mathematical problems associated with interpretation of the data have now been solved for a variety of geometries of restricting barriers[20,21].

The PFG method is in general better for studying restricted diffusion for the reasons just outlined and because in systems exhibiting such behaviour small values of D and T_2 are quite common thus giving this technique the advantages mentioned earlier. Stejskal *et al.* have for example observed restricted diffusion in systems such as hydrated vermiculite crystals[14,16], yeast cells[17], apple flesh[17], tobacco plant pith[17] and emulsions[17,22].

The experimental distinction between a system which is exhibiting restricted diffusion and one which is not is that the former shows values of $\ln R$ which are functions of two variables, $(\Delta - \frac{1}{3}\delta)$ and $\delta^2 g^2$ whereas the latter has values of $\ln R$ which are a function of the single variable $\delta^2 g^2 (\Delta - \frac{1}{3}\delta)$. In terms of equation (2) a system exhibiting restricted diffusion appears to have a diffusion coefficient which is a function of Δ for a given $\delta^2 g^2$.

The barrier-to-barrier distances which may be measured depend in detail on the geometry of the barriers, the self-diffusion coefficient of the diffusing species and the range of Δ available which in turn depends on the T_2 of the

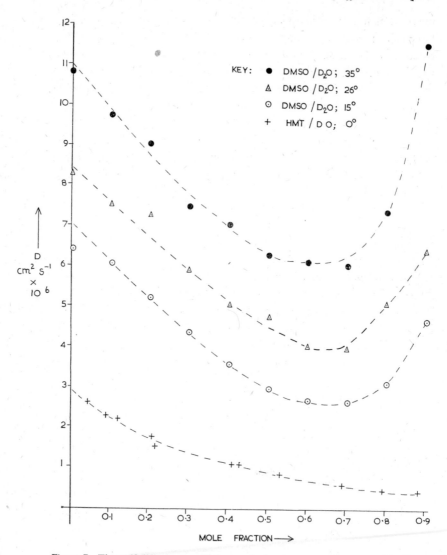

Figure 7 The self-diffusion coefficients of the proton containing molecules in dimethylsulphoxide (DMSO)/D_2O and hexamethylenetetramine (HMT)/D_2O mixtures as functions of composition and temperature. The concentration scale for the HMT/D_2O solutions is $0.1 \times$ the scale on the diagram

nuclei.* In general terms distances in the region of $1 \times 10^{-4} - 10^{-2}$ cm may be measured although in favourable cases the upper limit may be greater. Systems containing a gas as the diffusing entity would lead, given similar values of T_2, to barrier-to-barrier distances being accessible to measurement due to the larger diffusion coefficients.

The equations describing the behaviour of $\ln R$ as a function of Δ, g, δ and a barrier-to-barrier distance, for systems exhibiting restricted diffusion, are complicated and in general experimental data must be fitted to a particular model by a computer programme. Equations have been derived for $\ln R$ for the case of plane parallel barriers[13,17,20,21], cylindrical barriers[21], spherical barriers[17,21] and for diffusion near an attractive centre[13,17]. In some cases a simple distribution of barrier distances has been used[22].

B   Examples of Studies Using the PFG Technique

Fig. 7 shows values of the self-diffusion coefficients of dimethylsulphoxide (DMSO) molecules in $DMSO/D_2O$ mixtures and of hexamethylenetetramine (HMT) molecules in HMT/D_2O mixtures as functions of composition and temperature. These measurements were made on 0.1 cm^{-3} samples in a matter of some 80 hours of spectrometer time. The lowest concentration of protons studied was 10^{20} protons cm^{-3}.

Boss, Stejskal and Ferry[15] have used the PFG technique near its limit to measure the diffusion coefficient of polyisobutylene molecules in benzene solutions. At volume fractions of benzene of 0.978 and 0.903, D was found to be $(2.2 \pm 0.2) \times 10^{-7}$ cm^2 s^{-1} and $(2.7 \pm 1) \times 10^{-9}$ cm^2 s^{-1} respectively. This large change over a small concentration range was interpreted as due to the onset of entanglements between the polymer chains at the higher polymer concentration.

An excellent example of the use of the PFG method to study restricted and anisotropic diffusion is the work of Boss and Stejskal on water diffusion in swollen hydrated vermiculite crystals[14,16]. In these systems evidence was found for both restricted and anisotropic diffusion as well as multiphase and anisotropic transverse nuclear magnetic relaxation. Analysis of the data showed that diffusion of the water protons parallel to the silicate layers was unrestricted and similar to that in liquid water, except for some behaviour

---

* A modification of the PFG experiment has been devised[23], based on the Carr-Purcell/Gill-Meiboom pulse sequence, which reduces certain contributions to T_2 thus making longer values of Δ available.

consistent with slight buckling of the layers. Diffusion perpendicular to the layers was unrestricted for up to half the protons but was severely restricted for the rest. The "free" behaviour was attributed to water molecules diffusing in large faults in the crystal structure whilst the restrictions were identified as pores having sizes in the range 200–1800 Å in the unswollen material.

Fig. 8 shows a selection of ln R values plotted against $(\Delta - \frac{1}{3}\delta)g^2\delta^2$

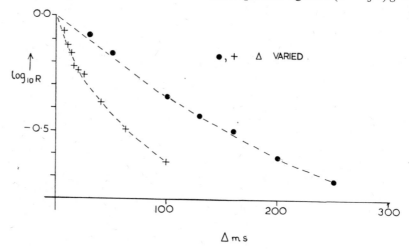

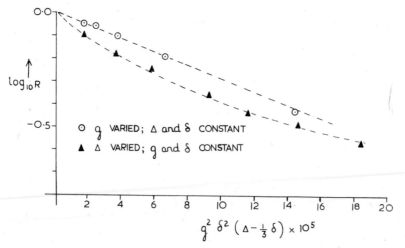

Figure 8 log R values as functions of Δ and $(\Delta - \frac{1}{3}\delta)\delta^2g^2$ for systems showing restricted diffusion. Top graph: Two samples of cucumber flesh cells; Bottom graph: A sample of apple cells

or Δ for two systems exhibiting restricted diffusion. The upper graph shows measurements made on two separate samples of cucumber flesh cells taken from different places in the same cucumber. The lower of these two curves was obtained using a gradient twice that used for the measurements in the upper curve. These results have not yet been analysed in terms of a specific model and the dotted lines are just to emphasise the curvature. Since examination of the cells by microscope indicated an average cell diameter of 3.3 $\times 10^{-3}$ cm and since the largest value of Δ (250 ms) corresponds to an average (rms) molecular displacement of $\sim 4.8 \times 10^{-3}$ cm (assuming a D of 1.5×10^{-5} cm^2 s^{-1}), the fact that $\log_{10} R$ does not appear to approach a constant value at large Δ might indicate that the walls of these cucumber cells are fairly permeable to the cell fluid.

The lower graph in Fig. 8 shows measurements of $\log_{10} R$ on a single sample of the cells from an apple. The upper straight line graph was obtained at constant Δ by varying g whilst the lower curved line was determined by variation of Δ. These two curves illustrate the dependence of $\log_{10} R$ on Δ and $\delta^2 g^2$ as independent variables for a system showing restricted diffusion as did similar measurements on apple cells made by Stejskal and Tanner[17,22].

SUMMARY

The PFG spin echo technique is a versatile addition to the methods available for studying self-diffusion in fluids. It can be used for measurements of D in systems showing both unrestricted, restricted and anisotropic diffusion. In suitable cases where restricted diffusion effects are observed the internal structure of the system, as seen by the diffusing particles, may be studied.

At the present time the authors are studying a variety of systems where the capabilities of this technique are likely to be of use. For example we are studying a variety of emulsions prepared under a range of conditions to see if it is possible to measure the rate of coalescence of the droplet phase, a property it is often difficult to distinguish from the flocculation rate. We are also studying liquid crystal systems with and without solutes to investigate the effects of magnetic ordering on the isotropy of the diffusion processes.

Other areas where this technique could be of great value are, for example, in studying the nature of the diffusion in two component systems near upper or lower consolute temperatures and in studying the binding of counterions by polyelectrolytes.

ACKNOWLEDGEMENTS

We wish to thank Mr. A.J.Strike for assistance with electronics, Mr. W.Plumbley for engineering work, the SRC for maintenance grants (C.R. and D.J.T.) and a grant for the purchase of equipment and Dr. C.H.Neumann for permission to refer to his unpublished work.

REFERENCES

1. E.L.Hahn, *Phys. Rev.*, **80**, 580 (1950).
2. H.Y.Carr and E.M.Purcell, *Phys. Rev.*, **94**, 630 (1954).
3. D.C.Douglass and D.W.McCall, *J. Phys. Chem.*, **62**, 1102 (1958).
4. R.L.Garwin and H.A.Reich, *Phys. Rev.*, **115**, 1478 (1959).
5. K.A.Valiev and M.I.Emel'yanov, *Zhur. Strukt. Khim.*, **5**, 7 (1964).
6. D.W.McCall and D.C.Douglass, *J. Phys. Chem.*, **69**, 2001 (1965).
7. R.Hausser, G.Maier and F.Noack, *Z. Naturforsch.*, **21a**, 1410 (1966).
8. D.W.McCall and D.C.Douglass, *J. Phys. Chem.*, **71**, 987 (1967).
9. D.Kessler, A.Weiss and H.Witte, *Ber. Bunsenges. Phys. Chem.*, **71**, 3 (1967).
10. C.J.Clemett, *J. Chem. Soc. (A)*, 458 (1969).
11. E.O.Stejskal and J.E.Tanner, *J. Chem. Phys.*, **42**, 288 (1965).
12. J.E.Tanner, *Rev. Sci. Inst.*, **36**, 1086 (1965).
13. E.O.Stejskal, *J. Chem. Phys.*, **43**, 3597 (1965).
14. B.D.Boss and E.O.Stejskal, *J. Chem. Phys.*, **43**, 1068 (1965).
15. B.D.Boss, E.O.Stejskal and J.D.Ferry, *J. Phys. Chem.*, **71**, 1501 (1967).
16. B.D.Boss and E.O.Stejskal, *J. Colloid Interfac. Sci.*, **26**, 271 (1968).
17. J.E.Tanner and E.O.Stejskal, *J. Chem. Phys.*, **49**, 1768 (1968).
18. R.C.Wayne and R.M.Cotts, *Phys. Rev.*, **151**, 264 (1966).
19. D.E.Woessner, *J. Phys. Chem.*, **67**, 1365 (1963).
20. B.Robertson, *Phys. Rev.*, **151**, 273 (1966).
21. C.H.Neumann, Private communication.
22. J.E.Tanner, Ph.D.Thesis, University of Wisconsin, U.S.A. (1966).
23. K.J.Packer, C.Rees and D.J.Tomlinson, *Mol. Phys.*, **18,** 421 (1970).

2.4

Line broadening techniques for measuring diffusion coefficients

A. J. HYDE

Department of Pure and Applied Chemistry,
University of Strathclyde, Glasgow

ABSTRACT

The advent of light sources with extremely small linewidths (LASERS) has brought about the possibility of measuring diffusion coefficients from the spectral broadening of the line after scattering of the light by a solution.

The factors governing the broadening will be discussed and the broad outline of the theory presented.

Experimental results obtained by various authors using the above technique will be presented and compared both with former measurements on the systems and with one another where this is possible. Recent measurements of rotational diffusion coefficients by a similar method will be mentioned.

The advent of the laser as a source of radiation has led to advances in many fields of science and has given rise to a new method of measuring diffusion coefficients of large molecules in solution.

If light of a single wavelength (or as narrow a line as possible) is passed into a solution of macromolecules, and the scattered light of 'unchanged' frequency is observed under high enough spectral resolution it is found that the line is broadened compared with the incident light. In addition there will be two lines of distinctly different frequency (the Brillouin lines) and a number of lines of very different frequency (the Raman lines). This broadening arises from Döppler shifts from the macromolecules undergoing Brownian motion. (The analysis is similar to that put forward by Rayleigh[1] concerning the natural (Döppler) width of spectral lines.)

Analyses due to Pecora[2] and Benedek[3] give the following expression for the power spectrum:

$$S\left(v, \Gamma_D\right) \sim \left[\frac{\Gamma_D/2\pi}{(v - v_0)^2 + (\Gamma_D/2\pi)^2}\right]$$

i.e. a Lorentzian of half width at half height given by $\Gamma_D/2\pi$ where

$$\Gamma_D = D\left[\frac{4\pi \sin \theta/2}{(\lambda_0/n)}\right]^2.$$

D being the diffusion coefficient of the solute molecule, θ of the angle of observation, λ_0 the wavelength of the incident radiation in vacuum and n the refractive index of the solution.

Since $D \sim 10^{-8} \rightarrow 10^{-6}$ cm^2 sec^{-1} for most macromolecules, the expected half widths due to this effect range from ca. $100 \rightarrow 10^4$ Hz.

This broadening is very small compared with the frequency of the exciting line (1 part in 10^{11} or less), is far beyond the resolving power of a conventional spectrometer and even beyond the resolution of the conventional interfero-meter (Fabry-Perot). In order to resolve this small broadening it is necessary to use either a heterodyne[4] or homodyne[5] beating technique.

The former technique was the first to be used but has been almost entirely abandoned in favour of the latter, which is experimentally much simpler. Light from a very monochromatic source (ac.w. laser of some type) illu-minates the appropriate solution contained in a cell, the light scattered at some suitable angle falls on the photocathode of a photomultiplier the out-put from which is fed to a wave analyser, the output of which is plotted on an X–Y recorder.

The light which reaches the photomultiplier consists partly of light scat-tered from the moving macromolecules in solution (frequency v) and partly of stray light scattered from stationary surfaces (frequency v_0). Beats will therefore occur between the different scattered rays and one set will have frequency $v_0 - v$. This frequency is in the kilocycle range and can be handled by conventional electronic techniques and so the X–Y plot from the spectrum analyser is the square root of the intensity spectrum of the broadened line which can be used as it stands or squared by an analogue squarer. In Table 1 values are given of diffusion coefficients determined by

Table 1

Material	$10^7 D$ from line broadening $(\text{cm}^2\text{sec}^{-1})$	$10^7 D$ other measurements $(\text{cm}^2\text{sec}^{-1})$	
Bovine serum albumin	6.7[a]	6.7	
Ovalbumin	7.1[a]	8.3	
Lysozyme	11.5[a]	11.6	
Tobacco mosaic virus	0.4[a]	0.3	
	0.28[b]		
D.N.A.	0.2[a]	0.13	
Polystyrene 440 Å	0.59[a]	0.56	
Spheres 630 Å	0.368[a]	0.38	calculated
1830 Å	0.134[a]	0.134	

a. From reference 3
b. From reference 6

the line broadening technique as compared with values determined by conventional methods or by calculation in the case of spheres.

It is clear that in the cases where the materials are completely characterised the values obtained are in close agreement. The present method has however the enormous advantage of determining D at a unique concentration and is therefore invaluable as a method for studying $D = f(c)$.

In addition to the translational movement of the molecule there also exists rotational motion which also affects the spectrum in the case of anisotropic molecules (or particles). The presence of this rotational motion of anisotropic molecules also gives rise to a broadening, the half width at half height being $6\Theta^2/\pi$, where Θ is the rotational diffusion coefficient (and also to a cross term). It is necessary therefore to fit the broadened line with two or more Lorentzians characteristic of the different processes. Fitting is helped by the fact that the translational broadening tends to zero with the angle of observation. This method has been used to find both D and Θ for tobacco mosaic virus.[6]

Recently a modification of the above method has been used for measuring Θ without interference from translational diffusion.[7] In this method, laser light polarised say vertically, is passed into a long cell containing the macromolecular solution and the light scattered at 0° along with the transmitted beam is passed through an analyser (Nicol prism) set to allow the passage of horizontally polarised light. The light transmitted by the analyser consists in the main of two parts: (1) the small amount of horizontally polarised light scattered from the anisotropic macromolecules (depolarised light) and (2) a small amount of vertically polarised light trasmitted because of the imperfection of the analyser. The light scattered from the molecules is broadened due to rotational diffusion (and also due to translational but since the light scattered is observed almost at zero angle this broadening is small) whereas the greater part of the transmitted vertically polarised light is the unscattered incident beam. As in the previous case beating occurs between the different types of light and the spectrum of the broadened line is available from the photomultiplier output. A value of $\Theta = 350 \text{ rad}^2 \text{sec}^{-1}$ is obtained for T.M.V. as compared with $\Theta = 356 \text{ rad}^2 \text{sec}^{-1}$ obtained from decomposition of the sum of Lorentzians.

This new method of determining diffusion coefficients is thus seen to be a powerful technique for studying concentration variations of D and Θ. (In pure fluids the same measurement gives the thermal diffusivity of the fluid instead of the diffusion coefficient.)

REFERENCES

1. Lord Rayleigh, *Nature*, **8**, 474 (1873), *Phil. Mag.*, **27**, 298 (1889), *Proc. Roy. Soc.*, A 76, 440 (1905).
2. R. Pecora, *J. Chem. Phys.*, **40**, 1604 (1964).
3. S. B. Dubin, J. H. Lunacek, G. B. Benedek, *Proc. Nat. Acad. Sci.*, **57**, 1164 (1967).
4. H. Z. Cummins, N. Knable, Y. Yeh, *Phys. Rev. Letters*, **12**, 150 (1964).
5. N. C. Ford, G. B. Benedek, *Critical Phenomena*, p. 150, N.B.S., 1966.
6. J. J. Herbert, F. D. Carlson, H. Z. Cummins, *Biophys. J.*, **8**, A 95 (1968).
7. A. Wada, N. Suda, T. Tsuda, K. Soda, *J. Chem. Phys.*, **50**, 31 (1969).

2.5

Diffusion near the lower consolute temperature point

M. M. BREUER and M. E. BURNEL

Unilever Research Laboratory
Isleworth, Middlesex

ABSTRACT

Thermodynamic considerations suggest that the diffusion coefficient of binary systems approaches zero near the lower consolute temperature point of the system. Some experimental results confirming these theoretical predictions have been obtained by Russian workers.

In a study of an aqueous glycol ether system, we have extended the experimental support of this effect. Retarded diffusion has been noted over a wide composition range corresponding to a broad 'minimum' in the temperature-composition profile. The effect seems to extend to temperatures several degrees below the consolute point.

The results have been correlated with existing theories of diffusion and solution thermodynamics.

INTRODUCTION

The mutual diffusion coefficient, D, of a binary liquid mixture gradually diminishes as the system approaches its critical (consolute) point.[1,2] The phenomenon has been generally attributed to the fact that the concentration derivatives of the chemical potentials, $(\partial\mu/\partial c)_{T,P}$, vanish at the critical point, as pointed out by J. W. Gibbs, D being proportional to $(\partial\mu/\partial c)_{T,P}$. In addition, Anisimov and Perelman[3] have expressed the view that the observed slowing down of the mass transfer in these systems is due to the fact that diffusional processes do not follow Fick's law in the vicinity of the critical point, as the material fluxes become non-linear high-order functions of the concentration gradients. On the basis of several approximations they derived the following diffusion equation for use in the near critical region for a one-dimensional infinite slab

$$C(x, t) = \left(\frac{D_1}{4Do^2t}\right)^{1/4} \cdot f(\xi) \tag{1}$$

where $\xi = x/(4D_1t)^{1/4}$, x representing distance and t, time.

Do is related to the conventional diffusion coefficient, D by $D = Do$ (C-Ccrit.).[2] Ccrit. is the concentration at the critical point.

D_1 is the only additional coefficient retained in Anisimov and Perelman's approximation to a higher order flux equation.

According to these authors, a consequence of equation (1) is that the law of motion for points of equal concentration follows the proportionality $x \propto t^{1/4}$ and not $x \propto t^{1/2}$ as was postulated by Boltzmann. In spite of the great theoretical importance of this prediction, the theory of Anisimov and Perelman has not received experimental confirmation. We decided therefore to check the validity of their deductions.

The second aspect of this paper is an attempt at resolving the vexed question of how much the slowing down of diffusion in the critical region is due to the vanishing thermodynamic driving force, $\partial\mu/\partial c$, and how much to the greater frictional resistance which an increased viscosity represents. We were also interested to find out the extensiveness of the retarded diffusion phenomenon in terms of the temperature and composition ranges over which the effect is noticeable.

We have measured mutual diffusion coefficients and viscosities of the system monoamyl diethyleneglycol ether $(C_{5,2})$-water at various concentrations both at 39 and 25°C. The lower consolute point of the system (our determination) is at 39.5 ± 0.4°C and 10% $C_{5,2}$ by weight.

EXPERIMENTAL

Materials

The glycol ether $C_{5,2}$ was prepared according to a slightly modified form of the procedure of Chakhovskoy.[4] The purity of the material was better than 99.5% as determined by GLC.

Highly purified water (conductivity $<5 \times 10^{-8}$ mho) was employed.

Methods

A Beckman-Spinco Model H convergent light diffusiometer was used with Gouy optics according to the methods developed by Gosting and his associates.[5]

Densities (pyknometer), flow times (Ostwald viscometer) and refractive indices (Abbé instrument) were determined for the same range of compositions over which the diffusion was studied.

Flow times (typically several minutes) were reproducible to within 1% in the PSL type BS/IP/SL, No. 2 viscometer used. In a few experiments flow times were measured only after a solution had been left standing in the top bulb of the viscometer for several hours. No significant difference was noted between the flow times thus obtained and those for freshly loaded solutions.

Densities were measured to four decimal places.

Refractive indices were measured to ± 0.0005.

Diffusion coefficients quoted were mostly the average of several measurements. The maximum departure of an experimental value from any such mean was $\pm 0.25 \times 10^{-4} \text{ m}^2\text{s}^{-1}$.

Results

The refractive index was found proportional to the concentration as required by the interferometric method for studying diffusion.

The macroscopic dynamic viscosity, η, varied linearly with concentration at the temperatures of interest (Fig. 1) except for the slightly sigmoidal char-

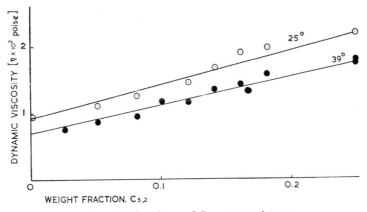

Figure 1   Viscosity, η of $C_{5,2}$-water mixtures

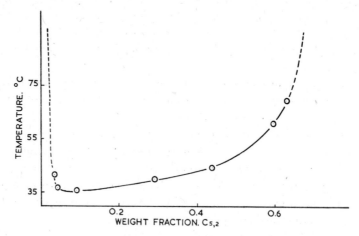

Figure 2 Phase separation temperature curve for the system $C_{5,2}$-water
(taken from N. Chakhovskoy[4])

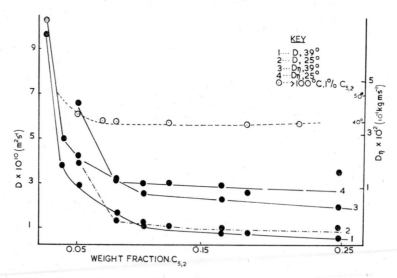

Figure 3 ●, diffusion coefficient and D–η product for $C_{5,2}$-water mixtures at 25 and 39 °C. ⊙ Phase separation temperature curve, our determination

acter of both curves. No evidence was found for any unusual increase in bulk viscosity near the critical point.

The phase separation temperature curve for the $C_{5,2}$-water system agreed substantially with that reported by Chakhovskoy[6] (Fig. 2) as far as we determined it (uppermost curve of Fig. 3).

The binary diffusion data recorded at 25° and 39°C are also represented in Fig. 3 along with the corresponding products, $D \cdot \eta$.

The form of the plots of both D and $D \cdot \eta$ versus composition followed closely the phase separation line, the resemblance being strong even at 25°C, 14° below the consolute point. At 39°C a 10% aqueous solution by weight of $C_{5,2}$ is only a fraction of a degree below the critical demixing temperature.

Discussion

In order to clarify the question posed by the work of Anisimov and Perelman, namely whether or not the diffusion is Fickian in the vicinity of the critical point, we proceeded as follows:

We plotted Ym, the distance of the mth (extreme) interference fringe from the optical axis at the image plane of the diffusometer as a function of $t^{-1/2}$ and $t^{-1/4}$ (t being the time into a diffusion run). According to the theory of the Gouy interferometer[7]

$$Ym = ab \frac{\Delta n}{2} - \text{'optical path difference'} \qquad (2)$$

where a and b are the cell thickness and the total optical path between the cell and the photographic plate respectively. Δn is the refractive index difference between two chosen planes of a certain concentration instantaneously at a displacement x from the centre of the cell and from which arises the mth fringe. Since the same interference fringe corresponds to the same path difference and as a, b and Δn can be considered constant, the change in the value of Ym with time will be inversely proportional to x and a direct measure of the advancement of the diffusion front.

The results of our experiments are shown in Figs. 4, 5 and 6. Better agreement is obtained both in the near critical and the noncritical region if the $t^{-1/2}$ rule is assumed. The sucrose data are included for the sake of comparison, since the validity of Fick's law for this system has been established in the past. We conclude that our experimental evidence does not support Anisimov and Perelman's contention and Fick's law appears to be valid even when diffusion occurs very close to the critical point (Fig. 6).

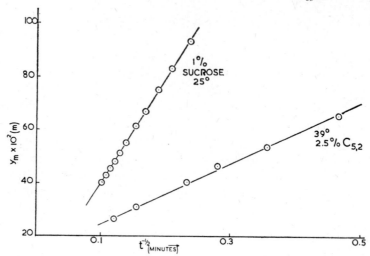

Figure 4 Fringe displacement as a function of the inverse square root of time

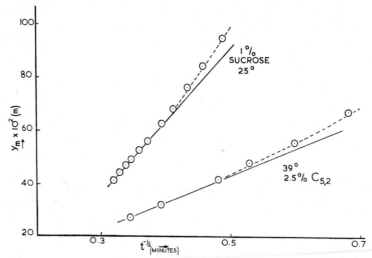

Figure 5 Fringe displacement as a function of the inverse fourth root of time

The binary diffusion coefficient can be expressed as

$$D = \frac{1}{f} \cdot \frac{\partial \mu_2}{\partial C_2} \tag{3}$$

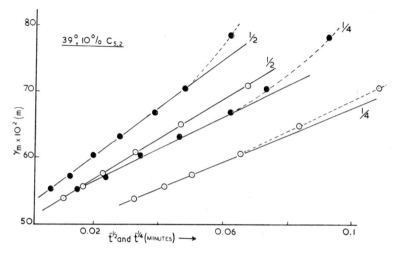

Figure 6 Fringe displacement as a function of the inverse square and inverse fourth root of time. $C_{5,2}$-water mixture close to the lower consolute point. Two experiments

where f is the molecular frictional coefficient, μ_2 the chemical potential and C_2 the concentration of the solute.

It is of considerable interest to establish to what relative extent the frictional and the thermodynamic terms in equation (3) govern the value of D

in the critical region. Since activity coefficient data for our system are not yet available, we have assumed that the thermodynamic behaviour of $C_{5,2}$-water solutions can be described by the Flory-Huggins theory. Accordingly the chemical potential of the solute, relative to that of the pure liquid as a standard state, is given by[8]

$$\mu_2 = \mu_2^0 + RT\left[\ln v_2 - (x-1)(1-v_2) + \chi_1 x (1-v_2)^2\right] \qquad (4)$$

where $v_2 = xV_1C/M = C\bar{v}$, the volume fraction of the solute,
C is the solute concentration in kg l^{-1},
R is the gas constant and T the absolute temperature,
x is the molar volume ratio of solute to solvent,
M is the molecular weight of the solute,
V_1 is the molar volume of the solvent,
$\bar{v}$ is the partial specific volume of the solute, and
χ is the solute-solvent interaction constant.

The parameters x and χ need to be evaluated before numerical calculations can be carried out with equation (4). According to the Flory-Huggins theory, this can be achieved using Cc, the value of the solute concentration at the critical point (page 544 of reference 7).

Thus, using the critical conditions $(\partial\mu_2/\partial v_2)_{T,P} = 0$,

$$Cc = \frac{M}{V_1 x} \cdot \frac{1}{(1 + x^{1/2})} \tag{5}$$

and the corresponding χ_{1c} emerges as

$$\chi_{1c} = \frac{(1 + x^{1/2})^2}{2}.$$

Calculations give the following values for our system

$$x = 18.53$$

$$\chi_{1c} = 0.759.$$

We have used these data in conjunction with equation (3) and the differentiated form of equation (4) to evaluate the frictional coefficient f and the

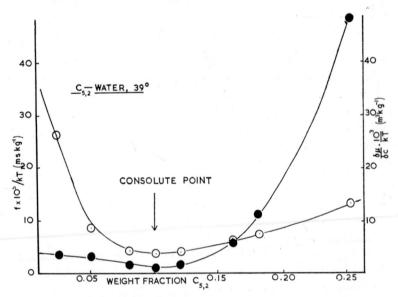

Figure 7   Frictional and thermodynamic terms of equation (4) from Flory-Huggins treatment of data for $C_{5,2}$-water mixtures at 39 °C

thermodynamic term $\partial\mu/\partial C$ over a range of compositions including the critical region. The results are shown in Fig. 7.

On the low concentration side of the critical region the diffusion coefficient seems to be controlled more by the thermodynamic term. The two curves in the diagram run close to and very nearly parallel to the composition axis near the critical point. This reflects the marked degree of concentration independence of the diffusion coefficient in this region. At high $C_{5,2}$ concentrations the frictional factor assumes predominance and D stays at a low value (Fig. 3).

It is debatable whether the concise form of the Flory-Huggins equation used here is valid for a polar situation[8] such as the $C_{5,2}$-water system provides, but in the absence of supporting activity data it is considered to give a worthwhile, albeit qualitative, insight into the diffusion anomaly near the consolute point.

Some recent NMR results[9,10] are relevant to this work. Spin echo measurements of self-diffusion coefficients, D, for some hydrocarbon systems revealed no abrupt abnormality in the $\ln D$ vs. reciprocal temperature curve in the critical region. This is further evidence that the thermodynamic term overrides the frictional contribution in the determination of diffusion behaviour near the critical point.

On the basis of our results we can draw the following conclusions:

1 Fick's law is adhered to even near a critical solution point.

2 No definite decision can be reached regarding the relative importance of the thermodynamic and frictional terms in the determination of the value of D without obtaining further accurate thermodynamic data. It appears however that the change in the thermodynamic term plays the larger part.

REFERENCES

1. I.R.Krichevskii and Yu.V.Tsekhanskaya, *J. Phys. Chem. (U.S.S.R.)*, **30**, 2315 (1956).
2. R.Haase and M.Siry, *Z. Physik. Chem.*, **57**, 56 (1968).
3. S.I.Anisimov and T.L.Perelman, *Int. J. Heat Mass Transfer*, **9**, 1279 (1966).
4. N.Chakhovskoy *et al.*, *Bull. Soc. Chim. Belg.*, **65**, 453 (1956).
5. L.J.Gosting and M.S.Morris, *J. Amer. Chem. Soc.*, **71**, 1998 (1949).
6. N.Chakhovskoy, *Bull. Soc. Chim. Belg.*, **65**, 474 (1956).
7. H.J.V.Tyrrell, *Diffusion and Heat Flow in Liquids*, Butterworths, 1961, p. 106.
8. P.J.Flory, *Principles of Polymer Chemistry*, Cornell University Press, 1953, p. 513.
9. G.M.Panchenkov *et al.*, *Russ. J. Phys. Chem.*, **43**, 421 (1969).
10. J.E.Anderson, *J. Chem. Phys.*, **50**, 1474 (1969).

2.6

Isothermal diffusion measurements on the system H₂O—Tetraethylammonium chloride—KCl at one composition

A. A. DESHMUKH and R. FLEMING

Chemistry Dept., The School of Pharmacy, University of London

ABSTRACT

Isothermal diffusion and density measurements have been made on the binary system H_2O—$(C_2H_5)_4NCl \cdot H_2O$ over the concentration range 0.01 to 0.02 g cc^{-1}, and on the ternary system H_2O—$(C_2H_5)_4NCl \cdot H_2O$—KCl at one composition where the mean concentration of both solutes was 0.05 g cc^{-1}. The two fixed volume cross-diffusion coefficients were found to be relatively large, and according to the measured $(D_{ij})_v$ the ternary system should have been gravitationally stable at $\alpha_1 = 0.75$. It was found that localised mixing occurred in the region of the initial boundary soon after the start of diffusion.

INTRODUCTION

In the system H_2O-tetraethylammonium chloride-KCl the flow of $(C_2H_5)_4NCl$ and KCl can be described by the equations[1,2]

$$(J_1)_v = -D_{11}\frac{\partial \varrho_1}{\partial x} \quad -D_{12}\frac{\partial \varrho_2}{\partial x} \tag{1}$$

$$(J_2)_v = -D_{21}\frac{\partial \varrho_1}{\partial x} \quad -D_{22}\frac{\partial \varrho_2}{\partial x} \tag{2}$$

where the subscript 1, here and throughout this paper refers to the quaternary ammonium compound and the subscript 2 to the potassium chloride. For the small concentration differences used in these experiments the volume-fixed frame of reference (indicated by the subscript v) may be considered to be identical with the cell (or apparatus)—fixed reference frame[3,4]. The flows J_i have the units of $g \, cm^{-2} sec^{-1}$, the four (D_{ij}) diffusion coefficients have the units of $cm^2 sec^{-1}$, and the concentrations of the solutes ϱ_i are expressed in $g \, cc^{-1}$.

EXPERIMENTAL

The diffusion measurements were made on a recently built Gouy diffusiometer, whose optical lever arm length, b, was 260.097 cm, and the internal width of the centre section of the Tiselius cell, a, along the optic axis was 2.4996 cm. Monochromatic light of wavelength 5460.7 Å was obtained from a 125 watt water-cooled Mazda MB/D mercury vapour lamp using a Wratten 77A filter and a collimating device which focussed the light onto a Hilger and Watts source slit (1 div. = 0.005 mm). The image of the slit was focussed on a photographic plate by a 101.5 cm focal length achromatic doublet objective lens, 11 cm in diameter and corrected to 1/8th, wavelength at 5461 Å. The temperature of the water bath was maintained at $25° \pm 0.001°$ during the diffusion run using a mercury-in-glass contact thermometer connected to a variable-load heat controller designed by Skerrett[5]. The temperature of the bath was frequently read during the course of the experiment using an N.P.L. calibrated thermometer. The bath windows were 2 cm thick, 11 cm in diameter and polished flat to 1/20th, wavelength at 5460.7 Å. The cell frame was similar to that described by Gosting, Hanson et al.[6], and Dunlop[7].

Photographs of the interference fringes were taken on Kodak Ortho-800 plates and were measured using a Gaertner M2001P toolmakers' microscope.

The density of each solution was determined using a 34 cc capacity single-neck pycnometer and was weighed against a tare of similar shape and volume, and which contained doubly-distilled air-free water whose density was taken to be 0.997048 $g \, cc^{-1}$ at 25°.

MATERIALS

Analar potassium chloride (99.8% pure after ignition) purchased from British Drug Houses Ltd., was dried in an oven overnight at 120° and cooled in a desiccator prior to weighing. Tetraethylammonium chloride was purchased in the monohydrate form from British Drug Houses Ltd., and from Eastman-Kodak Organic Chemicals Dept., of Distillation Product Industries. Before making up the solutions of quaternary ammonium salt the solid was dried overnight in a vacuum desiccator containing magnesium perchlorate and then assayed for chloride content by potentiometric titration against Analar silver nitrate solution. The molecular weights, M_i, of the tetraethyl-ammonium chloride monohydrate, KCl, and water were taken to be 183.72, 74.56, and 18.016 respectively.

SOLUTIONS

For each ternary experiment the mean solute concentration $\bar{\varrho}_i$, is defined by the expression

$$\bar{\varrho}_i = \frac{(\varrho_i)_A + (\varrho_i)_B}{2} \quad (i = 1, 2) \tag{3}$$

where $(\varrho_i)_A$ is the initial concentration of the solute in the upper solution, and $(\varrho_i)_B$ is the initial concentration of the solute in the lower solution in the Tiselius cell. The initial concentration difference is given by

$$\Delta\varrho = (\varrho_i)_B - (\varrho_i)_A \quad (i = 1, 2). \tag{4}$$

All solutions were prepared by weight and were corrected for weight in vacuo using the values 1.1115 and 1.984 g cc^{-1} for the densities of $(C_2H_5)_4NCl \cdot H_2O^8$ and KCl^9 respectively and the experimentally determined solution density. Tetraethylammonium chloride monohydrate contained 19.2979% Cl and the difference between this value and the chloride content for the salt determined experimentally by potentiometric halide titration was assumed to be water and the appropriate corrections were made to the weight of salt in air. Experiments on the binary system H_2O—$(C_2H_5)_4NCl \cdot H_2O$ showed that the density-concentration curve had a minimum in the region of 0.02 g cc^{-1}. The following cubic equation fits the experimental density results ($\pm 0.006\%$) and was obtained by the method of least squares

$$d = 0.997048 - 0.006134\varrho + 0.11697\varrho^2 - 0.0646\varrho^3 \quad (0 < \varrho < 0.2) \tag{5}$$

To prepare a ternary solution at a predetermined concentration it was necessary to be able to predict the density of the solution within a few parts in a hundred thousand, and to do this the procedure of Wolf et al.,^{10} was adopted.

Table 1[a]   Diffusion data for the binary system H_2O—$(C_2H_5)_4NCl \cdot H_2O$

Expt. No.	5	11
$\bar{\varrho}$	0.00989$_1$	0.04755$_8$
ϱ_A	0.0197820	0.039069$_9$
ϱ_B	0.0	0.056125$_4$
$\Delta\varrho$	0.0197820	0.017055$_5$
d_A	0.996948	0.997001
d_B	0.997048	0.997080
δ	0.0	1.$_3$
J_m	130.96	114.44
$D_a \times 10^5$	1.052$_9$	0.9768$_2$
ΔT	32.76	41.63

[a] Units: concentrations and densities, $g\,cc^{-1}$;

δ,   microns
D_a, cm^2sec^{-1}
ΔT, sec

Using the expression

$$d = d_0 + \sum_{i=1}^{2} \left(\frac{C_i}{1000}\right)(M_i - d_0\phi_i) \qquad (6)$$

a preliminary value for the density of the solution was computed at a pre-determined molarity C_i. Values of ϕ_1 were calculated using the expression

$$\phi_1 = \frac{M_1}{d_0} - 1000\left(\frac{d - d_0}{C_1 d_0}\right) \qquad (7)$$

and the binary density results obtained at concentration C_1. The apparent molar volumes of KCl were calculated from the expression[11]

$$\phi_2 = 26.742 + 2.000C_2^{1/2} + 0.1110C_2. \qquad (8)$$

The predicted value of the density of the ternary solution was tested experimentally on a solution in which $(C_2H_5)_4NCl \cdot H_2O$ and KCl both had concentrations of 0.05 g cc^{-1} and a value of the coefficient $k_1 = 2.5 \times 10^{-4}$ of the correction term $k_1 C_1 C_2$ was evaluated. This correction term was subtracted from the densities predicted by equation (6), when preparing solutions for the first ternary experiment. Multiple linear regression analysis on the density data from all the ternary diffusion experiments yielded the relation

$$d = 0.997451 + 0.00248_4\varrho_1 + 0.60428_3\varrho_2 \quad (\pm0.007\%). \qquad (9)$$

The partial specific volumes of solvent and solutes were calculated using the equations[12]

$$\bar{v} = \frac{1 - H_i}{d - (H_1 \varrho_1 + H_2 \varrho_2)} \tag{10}$$

$$d = \varrho_0 + \varrho_1 + \varrho_2 \tag{11}$$

$$\bar{v}_0 \varrho_0 + \bar{v}_1 \varrho_1 + \bar{v}_2 \varrho_2 = 1 \tag{12}$$

where the subscript 0 refers to the solvent and where $H_1 = 0.00248_4$ and $H_2 = 0.60428_3$, being the coefficients of ϱ_1 and ϱ_2 in equation (9). The results are given in Table 2.

Table 2[a]   Diffusion data for the ternary system H_2O—$(C_2H_5)_4NCl \cdot H_2O$—KCl $\bar{\varrho}_1$
         $= \bar{\varrho}_2 = 0.05$ $(0 = H_2O, 1 = (C_2H_5)_4NCl \cdot H_2O, 2 = KCl)$

1	Expt. No.	6	7	8	13
2	J_{expt}	104.99	112.38	108.70	114.09
3	J_{calc}	104.96	112.37	108.76	114.07
4	$Q_{expt} \times 10^4$	-73.05	65.76	25.43	71.82
5	$D_a \times 10^5$	1.839_9	1.238_1	1.479_5	1.184_4
6	$D_{a\,calc} \times 10^5$	1.838_3	1.235_7	1.482_3	1.185_5
7	α_1	0.0002	0.5341	0.2764	0.5962
8	R_1	0.14736			
9	R_2	0.12741			
10	$(D_{11})_v \times 10^5$	$0.852_0 \mp 0.001_9$			
11	$(D_{12})_v \times 10^5$	$0.140_5 \mp 0.001_8$			
12	$(D_{21})_v \times 10^5$	$0.141_4 \pm 0.003_8$			
13	$(D_{22})_v \times 10^5$	$1.567_4 \pm 0.003_5$			

[a] Units: Refractive index derivatives R_i, ccg^{-1}
   Reduced height: area ratio D_a, cm^{2}sec^{-1}
   Diffusion coefficients $(D_{ij})_v$, cm^{2}sec^{-1}

DIFFUSION EXPERIMENTS

Woolf, Miller and Gosting[10] have described in detail the method of obtaining coefficients from Gouy interference fringes and only a brief description is given here.

The two solutions filling the Tilselius were allowed to equilibrate for 45 minutes before a set of six δ-pictures were taken; these had values which varied from -3 to $+3$ microns for all the experiments reported in this paper.

When the top section of the cell was open a crude boundary was formed between the upper, less dense solution A, and the lower, more dense solution B. A stainless steel capillary was inserted into the cell so that its tip lay on the optic axis and siphoning was commenced. This resulted in the crude boundary being brought to the optic axis, and was sharpened by drawing off approximately 80 cc of solution at the rate of 1 to 1.6 cc min^{-1}. δ'-pictures were taken with horizontal Rayleigh slits during the last 15 cc's of siphoning. The siphoning flow was stopped ($t' = 0$) and the top and bottom sections of the cell were closed relative to the centre section. Eight to ten photographs were taken of the Gouy interference fringes at various times t' after the start of diffusion.

CALCULATION OF $(D_{ij})_v$

For each interference pattern at time t' values of the ratio $Y_j/\exp(\zeta_j^2)$ were obtained for the fringes 0 $\cdots$ 6, 10, 15 $\cdots$ 30, 40, 50 $\cdots$ and were then plotted against $Z_j^{2/3}$; extrapolation to $Z_j^{2/3} = 0$ yielded C_t, the maximum displacement of light according to ray optics. The reduced height area diffusion coefficient D_a' at time t' was then calculated according to the expression

$$D_a' = (J\lambda b)^2/(4\pi C_t^2 t').$$ (13)

The diffusion coefficient D_a and Δt were calculated from the values of D_a' and $1/t'$ using a linear least squares method. An average value of the reduced fringe deviation Λ_j was obtained and plotted against the corresponding reduced fringe number $f(\zeta_j)$. The area of this graph, Q, is a measure of the deviation of the refractive index curve from the Gaussian shape, and was obtained by numerical integration using Simpson's $\frac{1}{3}$ rule. Using the values of D_a, J and Q, the $(D_{ij})_v$ were calculated according to the method of Fujita and Gosting[13] from four experiments in which α_1, the solute fraction, was varied from 0.0 to 0.6.

RESULTS

Two binary diffusion experiments were made in which $\bar{\varrho}$ were 0.01 (Expt. 5) and 0.05 (Expt. 11), with $\Delta\varrho$ approximately 0.018 in each case ($i = 1$, equations (3) and (4)). In experiment 5 the solution of quaternary ammonium salt was layered on top of distilled water which filled the lower part of the Tiselius cell, because in this concentration region, the salt solution is less dense than pure water. This is illustrated by the density-concentration curve in Fig. 1 which exhibits a minimum in the region of 0.02_3 g cc^{-1}. The cubic equation (14) given below fits the density-concentration results, and in (14) the concentration, ω, is expressed in g of $(C_2H_5)_4NCl \cdot H_2O$ per g of H_2O.

$$d = 0.997047 - 0.05596\omega + 0.0997\omega^2 + 0.1352\omega^3$$

$$(0 < \omega < 0.25). \tag{14}$$

Details of the binary diffusion experiments are given in Table 1. The starting time corrections, ΔT, were relatively long, but this may be due to the small density difference between the initial solutions making it difficult to form a very sharp boundary. The fringe deviation graph in experiment (11) had an area, Q, of -2.65×10^{-4} units which is just slightly greater than the expected experimental error ($= 1 \times 10^{-4}$) which implies a slight concentration dependence on the part of the solute. Apparatus errors had been ruled out as the apparatus has only just recently been calibrated with sucrose solutions. The above value of Q was subtracted algebraically from the areas of the fringe deviation graphs obtained in the ternary experiments.

The first ternary experiment in which $\alpha_1 = 1.0$ failed due to convective mixing which occured in the diffusion cell, and with prior knowledge of the $(D_{ij})_v$ this instability could have been predicted. The four experiments in which the value of α_1 was 0.0, 0.28, 0.53, 0.60 respectively were all stable, and the $(D_{ij})_v$ were calculated according to the method of Gosting and Fujita which utilises the whole area of the fringe deviation graph Q and the results of these calculations are shown in Table II. Because the solute concentrations were always close to the mean concentration, $\bar{\varrho}_1 \simeq \varrho_2 \simeq 0.05$, the refractive index of the three component solution is given by

$$n(\varrho_1, \varrho_2) = n(\bar{\varrho}_1, \bar{\varrho}_2) + R_1(\varrho_1 - \bar{\varrho}_1) + R_2(\varrho_2 - \bar{\varrho}_2) \tag{15}$$

and the values of R_1 and R_2 shown in Table II were calculated from the expression[14]

$$J\lambda/(a[\Delta\varrho_1 + \Delta\varrho_2]) = [\Delta\varrho_1(R_1 - R_2)/(\Delta\varrho_1 + \Delta\varrho_2)] + R_2 \tag{16}$$

by the method of least squares. The smoothed values of R_1 and R_2 were then used to calculate J_{calc} and it can be seen from Table II that the agreement between the experimental and calculated value of J is approximately 0.03%. This was considered to be satisfactory as the quaternary ammonium salt was deliquescent and batches from two different manufacturers were used to prepare the solutions. The solute fraction α_1 was calculated according to the equation

$$\alpha_1 = R_1 \, \Delta\varrho_1/(R_1 \, \Delta\varrho_1 + R_2 \, \Delta\varrho_2). \tag{17}$$

Experimental values of Q were altered by $\pm 1.0 \times 10^{-4}$, the expected experimental error[9] and the $(D_{ij})_v$ were calculated to give some idea of the accuracy of the $(D_{ij})_v$'s. The change in the $(D_{ij})_v$ caused by the change in Q is shown by the quantity with the $\pm$ sign after each $(D_{ij})_v$ and the upper sign in each case corresponds to the change caused by adding 1×10^{-4} to the value of Q.

Partial specific volume and density data are reported in Table 3. The density derivatives H_1 and H_2 were calculated by multilinear regression using the density data from all the three component experiments, including those which were not gravitationally stable.

Table 3[a] Partial specific volumes and density data for the system
H_2O—$(C_2H_5)_4NCl \cdot H_2O$—KCl

1	Expt. No.	6	7	8	13
2	$(\varrho_1)_A$	0.049534_0	0.045111_7	0.047329_4	0.044119_3
3	$(\varrho_2)_A$	0.040980_2	0.045513_1	0.043249_5	0.045972_4
4	d_A	1.02228_4	1.02504_4	1.02365_4	1.02541_1
5	$(\varrho_1)_B$	0.049537_2	0.054009_5	0.051784_6	0.054201_9
6	$(\varrho_2)_B$	0.058972_9	0.054488_8	0.056744_3	0.053870_3
7	d_B	1.03316_8	1.03048_7	1.03180_4	1.03019_9
8	$\bar\varrho_1$	0.049535_6	0.049563_2	0.049557_0	0.049158_3
9	$\bar\varrho_2$	0.049976_5	0.049998_0	0.049996_9	0.049921_3
10	H_1	0.00248_4			
11	H_2	0.60428_3			
12	$\bar v_0$	1.00256			
13	$\bar v_1$	1.00006			
14	$\bar v_2$	0.39673			

[a] Units: concentrations ϱ_i, $g\,cc^{-1}$
densities d, $g\,cc^{-1}$
partial specific volumes $\bar v_i$, $cc\,g^{-1}$

DISCUSSION

Since $(D_{12})_v$ and $(D_{21})_v$ are positive and significantly different from zero, a concentration gradient of either solute contributes markedly to the flow of the other solute. The conditions for gravitational stability, in terms of the

$(D_{ij})_v$ and $\varDelta\varrho_i$, of ternary solutions at all times $t > 0$ were derived from Wendt[15]. For a gravitationally stable diffusing liquid column, the following inequalities must hold, viz:

$$G_\alpha \geqslant 0 \quad \text{and} \quad G_\beta > 0 \tag{18}$$

where

$$G_\alpha = H_1 K_1^- + H_2 K_2^- \tag{19}$$

$$G_\beta = H_1 K_1^+ + H_2 K_2^+ + [\sigma_-/\sigma_+]^{1/2} (H_1 K_1^- + H_2 K_2^-) \tag{20}$$

$$K_1^{\pm} = \pm \frac{[(D_{22} - D_{11} \pm U)\varDelta\varrho_1 - 2D_{12}\varDelta\varrho_2]}{4U} \tag{21}$$

$$K_2^{\pm} = \pm \frac{[(D_{11} - D_{22} \pm U)\varDelta\varrho_2 - 2D_{21}\varDelta\varrho_1]}{4U} \tag{22}$$

$$\sigma_{\pm} = \frac{(D_{11} + D_{22} \pm U)}{2S^2} \tag{23}$$

and

$$U = (D_{11} - D_{22} + 4D_{12}D_{21})^{1/2} \tag{24}$$

$$S = (D_{11}D_{22} - D_{12}D_{21})^{1/2}. \tag{25}$$

The values of G_α and G_β are tabulated in Table 4. According to the $(D_{ij})_v$ listed in Table 2, both G_α and G_β are positive when $\alpha_1 = 0.75$, and the cal-

Table 4[a] Values of G_α and G_β

Expt. No.	6	7	8	13
α_1	0.0002	0.5341	0.2764	0.5962
$\varDelta\varrho_1$	0.000003_2	0.008897_8	0.004455_2	0.010082_6
$\varDelta\varrho_2$	0.017992_7	0.008975_7	0.013494_8	0.007897_9
G_α	0.005251_4	0.003114_3	0.004186_2	0.002865_6
G_β	0.003963_2	0.001849_4	0.002908_6	0.001595_0
Expt. No.	12	9		
α_1	0.7478	0.7422	1.0	
$\varDelta\varrho_1$	0.012807_6	0.012441_6	0.018	
$\varDelta\varrho_2$	0.004996_6	0.004998_4	0.0	
G_α	0.002170_4	0.002150_6	0.001000_9	
G_β	0.000916_8	0.000922_4	-0.000258_4	

[a] Units: Concentrations ϱ_i, g cc^{-1}
 G_α and G_β, g cc^{-1}

culated density difference between the two diffusing solutions, on the basis of previous experience with binary systems, is adequate to ensure that the columns are stable. However, it was found that a localised zone of mixing was observed within minutes of the start of diffusion which spread slowly with time (Expt. Nos. 9 and 12). It is suggested that the value of $G_\beta \simeq 0.0009$ is smaller than the "practical limit" at which these systems can be studied successfully. Slight vibrations transmitted from the mechanical stirrer or the manipulations necessary after closing the siphon-tap together with possible convective heat of mixing are apparently sufficiently large to cause this slight instability.

ACKNOWLEDGEMENTS

The authors are indebted to the Royal Society (Government Grant), the Science Research Council, and the Central Research Fund of the University of London for grants which defrayed the cost of the apparatus. One of use (A.A.D.) is grateful to the S.R.C. for the award of a postgraduate studenship.

REFERENCES

1. R.L.Baldwin, P.J.Dunlop, and L.J.Gosting, *J. An. Chem. Soc.*, **77**, 5235 (1955).
2. G.J.Hooyman, *Physica*, **22**, 751 (1956).
3. J.G.Kirkwood, R.L.Baldwin, P.J.Dunlop, L.J.Gosting, and G.Kegeles, *J. Chem. Phys.*, **33**, 1505 (1960).
4. G.J.Hooyman, H.Holtan, Jr., P.Mazur, and S.R. de Groot, *Physica*, **19**, 1095 (1953).
5. E.J.Skerrett, *Chem. and Ind.*, 1996 (1968).
6. L.J.Gosting, E.M.Hanson, G.Kegeles, and M.S.Morris, *Rev. Sci. Instr.*, **20**, 209 (1949).
7. P.J.Dunlop and L.J.Gosting, *J. Am. Chem. Soc.*, **75**, 5073 (1953).
8. *Dictionary of Organic Compounds*, Eyre and Spottiswoode, London, 1964.
9. G.Reinfelds and L.J.Gosting, *J. Phys. Chem.*, **68**, 2464 (1964).
10. L.A.Woolf, D.G.Miller, and L.J.Gosting, *J. Am. Chem. Soc.*, **84**, 317 (1962).
11. L.J.Gosting, *J. Am. Chem. Soc.*, **72**, 4418 (1950).
12. P.J.Dunlop and L.J.Gosting, *J. Phys. Chem.*, **63**, 86 (1959).
13. H.Fujita and L.J.Gosting, *J. Phys. Chem.*, **64**, 1256 (1960).
14. I.J.O'Donnell and L.J.Gosting, *The Structure of Electrolytic Solutions* (W.J.Hammer, ed.), Chapter 11, John Wiley and Sons, Inc., New York, N.Y., 1959.
15. R.P.Wendt, *J. Phys. Chem.*, **66**, 1740 (1962).

2.7

Diffusion in multicomponent systems

R. MOHAN

Weizmann Institute, Israel

ABSTRACT

A more general approach to the analysis of translational diffusion in multicomponent systems is attempted in this work to check the generalised Fickian mass flow equation and the corresponding irreversible formalism. Earlier works by Gosting and his school were primarily limited to electrostatically interacting systems with nearly equal main diffusion coefficients and small cross term coefficients i.e. to weakly interacting systems. A more complicated system of small and large molecules interacting hydrodynamically was studied (Bovine serum albumin–KCl–H_2O and Bovine serum albumin–Haemoglobin–KCl–H_2O without dialysis). The initial diffusion conditions were not restricted to small gradients as in Gosting, thereby providing ample opportunity to devise and compare various experimental methods and calculation procedures. In these experiments, in addition to the high concentration dependency of the cross coefficients, the question of the stability and the presence of mass flow terms other than those due to purely diffusional flow are important factors to be considered. Final analysis suggests the doubtful validity of the local thermodynamic equilibrium or stability condition required for the Onsager Reciprocal relation assumption.

INTRODUCTION

Twelve years ago, in the "Advances in Protein Chemistry", Gosting[1] reviewed the theory and practice of diffusion in solutions. The final section dealing with multicomponent systems (three components including the solvent) dealt only with strong electrolytes where the components were of approximately equal size. Gosting's experimental methods as well as his procedures for calculating diffusion coefficients D and connecting these coefficients to the thermodynamic properties of the system were to become standard practice in further works of analysis of multicomponent isothermal diffusion processes. The methods he evolved were mostly for weakly interacting systems which could be generally characterized as systems where the components have similar *main* diffusion coefficients and only small cross coefficients.

There are many assumptions involved in Gosting's methods. These are essentially due to the attempt to obtain some form of analytical solution, an the limitations of the type of interferometry chosen. His main assumptions and restrictions are:

i) The concentration of each component ion can be expressed as a function of a single variable, $xt^{-1/2}$.

ii) In order to obtain analytical solutions, at least one of the cross coefficients should be negligible compared with the main diffusion coefficients.

iii) To solve the set of simultaneous equations, all the four D's refer to the solutes, the third component being the solvent. Gosting meets these requirements using very weak concentration gradients. In order to solve for the four unknown D's four equations are required and Gosting obtains these from four experiments with differing concentrations in the two diffusion layers, but with the same average $\bar{c}_1$, $\bar{c}_2$ of the solutes.

Application of Gosting's methods of evaluating D's to systems with the type of strong interactions which could be expected from the hydrodynamic interaction of a fast-moving salt with a slow-moving protein could be erroneous, since in such systems the cross coefficients may be high and not negligible. In addition, there is always the possibility of mass flow due to non-diffusional terms, and an approach analogous to Gosting's becomes difficult. The present work is aimed at eliminating these restrictions and developing a method of determining the concentration dependence with the minimum number of experiments, i.e. by using high initial gradients in the components. To avoid the introduction of an additional flow term due to changes in specific volume on mixing in such a system, a direct analysis based on mass flow terms in the system is used. The assumption that c is a function of $xt^{-1/2}$ is not true in general in a multicomponent system and is thus avoided. The experiments were carried out with bovine serum albumin–KCl–H_2O and bovine serum albumin–haemoglobin–KCl–H_2O.

MATERIALS AND METHODS

Crystalline bovine serum albumin, B.S.A. (Armour), KCl (Analar) and human haemoglobin, Hb, (Pentax) were used. The required solutions were prepared in distilled water, the concentrations being measured in g/cc. Since temperature corrections would be required in calculating the densities of the various constituents, the temperature at which the solutions was made was noted. The compositions of the two layers in the two experiments were:

	BSA	KCl	
BSA–KCl–H$_2$O:	0.6001%	1.4998%	
vs.	0.1021%	0.1002%	in water at 20°C

and

	BSA	KCl	Hb	KCl
BSA–Hb–KCl–H$_2$O:	0.6016%	1.5002%	vs. 0.1008%	0.1006%
				in water at 20°C

Diffusion Cell

The diffusion cell for these experiments was the conventional Rayleigh cell 7 cm long, 2.5 cm through the optical path and 0.4 cm thick. A sharp boundary between the two layers was achieved by the method of Kahn and Polson[2]. This method, which has not been tested previously for such systems was found to be perfectly satisfactory. A long straight platinum coated needle (No. 719) with bevelled edge was used. A rate of flow has to be adjusted in order to get a sharp boundary for the salt as well as for the protein. The multiple diffraction Rayleigh interferometer used was highly suitable for ascertaining boundary sharpness.

Optical System

The optical system was of the Schlieren type with improved optical components for interferometry to give $\lambda/100$ accuracy. The Schlieren lenses of focal length 170 cm are corrected for the red and green. The achromatic lens and the objective lens were used at unit magnification. The optical alignment and resolution were tested by modulation transfer functions for the lenses by using the photoelectric scanner in addition to the usual geometrical optical methods.

The optical system was designed for both Rayleigh and Gouy interferometry as well as for light absorption scanning. The Rayleigh technique with a single-slit light source required the slit and light source to be vertical, while the Gouy method required them to be horizontal. The problem was solved by using a horizontal slit and horizontal light source, with a raster containing vertical lines superimposed on the horizontal slit. The raster could be easily taken out of the optical path and quickly and accurately brought back into position. In this way, by using a phase plate, both the Schlieren gradient curve and the Rayleigh fringes could be photographed simultaneously. The rasters were made by ruling lines on the gelatin of blackened and hardened high contrast photographic plates. The performance of these rasters was as good as that of the expensive ones ruled on glass, or replicas. Rasters with different line spacing were required for each wavelength. The horizontal light source slit could be accurately controlled with a vernier. Without the raster, the light source positioning serves both Gouy and light absorption scanning.

The cylindrical lens was fitted with a special holder which permitted easy removal of the lens from the optical path for light absorption scanning. The holder had two fine adjustment screws for correct horizontal and vertical positioning for Rayleigh and Gouy interferometry respectively.

Special masking devices were incorporated on the diffusion cell carrier, so that the choice for Gouy, Rayleigh or light absorption could be made easily from outside the thermostat without disturbing the diffusion cell. Similar maskings were employed at the image plane. The Rayleigh masking slit width on the diffusion cell was set at 2 mm, which was also sufficient for light absorption scanning by photoelectric means with slit widths ranging from 0.015 mm to 0.5 mm. Different scanning speeds and slit widths were used depending on the smoothness of the concentration profiles.

The photoelectric cell for scanning the Gouy fringes employed a 7μ slit. The photocell carrier could be displaced easily from the image plane to permit photographs to be taken. The scanner screw used was carefully checked to see that there was no backlash etc. The photocell housing incorporated a microscope objective to spread the light uniformly over the photosensitive cathode of the photomultiplier.

The various methods mentioned were tested on the simple two-component systems $KCl–H_2O$ and $Hb–H_2O$. The correct diffusion coefficients were obtained in all cases.

Methods of Calculation

*1   Evaluation of the concentration of BSA and KCl in the system
BSA–KCl–H$_2$O from the Rayleigh fringes by the two-wavelength method*

Rayleigh fringes were photographed at the Hg violet and Cd red wavelengths.
To obtain the concentrations of BSA and KCl at any point in the diffusion
cell from these photographs accurate specific refractive index increments data
for BSA and KCl at these wavelengths are required. While the specific refractive index increments for BSA are constant at each wavelength, for KCl they
are proportional to the square root of the salt concentration. However, the
ratio of the specific refractive increments for KCl at the two wavelengths is a
constant independent of the salt concentrations. Utilising these two parameters and the total number of fringes at the two wavelengths at any position in the diffusion cell, the corresponding concentrations of BSA and KCl
could be evaluated from the two simultaneous equations.

2

In the BSA–Hb–KCl–H$_2$O system, the haemoglobin concentrations obtained from the light absorption scanning were converted into fringe numbers utilising specific refractive index increment data for Hb. This number
of fringes was added or subtracted to the total number of fringes obtained by
interferometry, to give the fringe number corresponding to the concentration
of BSA and KCl only. The individual concentrations of BSA and KCl were
obtained in the manner indicated in paragraph 1.

3   Gouy

To check the accuracy and sensitivity of the Gouy photoelectric scanning
method for multicomponent systems, the results were compared with Rayleigh fringes photographed immediately prior to the Gouy scanning. The
time required to convert the apparatus was one second, during which there
was negligible change in the diffusion pattern. From each Rayleigh photo-
graph, which gives the refractive index profile in the diffusion cell directly,
the corresponding positions and intensities of the Gouy fringes at the same
wavelength were calculated on the basis of Kirchhoff's diffraction theory.
Comparison of the results with experimental Gouy fringe positions and
intensities showed very good agreement, confirming the reliability of the
fringe scanning system used. Since the refractive index profile of the BSA–
KCl–H$_2$O system is not smooth, this test is quite severe. The high resolution

and distortion-free scanning of the fringes gave a fringe measuring accuracy of $\lambda/100$, which is ten times higher than that which could be obtained by the Rayleigh method. This is also the limit of the optical tolerance of the basic Schlieren optical system. The reverse operation of using the Gouy method to find the refractive index profile in the diffusion cell, for the calculation of the diffusion coefficients is more tedious and would require refinement of the computation methods. Such refinements and the asymptotic improvement in the calculation accuracy could be tested only with a digital computer.

4 Calculation of diffusion coefficients for the system BSA–KCl–H$_2$O

From the detailed knowledge of the concentration profiles of BSA and KCl photographed at different times '*t*' during the diffusion process, the amount of mass flow J and the concentration gradients at any point in the diffusion cell could be calculated. This permits direct solution of the set of algebraic equations for the various concentration-dependent diffusion coefficients:

$$-J_1 = D_{11} \, (\partial c_1/\partial x) + D_{12} \, (\partial c_2/\partial x) \tag{1}$$

$$-J_2 = D_{21} \, (\partial c_1/\partial x) + D_{22} \, (\partial c_2/\partial x) \tag{2}$$

where D_{11} and D_{22} are the main diffusion coefficients of KCl and BSA respectively, D_{12} is the cross diffusion coefficient of KCl under the influence of the concentration gradient in BSA, and D_{21} is the cross coefficient of BSA under the influence of KCl. This set of equations can be solved provided D_{11} and D_{22} are known. Alternatively two further equations could be added to the above two, if, in the entire series of c–x–t diagrams another point in the diffusion could be found where the concentrations of BSA and KCl are exactly the same as in the above set of equations. To obtain such points from the same experiment a large number of photographs of the diffusion process would be needed. In the absence of such a pair of points from which a set of four simultaneous equations could be solved to obtain all the four diffusion coefficients, we have substituted the values of D_{11} and D_{22} obtained from the corresponding two-component systems. For D_{11} we used the diffusion coefficient corresponding to the same concentration of KCl in the KCl–H$_2$O system, and for D_{22} we used the diffusion coefficient of the same concentration of BSA in water, as the concentration of the supporting electrolyte goes to zero.

SUMMARY OF RESULTS

A detailed account of the calculation procedures and experimental results are to be given elsewhere[3]. The pertinent trend of results are noted below.

The diffusion coefficients of the system BSA–KCl–H_2O at 20 °C were corrected to 25 °C as most data are given for this temperature. The temperature correction was made by using the viscosity coefficient of water at these two temperatures. Further, since there is a volume flow due to changes in the specific volumes of the components during diffusional mixing, a volume correction was made based on the specific volume of aqueous solutions of BSA and KCl at different concentrations.

The main diffusion coefficients of KCl and BSA differ by a factor of twenty and are much less concentration-dependent than the cross coefficients. The values of the cross coefficients lie between the values of the two main diffusion coefficients. While the cross diffusion coefficients of the salt, D_{12}, show an increasing trend with increasing salt and protein concentrations, the cross diffusion coefficient of the protein fluctuates wildly. The cross coefficient of the protein is of the same order of magnitude as that of the salt.

Thermodynamic Formulation of the Diffusion Coefficients

On the assumption that an equivalent representation of Fick's law in the thermodynamic formalism would require replacement of the concentration gradients by chemical potential gradients, the Fickian D-coefficients would be converted into thermodynamic L-coefficients. The essential interest is in the L cross coefficients. Since irreversible thermodynamics asserts that $L_{ij} = L_{ji}$, its validity to the system under study would prove that a sound approach has been adopted in the formulation of the flow equations, the calculation of the diffusion coefficients and the design of the experimental methods.

To obtain the L_{ij} coefficients from the diffusion coefficients, data on the partial chemical potentials for the system BSA–KCl–H_2O are required in the concentration range of the experiment. These partial chemical potential values could be evaluated from the activity coefficients reported by Scatchard[4] and by Harned and Owen.[5] Since in correcting the diffusion coefficients for specific volume changes, the diffusion coefficients were corrected with terms including the density, the derivatives of chemical potentials with respect to molality are expressed in terms of partial mass densities. The molalities corresponding to these concentrations are in the range 0.015 to

0.2 for KCl and 0.2 to 0.6×10^{-6} for BSA. $(\partial\mu_i/\partial\varrho_j)$ is not equal to $(\partial\mu_j/\partial\varrho_i)$ while $(\partial\mu_i/\partial m_j)$ equals $(\partial\mu_j/\partial m_i)$ where chemical potentials are expressed in terms of molalities.

The cross L coefficients calculated on this basis were not equal to one another. The discrepancies were rather large and sometimes even the sign was wrong.

DISCUSSION

Our main aim in setting up the thermodynamic formulation of the diffusion problem in multicomponent systems was to check the applicability of the so-called generalized Fick equation, by examining its equivalent representation in terms of Onsager's methodology of irreversible thermodynamics. This would also serve to test the validity of splitting the diffusion coefficients in multicomponent systems into *main* and *cross* coefficients.

The test for the validity of the Onsager reciprocal relation is whether $L_{ij} = L_{ji}$. Our results show that this is not so, and thus the Onsager reciprocal relation is not valid. Broadly, the failure could be due to (a) incorrect diffusion coefficient, (b) incorrect thermodynamic data or (c) severe limitations in the applicability of irreversible thermodynamics.

1 Experimental Accuracy

In view of the fact that the cross coefficients are large, the instrumental sensitivity required to obtain accurate diffusion coefficients is not as extreme as in the simple multicomponent electrolyte systems studied by Gosting[1]. Nevertheless our instrumental accuracy was high, giving an uncertainty of 0.1% in the diffusion coefficients in the two-component systems.

The errors in the evaluation of the individual concentrations of KCl and BSA stem from the errors in the fringe number read out by the microcomparator used. The microcomparator could read the position of the fringes to an accuracy of 5μ entailing an error in the concentrations of not more than 5%. These errors are usually smoothed out by the integration procedures used to obtain the diffusion coefficients, hence the overall error in the cross coefficients is no more than approximately 10% whereas the deviation from the Onsager reciprocal relation is much higher than this.

2 Errors in the Main Diffusion Coefficients

The main diffusion coefficients D_{11} and D_{22}, unlike the cross coefficients were not obtained directly from the multicomponent system but from the two-component system $KCl-H_2O$ and $BSA-H_2O$, on the assumption that the main diffusion coefficients could be considered independent of the concentration of the second solute. However, on the basis of the thermodynamic formulation this need not be true. In fact the thermodynamic formulation required that not only the cross coefficients but also the main diffusion coefficients be dependent on the concentration of both the solutes. It would be difficult to derive the magnitude of this concentration dependency on theoretical grounds, since the concentration dependence of the corresponding L coefficients has not been studied thoroughly either theoretically or experimentally. However, to obtain $L_{ij} = L_{ji}$ it would be necessary that the main diffusion coefficients of the protein should be less than the values used and that of the salt should be more by factors of ten to hundred. Further, to obtain $L_{ij} = L_{ji}$ the main diffusion coefficient of the protein should be strongly dependent on the salt concentration. These requirements do not seem reasonable from the thermodynamic point of view, since it can be shown that within the concentration range of the mixture used,

$$D_{11} = L_{11} \left(\partial \mu_1 / \partial \varrho_1 \right), \quad \text{and} \quad D_{22} = L_{22} \left(\partial \mu_2 / \partial \varrho_2 \right).$$

Under these conditions the values used for the main diffusion coefficients may not be a bad approximation.

3 Possibility of the presence of Additional Components

In the diffusional and thermodynamic treatments of the $BSA-KCl-H_2O$ system, the solvent as the third diffusing component has been eliminated, leaving only four diffusion coefficients to be measured. Nevertheless, it is possible that additional components are present due to chloride binding by the protein, isomerization of the protein or the presence of a pH gradient. In the present experiments the initial composition of $BSA-KCl-H_2O$ in the two layers was chosen such that additional components were negligible and could not be the main cause for the failure of the Onsager reciprocal relation.

4 Extension of Fick's Law

Apart from the way in which the diffusion coefficients are split into main and cross coefficients in the Fick equation, it is also assumed that the diffu-

sion coefficients are dependent only on the concentrations of the various
solutes. However, if the diffusion coefficients depend also on the concentration gradients, then even on theoretical grounds alone the flows are not
linear in the thermodynamic forces, and hence the present form of irreversible
thermodynamics does not apply. In such dilute systems, it would be difficult to visualize other possibilities such as time-dependence of the diffusion
coefficients or some such memory processes, which could still be within the
limits of irreversible thermodynamics.

5 Presence of Mass Flow Terms Other Than the Diffusional

We have already indicated that the diffusion coefficients were corrected for
volume flow due to changes in the specific volume of the mixtures as the
diffusion proceeds. It is generally assumed that in most diffusion processes
where an initial concentration gradient is set up between two layers of
liquid, only mass flow occurs, and this is purely diffusional flow. An acceleration process, if present initially due to the high concentration gradient, dies
out very quickly. A direct experimental test for this statement in a properly
devised diffusion experiment would be worthwhile. In addition, in a system
like the one we have chosen where there is a strong interaction between a
slow-moving component and a fast-moving one, the problem of momentum
flow or poorly maintained mechanical equilibrium with the chance of instability is as yet unexplored.

A detailed examination of the Rayleigh fringe photographs and the Gouy
fringe profile diagrams for the system BSA–KCl–H_2O does not show any
undue convection spots. The oscillations observed in haemoglobin concentration profiles in the system BSA–Hb–KCl–H_2O are in keeping with similar
oscillations in the BSA concentrations in the BSA–KCl–H_2O experiment.
However, the total concentration gradient or density gradient is a smooth
function. Hence if there was any momentum flow it is not evident experimentally. Using light absorption scanning where a small amount of strongly
absorbing coloured material has been introduced into the diffusion system,
and the sensitive Gouy photoelectric scanning methods, it should be possible
to detect momentum flow of the kind described above. However the point
under discussion here is whether momentum flow could have given spurious
diffusion coefficients or could have made the system unsuitable for testing
the Onsager reciprocal relation. In view of the fact that it was negligible it
could not have interfered with the measurements.

Having thus established the reliability of our procedure and having eliminated the possibility of disturbance of the local mechanical equilibrium, we are led to the conclusion that the disturbance of the local thermodynamic equilibrium during diffusion is the only process that could account for the failure of irreversible thermodynamics.

It would be worthwhile to make a further study of local thermodynamic equilibrium, and the question of mechanical equilibrium in multicomponent systems, in relation to generalized Fick's equation and Onsager's reciprocal relation.

ACKNOWLEDGEMENT

This work was done in partial fulfilment of the Ph.D. at Weizmann Institute of Science, under the supervision of Prof. A.Katchalsky and Dr. N.Shavit.

REFERENCES

1. R.J.Gosting, *Advances in Protein Chemistry*, Vol. 11 (1956).
2. D.S.Kahn and A.Polsen, *J. Phys. and Colloid Chem.*, **51**, 816 (1943).
3. Ram Mohan and N.Shavit (to be published).
4. G.Scatchard, *J. Amer. Chem. Soc.*, **68**, 2610 (1946).
5. H.S.Harned and B.B.Owen, *The Physical Chemistry of Electrolyte Solutions*, Reinold N.Y. 1958.

Mass and other effects
on diffusion processes

Isotope effects in self-diffusion

A. D. LE CLAIRE

Solid State Division,
Atomic Energy Research Establishment,
Harwell, Didcot, Berks

INTRODUCTION

Self-diffusion, or the diffusion of one set of molecules into another identical set, is a concept that could only have followed on the firm establishment of the molecular theory of matter. This was not of course achieved until the time of Maxwell and Clausius so it is not too surprising that Graham's extensive writings on diffusion contain no mention of the idea of self-diffusion. It seems first to have been discussed by Maxwell and while he recognised it as being the ideally simplest type of diffusion to consider from a theoretical point of view, he recognised too that it was not a property that, strictly, could ever be studied experimentally. The very notion implies that one can distinguish the one set of the molecules from the other, and this one can never do, other than conceptually: if they are distinguishable, through some physical attribute, they can no longer be physically identical.

However, the use of radioactive or stable isotopes allows one to approach very closely indeed to the condition of what we may call "ideal self-diffusion". Isotopes are identical chemically and the small difference in mass between two interdiffusing isotopic species can often be ignored. Nevertheless, diffusion rates are determined in part by mass and there are measurable differences in the diffusion coefficients of two isotopes, diffusing in a common host, when they have different masses. These measurable differences are what we mean by an "isotope effect". By switching from one isotope to another in the same measurement we change in a known way just one of the many factors determining the diffusing rate and the resulting differences in D can often provide valuable information on the nature of the diffusion process operating. Such "isotope effect" measurements have proved especially valuable in solids.[1]

Apart from their use in measuring self-diffusion coefficients, the other great virtue of radioactive tracers is that they can be very accurately assayed even at very low concentrations. This means that one need work only with extremely small concentrations of tracer, so small in fact that even if diffusion is into some chemically different substance, the chemical concentration gradients are so small that diffusion occurs essentially in a chemically homogeneous system and therefore under near equilibrium conditions. Thus one can measure self-diffusion coefficients of the constituents of an alloy or a compound at a well defined composition. As a special case, one can measure the so-called Impurity Diffusion Coefficient of a foreign species of atom in a pure material at effectively vanishingly small concentrations.

All these measurements lend themselves to mass effect determinations, provided only that two or more suitable isotopes are available for the system under study.

Let us now see what part the mass of a diffusing isotope plays in determining its diffusion rate.

MASS DEPENDENCE OF D

Because we are generally dealing in isotope effect measurements with homogeneous systems we can conveniently start from the quite general relation for D

$$D = (1/6t) \langle R^2 \rangle \tag{1}$$

in terms of the mean square displacement, $\langle R^2 \rangle$, of a diffusing atom in time t. R is the vector sum of individual atom displacements r_i, so if there are N displacements in time t, and N is very large,

$$\langle R^2 \rangle = \langle (\Sigma r_i)^2 \rangle$$
$$= N \langle r_1^2 \rangle + 2N \{ \langle r_i \cdot r_{i+1} \rangle + \langle r_i \cdot r_{i+2} \rangle + \cdots \}$$
$$= N (\langle r_1^2 \rangle + 2 \langle r_i \rangle^2 \{ \langle \cos \theta_1 \rangle + \langle \cos \theta_2 \rangle + \cdots \}) \tag{2}$$

$\langle \cos \theta_1 \rangle$ is the mean value of the cosine of the angle between one displacement and the next, $\langle \cos \theta_2 \rangle$ the same between one displacement and the next but one, and so on. (In writing the last line of (2) we have assumed displacement lengths are independent of displacement directions, which will be true for the case we shall consider).

a Gases

If we are dealing with diffusion in a gas, the r_i are simply the flight paths between successive molecular collisions. Then $\langle r_i \rangle = \lambda$, the mean free path, and $\langle r_i^2 \rangle = 2\lambda^2$, as follows from the probability for a path of length $r - p(r) = (1/\lambda) \exp - r/\lambda$. Also, it is not difficult to show that for molecular collisions $\langle \text{Cos } \theta_j \rangle = \langle \text{Cos } \theta_1 \rangle^j$. The series for $\langle R^2 \rangle$ is now easily summed to give

$$\langle R^2 \rangle = \frac{2N\lambda^2}{1 - \langle \text{Cos } \theta_1 \rangle}. \tag{3}$$

If c is the mean molecular velocity, $N\lambda = ct$, so that

$$D = \frac{1}{3} \cdot \frac{c\lambda}{1 - \langle \text{Cos } \theta_1 \rangle}. \tag{4}$$

Elementary treatments assume that collisions deflect an atom in a random way so that $\langle \text{Cos } \theta_1 \rangle$ is zero. But this is not true. After a collision the velocity of a molecule is more likely than random to have a component along the direction of its motion *before* the collision—what is known as the "persistence of velocities" effect—so that $\langle \text{Cos } \theta_1 \rangle$ has some positive value.[2,3]

For molecules of mass m_1 diffusing at very low concentration in a gas of molecular mass m_2, the expression for c, λ and $\langle \text{Cos } \theta_1 \rangle$ are, for an elastic sphere model,[2]

$$c_1 = 2\,(2kT/\pi m_1)^{1/2} \tag{5}$$

$$\lambda = 1/\pi \varrho S^2 \left(1 + \frac{m_1}{m_2}\right)^{1/2} \tag{6}$$

and

$$\langle \text{Cos } \theta_1 \rangle = (m_1 - \delta m_2)/(m_1 + m_2). \tag{7}$$

S is a cross-section for the collision of the two molecular species and ϱ the number density of molecules. δ is a slowly varying function of m_1/m_2 lying between 0 for $m_1/m_2 \to 0$ and $\frac{1}{3}$ for $m_1/m_2 \to \infty$. $\delta = 0.188$ for $m_1 = m_2$.

Equation (4) for D_1 is therefore

$$D_1 = \frac{2\,(2kT)^{1/2}}{3\pi^{3/2}\varrho S^2\,(1 + \delta)} \cdot \left(\frac{m_1 + m_2}{m_1 m_2}\right)^{1/2}, \tag{8}$$

from which we can define a mass-effect as the ratio of the D's for two isotopes of the same species having masses $m_1 = m^\alpha$ or m^β,

$$\frac{D^\alpha}{D^\beta} = \left(\frac{m^\beta}{m^\alpha}\right)^{1/2} \left(\frac{m^\alpha + m_2}{m^\beta + m_2}\right)^{1/2} \left(\frac{1 + \delta_\beta}{1 + \delta_\alpha}\right). \tag{9}$$

There are more sophisticated expressions for D than that given by equation (8) but all calculations agree in giving an equation like (9) for the mass effect.[2,3] Usually the last term, with the δ's, can be omitted.

It is interesting to note how important it is to include proper account of the non-random aspects of the collision process—the persistence of velocities term, $\langle \text{Cos } \theta_1 \rangle$. Without it, there would be quite a different type of mass dependence.

We may comment too on the difference in form between equation (9) and Graham's laws of diffusion. The first factor in equation (9), $(m^\beta/m^\alpha)^{1/2}$, derives from the mass dependence of c. This same factor, alone, occurs in Graham's laws because the two phenomena to which they refer are both governed only by the magnitude of c. (These are (1) effusion through a hole of dimensions $\leq \lambda$, the rate of which is proportional to c, and (2) counter diffusion through a porous plug between two gases at equal pressure, where the rate of flow of each gas is inversely proportional to the momentum mc).[4] The diffusion we are considering is governed not only by c but also by collision between dissimilar molecules, the relevant parameters of which are λ and $\langle \text{Cos } \theta_1 \rangle$. The mass dependence of these two quantities (eqs. 6 and 7) combine to give the other mass factors in equation (9).

The kinetic theory of gases would appear to be sufficiently well developed for the role of mass in the various transport processes to be fully understood, and indeed, eq. (9) contains nothing but mass terms. It follows that there is little to be learned in general from measurements of the isotope effect and few such measurements seem to have been made. As an isolated example we may quote measurements of D^α/D^β for H and D diffusing in Tritium.[5,6] The results differed slightly from equation (9) and this was taken to indicate a difference in the cross-sections, or effective radii, of D and H molecules.

The main use of equation (9) is for correcting measured tracer diffusion coefficients into true self-diffusion coefficients, when the accuracy warrants it.

The situation is quite different with liquids and solids. In solids especially isotope effect measurements can be very informative and in recent years this has become a very fashionable area of investigation.

b   Solids

Returning to equation (2) for $\langle R^2 \rangle$, in solids the displacements r_i are the vectors representing atom jumps between lattice sites. In most simple solids these are all of the same magnitude, so if a is the interatomic spacing, $\langle r_i \rangle = a$ and $\langle r_i^2 \rangle = a^{2*}$. If each atom makes Γ displacements in unit time $N = \Gamma t$, and from the resulting equation for $\langle R^2 \rangle$ we obtain

$$D = \tfrac{1}{6} \Gamma a^2 \cdot f \tag{10}$$

where the quantity f is

$$f = 1 + 2 \sum_j \langle \text{Cos } \theta_j \rangle \tag{11}$$

and is called the "correlation factor".

Again, elementary treatments assume that successive atom jumps in a crystal occur at random so that the $\langle \text{Cos } \theta_j \rangle$ are zero and $f = 1$. But this is not necessarily so, as we shall see, and the correlation factor carries, just as with gases, an important dependence on mass. It is therefore essential to consider it in discussing isotope effects. In solids, unlike the case of gases, there are several different mechanisms by which atoms migrate. In general, each has a different value of f and this contributes to each having a different mass dependence of D.

It is quite easy to see why for many mechanisms of atomic migration the successive jump directions are not at random.

Consider vacancy diffusion. Suppose an atom has just made a jump from site (B) to site (A) (Fig. 1) so that there is a vacancy now at (B). If atom jumps really occurred at random the next jump would be as likely to (C) as back again to (B). But the probability will be greater than random that the next jump will be back to (B), simply because there is a vacancy already there to effect such a jump, and less than random that the next jump will be to C, for this would require the prior migration of the vacancy around to C. Successive jump directions are therefore correlated with one another and the $\langle \text{Cos } \theta \rangle$ cannot be zero. Since the tendency is for consecutive jumps to leave an atom undisplaced the net effect of this correlation must be to reduce D below the value calculated assuming jumps occur randomly, so that f is always less than one. (By contrast, in gases, f is always greater than one).

---

* More general cases are readily treated[1] but we shall not deal with these here.

X   X   X

X   ⊗   ◯
C   A   B

X   X   X

X   Non Tracer Atoms
⊗   Tracer Atom
◯   Vacancy

Figure 1   Vacancy diffusion

In many simple cases the relation $\langle \cos \theta_j \rangle = \langle \cos \theta_1 \rangle^j$ holds too for vacancy diffusion*. Then the series for f can be summed to give

$$f = \frac{1 + \langle \cos \theta_1 \rangle}{1 - \langle \cos \theta_1 \rangle}. \tag{12}$$

There is a very similar effect with interstitialcy diffusion—see Fig. 2. In an interstitialcy jump an interstitial atom, shown at (a), moves onto a lattice site and the atom on that site is displaced into an interstitial site, as at (b). The next jump of the atom must be back to an interstitial site again and, as before, there is a tendency on average for it to be in the reverse direction of the previous jump because, in this case, there is an interstitial atom available to cooperate in a jump in this direction. Obviously here too $f < 1$.

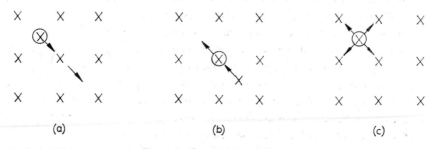

(a)   (b)   (c)

Figure 2   Interstitialcy diffusion

---

* This relation is valid provided there is at least two-fold symmetry around the jump vector. It does not therefore hold, e.g. for diffusion by divacancy pairs. Such cases are readily treated by more general methods[1], but we shall not deal with them here.

But now there is a difference from vacancy diffusion. The *third* jump of the atom from the interstitial site it occupies after the second jump will clearly occur with equal probability in any of the allowed directions (Fig. 2c). There is nothing to predispose it to jump preferentially in any particular direction. Thus while the first and second jump directions are correlated, the third jump direction is *not* correlated with the second, nor therefore with the first. This means that $\langle \text{Cos } \theta_j \rangle = 0$ for $j \geq 2$, so that f in this case reduces simply to

$$f = 1 + \langle \text{Cos } \theta_1 \rangle_{i-n-i}, \tag{13}$$

if the average is taken only for consecutive pairs of jumps that are correlated, i.e. interstitial to normal to interstitial pairs.

It will be clear now that for simple interstitial diffusion, where atoms diffuse by jumping between interstitial sites, there is no correlation between successive jump directions and $f = 1$. The same is true too for any migration by mutual exchange of positions of two or more atoms.

Calculations of f for self-diffusion have been made[7] for a number of cases and some results are shown in Table 1.* This illustrates how f depends on the

Table 1   Correlation factors for self-diffusion

	$\langle \cos \theta_1 \rangle$	f
Vacancy diffusion		
Diamond	$-1/3$	0.50
S. Cubic	-0.2098	0.6531
B.C.C.	-0.1579	0.7272
F.C.C.	-0.1227	0.7815
H.C.P.		$f_x = f_y = 0.7812$
		$f_z = 0.7815$
Divacancy diffusion		
F.C.C.	—	0.475
Interstitialcy diffusion		
B.C.C. (110) jumps	-0.333	0.666 (Direct)
	-0.068	0.932 (Indirect)
F.C.C. (100) jumps	-0.20	0.80 (Direct)
	0	1.0 (Indirect)
Exchange, ring or interstitial Diffusion		
Any Structure		1.0

* Methods of calculation are reviewed in reference 1.

crystal structure and also on the mechanisms of diffusion—e.g. for b.c.c. crystals $f = 0.73$ for the vacancy and 0.66 and 0.92 for the two types of interstitialcy jumps.

Now f is in these cases of self-diffusion a numerical constant and this is because there is only one jump frequency involved, all atoms and all jumps being identical. But this is not quite the situation in a tracer self-diffusion experiment. The difference in mass between the tracer isotope and the atoms of the host crystal will mean there will be differences in the jump frequencies of these two species and it is clearly vital for understanding the isotope effect to know how this affects the correlation factor.

There is an identical situation in considering the correlation factor for impurity diffusion—the diffusion of some foreign impurity tracer into a crystal—and tracer self-diffusion is in this connection only a specially simple case of impurity diffusion. So let us consider this and assume, say, a vacancy mechanism.

Fig. 3 shows an impurity atom (shaded) which we suppose has just arrived at this position by exchange with a vacancy which is now on a nearest neighbour site. We can distinguish several different jump frequencies for the

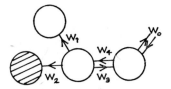

Figure 3   Impurity diffusion

vacancy. It may exchange with the impurity at a rate ω_2 or with a first nearest neighbour solvent atom (in f.c.c. crystals) at a rate ω_1, or it may dissociate and move to a second nearest neighbour at rate ω_3. ω_4 is the jump rate for the opposite association jump. We suppose all other vacancy jumps occur at the same rate ω_0 as in pure solvent, these occurring too far away to be affected by the impurity.

The correlation factor for this situation can be calculated[8,1] and turns out to be, for a f.c.c. crystal,

$$f = \frac{\omega_1 + F\omega_3}{\omega_2 + \omega_1 + F\omega_3} \tag{14}$$

where F is a slowly varying function of (ω_0/ω_4) lying between 1 and 7/2, the nature of which need not concern us, except to mention that $F(1) = 2.5756$.

It is easy to see why f should depend on the relative values of these jump frequencies by considering a couple of extreme cases.

Suppose that $\omega_2 \gg \omega_1, \omega_3$. Equation (14) shows that f will then be very small. This is because under these conditions one jump of the impurity is very likely indeed to be followed by a jump in the reverse direction by immediate exchange again with the vacancy. Successive jump directions are very highly correlated so that $\langle \text{Cos } \theta_1 \rangle$ approaches -1 and f is therefore very small.

On the other hand, suppose $\omega_2 \ll \omega_1, \omega_3$. After one jump of the impurity, a second jump will only occur on average after the vacancy has made many exchanges with solvent atoms. This will tend to randomise the direction from which the vacancy makes its second exchange with the impurity, so $\langle \text{Cos } \theta_1 \rangle$ will be near zero and f near one, as equation (14) confirms.

As we expect, equation (14) yields $f = 0.78146$, the value for pure self-diffusion, when we put $\omega_1 = \omega_2 = \omega_3 = \omega_4 = \omega_0$.

The most important feature of equation (14), because it leads as we shall see to a particularly simple result for the isotope effect, is that the same combination of solvent jump frequencies occurs in both numerator and denominator so that f is of the form

$$f = u/(\omega_2 + u) \tag{15}$$

where u contains solvent frequencies only. This, it turns out, is a fairly general form for f covering a large number of commonly occurring situations. There are exceptions[1], but we shall not discuss them here.

Now we can see how the correlation factors for two diffusing tracer isotopes will differ from one another, and from the value for pure self-diffusion. The mass differences make the ω_2's slightly different so for two isotopes α and β there are the two correlation factors

$$f^\alpha = u/(\omega_2^\alpha + u) \quad \text{and} \quad f^\beta = u/(\omega_2^\beta + u). \tag{16}$$

The u is the same in both because it contains only non-tracer atom jump frequencies.

The equation for D for each isotope tracer is, from equation (10)

$$D = \tfrac{1}{6} a^2 n_d (\omega_2 f) \tag{17}$$

where we have written for Γ, $\Gamma = \omega_2 n_d z$, where n_d is the fractional defect concentration and z the coordination number of jump directions available to an atom. The whole mass dependence of D resides in the bracket term,

11   Sherwood I (1426)

$(\omega_2 f)$ in (17), so

$$D^\alpha/D^\beta = \omega_2^\alpha f^\alpha/\omega_2^\beta f^\beta. \tag{18}$$

From (18) and (16) we can eliminate u and one of the f's to give

$$(D^\alpha/D^\beta) - 1 = f((\omega_2^\alpha/\omega_2^\beta) - 1) \tag{19}$$

where we have now dropped the distinction between f^α and f^β, their difference being insignificant in this equation. This is a very useful expression because we can now forget about the mass dependence of f and concentrate attention on its relation to mechanism. It remains now to consider the mass dependence of the jump frequency ratio $\omega_2^\alpha/\omega_2^\beta$.

Elementary classical mechanical derivations of ω give the result

$$\omega = \nu \exp\left(-Q/kT\right). \tag{20}$$

Q is the activation energy or potential barrier height for the atomic jump, which does not depend on isotope mass. ν is an average vibrational frequency of the migrating atom and vibrational frequencies of individual bound atoms vary inversely with the square root of their mass. Thus $\omega_2^\alpha/\omega_2^\beta = (m^\beta/m^\alpha)^{1/2}$ and therefore[9]

$$(D^\alpha/D^\beta) - 1 = f((m^\beta/m^\alpha)^{1/2} - 1). \tag{21}$$

This though is a very approximate result because equation (20) considers the jump process as involving only the atom that is diffusing. Any motional disturbance the rest of the lattice may undergo as a result of the atom jumping is completely ignored. This is clearly a very serious approximation in discussing mass effects, because the more atoms that are involved the less the effect will be on the dynamics of the jump process of changing the mass of one of them. In other words, if there is any appreciable coupling between the migrating atom and the rest of the lattice, inducing some motion in other atoms during a jump, we expect the mass effect to be reduced, by some fraction ΔK say, to give the result, with a little re-arrangement,[10,11]

$$E_\beta^\alpha(1) = \frac{(D^\alpha/D^\beta) - 1}{(m^\beta/m^\alpha)^{1/2} - 1} = f\Delta K. \quad 0 < \Delta K < 1 \tag{22}$$

The experimentally determinable quantity in the centre of this equation is called, in solid state diffusion, "the mass effect" and denoted by $E_\beta^\alpha(1)$.

We have been assuming so far that only one atom migrates at a time, as in vacancy diffusion. If there are two or more atoms migrating simultaneously, as in interstitialcy diffusion say, an extension of the argument gives

$$E_\beta^\alpha(n) = \frac{(D^\alpha/D^\beta) - 1}{\left(\dfrac{m^\beta + (n-1)\,m}{m^\alpha + (n-1)\,m}\right)^{1/2} - 1} = f \Delta K \qquad (23)$$

n is the number of atoms jumping simultaneously—$n = 2$ for interstitialcy—and m is the average mass of the host atoms. $E(n)$ is then "the mass effect" for an assumed n-atoms jump process. Obviously $E(n)$ increases with n, but since f and ΔK are both fractional $E(n)$ must be less than one. It follows that mass-effect measurements can be used to put an upper limit on the value of n.

Before discussing experimental results we need to say a little more about the factor ΔK. The way in which it was introduced as a fraction into equation (22) can be fully justified when an atomic jump is properly described and treated as a many-body process. The most convincing treatment is based on a model due to Vineyard.[12] With this, the isotope effect problem reduces to considering the effect of the mass change only on the total motion of the system, (migrating atom plus lattice) as the migrating atom begins to move from the saddle point to complete its jump—i.e., only on the motion leading to decomposition of the saddle-point configuration. There is a certain kinetic energy associated with this motion and it turns out that ΔK is just the fraction of it that is associated with the migrating atom itself. If only the migrating atom is moving (Fig. 4b) then clearly $\Delta K = 1$. If there is a concomitant motion of lattice atoms (Fig. 4c) then $\Delta K < 1$.

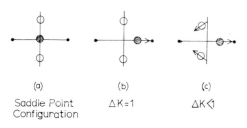

(a)
Saddle Point
Configuration

(b)
$\Delta K = 1$

(c)
$\Delta K \lessdot 1$

Figure 4

It seems very likely that ΔK will always be less than one if there is any relaxation of neighbouring atoms around a defect.[11,13] Figure 4c is drawn with this in mind. As the migrating atom moves from the saddle-point nearby atoms will at the same time begin to move towards their relaxed positions around, say, a vacancy, and the larger the ultimate amount of relaxation the smaller we might expect the fraction ΔK to be. One can in fact construct a very simple argument on these lines to give the result[11]

$$\Delta K \approx \frac{1}{1 + (N/3)\,(|1 - \Delta V_f|)} \qquad (24)$$

for ΔK in terms of the volume change ΔV_f on forming a defect. N is the number of saddle-point atoms that relax (~ 6 for b.c.c., ~ 4 for f.c.c.). If $\Delta V_f = 1$ atomic volume there is no relaxation and $\Delta K = 1$.

EXPERIMENTAL MEASUREMENT OF THE MASS EFFECT E

Measurements, usually with radioactive tracers, are always made by diffusing the two isotopes simultaneously into the host crystal from very thin layers of them (often only a few atomic layers) co-deposited onto the crystal surface. After diffusion, the sample is cut into slices and the activities c^α and c^β of the two isotopes determined in each slice. This requires that it must be possible to discriminate between the two tracers when present together, so their half-lives or the types or energies of the radiation they emit must be sufficiently different. The equations for c^α and c^β in a slice at distance x from the surface are

$$c^\alpha = (\text{const}/\sqrt{\pi D^\alpha t})\,\exp - x^2/4D^\alpha t$$

and

$$c^\beta = (\text{const}/\sqrt{\pi D^\beta t})\,\exp - x^2/4D^\beta t$$

from which it is easily found[14] that, because the times of anneal t are identical,

$$\ln(c^\alpha/c^\beta) = ((D^\alpha/D^\beta) - 1)\ln c^\alpha + \text{const.}$$

Thus, a plot of $\ln(c^\alpha/c^\beta)$ vs. $\ln c^\alpha$ gives a line of slope $(D^\alpha/D^\beta) - 1$, and to determine the mass effect it is only necessary to perform measurements of activity.

Fig. 5 shows a typical measurement[13] and illustrates the sort of precision that can be achieved. The slopes, and therefore the mass effect, can be determined to within a few per cent. This may seem surprising because it is

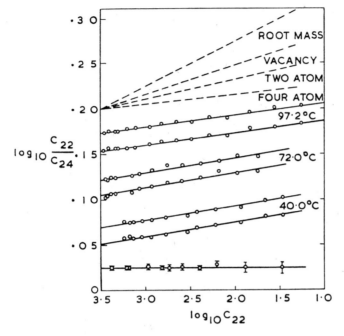

Figure 5 ^{22}Na and ^{24}Na diffusion in sodium

comparable with the accuracy of determination of a single D, but this accuracy is achieved largely because the technique avoids time and distance measurements and the usual errors associated with these.

RESULTS FROM SELF-DIFFUSION MEASUREMENTS

Table 2 shows the results of measurements of the isotope effect for self-diffusion in solids. From the experimental values of $D^x/D^\beta - 1$ the mass effect $E(1)$ has been calculated. These are all less than unity as the theory requires.

For the close-packed structures in the top half of the table the values of $E(2)$ (not shown) would all be greater than unity indicating the impossibility of any two-atom processes being responsible for their self-diffusion. In fact, of course, there is a deal of evidence that diffusion in these substances is by single vacancies so from the values of $E(1) = f\Delta K$, with $f = 0.782$ (Table 1), we obtain the values of ΔK in column 4. With the exception of Pd these are all

Table 2 Mass effect measurements—self-diffusion

Tracers	Solvent	$(D^\alpha/D^\beta) - 1$	$E^\alpha_\beta(1) = f\Delta K$	ΔK $\left(\begin{array}{l}f = 0.781 \text{ f.c.c.}\\ f = 0.727 \text{ b.c.c.}\end{array}\right)$	ΔK (Eq. 24)	Authors
Ag[105,111]	Ag	0.0184	0.67	0.86 ± 0.05	0.88	Barr and Peterson[20]
Cu	Cu	0.0154	0.68	0.87 ± 0.02		Rothman[21]
Pd[103,112]	Pd	0.0334	0.81	1.04 ± 0.05		Peterson[22]
Fe[52,59]	γ-Fe	0.0325	0.53	0.68 ± 0.02	0.77	Walter and Peterson[23]
Fe[55,59]	γ-Fe	0.0198	0.57	0.73 ± 0.08	0.77	Heumann and Imm[25]
Na[22,24]	NaCl	0.032	0.73	0.93 ± 0.14		Barr and Le Claire[14]
Zn[65,69]	Zn∥c	0.0211	0.70	ΔK_B = 0.92 ± 0.02		Peterson and
	⊥c	0.0225	0.75	ΔK_A = 0.95 ± 0.02		Rothman[24]
Na[22,24]	Na	0.0163	0.36	0.50 ± 0.02	0.51	Mundy, Barr and Smith[13]
Fe[52,59]	α-Fe	0.0272	0.43	0.59 ± 0.07		Walter and Peterson[23]
Fe[52,59]	δ-Fe	0.0206	0.34	0.46 ± 0.01		Walter and Peterson[23]
Ag[105,111]	AgBr	Vacancy	0.66			Barr and Peterson[20]
		D. interstitialcy	~1 ± 20%			
		I. interstitialcy	~0 ± 20%			

significantly less than unity (by ~ 10–20%) indicating a significant participation of non-migrating lattice atoms in the jump process.

In the case of Cu, measurements have been made over a range of temperature without indicating any significant variation in $E(1)$. This suggests that divacancy pairs, with their much smaller value of f (see Table 1) cannot be contributing much to the self-diffusion, a matter which is taken up later in the paper by Mehrer and Seeger.

For the non-close-packed metals Na and α and δ-Fe, $E(1)$ is quite noticeably less than for the close-packed materials. In fact, their values of $E(2)$ are also less than unity so this could mean that diffusion in these b.c.c. metals is by an interstitialcy process. However, there is strong evidence, from thermal expansion measurements[15], that the dominant defects in Na are vacancies so it is not unreasonable to conclude that diffusion is by a vacancy mechanism. (There is evidence[13] that E for Na shows a small but definite decrease with increase in temperature, which might indicate more than one mechanism.) Assuming a vacancy mechanism for both Fe and Na, we obtain the rather small values of $\sim \frac{1}{2}$ for ΔK.

The next column shows ΔK calculated from equation (24). This appears to give reasonable agreement with experiment and in particular provides an explanation for this low value of ΔK for Na as due to the large relaxation that is believed, from pressure effect measurements, to occur in this metal.

The problem with Na is still far from settled. There have recently been attempts, by Professor March at Sheffield and Professor Huntington at Troy, to calculate ΔK from detailed considerations of the saddle-point motion as described by the Vineyard model. Although the results seem very sensitive to the atom interaction potential used in the calculation, it may be possible, for vacancy diffusion in close-packed structures, to reproduce theoretically the experimental values of ΔK. It seems on the contrary quite impossible, with any reasonable choice of potential, to do the same for the low ΔK value for vacancy diffusion in Na. Assuming the adequacy of the Vineyard model, this result is a serious obstacle to belief in this mechanism. There is now a considerable body of information, both experimental and theoretical, on the diffusion and defect properties of Na and in a later paper this is to be assessed in an attempt to decide what the mechanism is in this metal. It does not seem possible yet to draw a final and fully convincing conclusion.

One way out of the difficulty with the ΔK for Na is to argue that the Vineyard model is inadequate in some respect. It does concentrate attention on the saddle-point configuration where the effects of coupling with the lattice are at a minimum and it may be more appropriate to calculate ΔK from the point of view proposed by Rice, that the jump rate is governed by fluctuations at the site from which the atom jumps and where the effects of coupling may be greater. Some consideration of mass-effects from this viewpoint has been given by Flynn.[16]

The final entry in Table 2 is results for Ag diffusion in AgBr. Here, there are three simultaneous mechanisms operating in proportions varying with the temperature, viz. the vacancy and two types of interstitialcy jumps.[36]

In consequence the analysis of the isotope effect measurements is more complex[37] and only the final ΔK values are quoted. They are interesting in being the only measurements for what is known to be interstitialcy motion. The difference between the ΔK values for the two types of jump is very striking, the zero value for the indirect interstitialcy reflecting a zero mass effect for this process and presumably therefore its very many body nature.

RESULTS FROM INTERSTITIAL DIFFUSION MEASUREMENT

Table 3 shows values for the mass effect in diffusion of solutes that dissolve interstitially in solids. Consider first the top block of results, Li in Si to He in quartz. There is little doubt that in these systems diffusion is by simple interstitial jumps, for which $f = 1$, so the value of $E(1)$ is the value of ΔK. There is an experimental error of 5–10% on most of these values so there is no significant departure of ΔK from unity. It follows that interstitial jumps of comparatively light interstitial atoms do not entail any significant disturbance of the host lattice, a result which is not unexpected, at least for the systems in Table 3.

For Na diffusion in soda glass $f\Delta K$ looks to be very small, especially in view of the larger value for the not too dissimilar system of Ne in quartz. This suggests that diffusion of Na in glass does not occur by simple interstitial jumps. In fact $f\Delta K$ calculated for a two atom process, i.e. $E(2)$, is still below the upper allowed limit of unity, so it seems Na might be diffusing by an interstitialcy process in glass.

The results for hydrogen in metals also show rather low values of $E(1)$. This is not because diffusion is not by simple interstitial jumps but most probably because hydrogen isotopes are so light that quantum effects are important. Their vibrational frequencies v are so high[17,18] that they have ground state energies hv comparable with kT at the temperatures at which diffusion measurements are made. Accordingly, the equations for ω (Eq. 20,

Table 3 Mass effect measurements — interstitial diffusion

Tracers $\alpha \beta$	Solvent	$\dfrac{D^\alpha}{D^\beta} - 1$	$E^\alpha_\beta(1)$	References
Li[6, 7]	Si	0.075	0.94	Pell[26]
Li[6, 7]	W	0.07	0.88	McCracken and Love[27]
C[12, 13]	Fe	0.044	1.07	Bosman[28]
Ne[20, 22]	Quartz	0.05	1.02	Frank, Swets and Lee[29]
He[3, 4]	Quartz	0.14	0.91	Jones[30]
Na[22, 24]	Soda Glass	0.11	0.24	Barr and Mundy[31]
H[1, 2]	Pd	0.3	0.73	Jost and Widman[32]
H[1, 2]	α Fe	0.2	0.48	Heumann and Primas[33]
H[1, 2]	γ Fe	0.28	0.68	Heumann and Primas[33]
H[1, 2]	Ni	0.19	0.46	Eichenauer, Löser and Witte[34]
H[1, 2]	Cu	0.19	0.46	Eichenauer, Löser and Witte[34]
H[1, 2]	Steel	0.37	0.89	Frank, Lee and Williams[35]

or Vineyard's equation), on which our mass effect expressions are based, must be rewritten with quantum instead of the classical ($h\nu \ll kT$) partition functions that feature in their derivation. The resulting expressions[11,19] for $E(1)$ account satisfactorily for the isotope effects shown in Table 3. They also account for the differences in diffusion activation energies for the two isotopes that many measurements reveal, a difference that is not catered for in the classical expressions.

MASS EFFECTS IN SUBSTITUTIONAL IMPURITY DIFFUSION

The other major area where isotope effect measurements in solids are proving of considerable value is in the study of impurity diffusion. It will be sufficient here merely to indicate the connection and we shall not attempt to review the results.[1]

The equations for the mass effect, (22) and (23) apply as well to impurity diffusion as to tracer self-diffusion. The only difference is that the correlation factor may now have any value between zero and unity, depending on the relative values of the several jump frequencies occurring in the expressions for f, like equation (14). We can obtain from a mass effect measurement a numerical value for f, and therefore information on the jump frequencies, if we have a value for ΔK. It is currently assumed that this is the same for impurity diffusion as the value directly measured for self-diffusion in the solvent. This is very probably quite a good approximation but the extent of its validity has not yet been investigated.

The information on jump frequencies contained in a knowledge of f is of particular value when it is combined with knowledge of other quantities like the self and the impurity-diffusion coefficients themselves and the dependence of the host self-diffusion coefficient on the concentration of solute impurity. These contain in different combinations the same set of jump frequency ratios that feature in f, so that quite detailed analyses can be made to yield individual values for many of them.

Because the frequencies occurring in f will in general have different activation energies, f itself will vary with temperature. This provides in many cases a significant contribution to the overall temperature variation of an impurity diffusion rate and measurements of f as a function of T are hence of considerable value in understanding the detailed features of this type of diffusion.

Mass effect measurements for impurity diffusion are usually less informative as to mechanism simply because, for the defect processes, the magnitude of f is not unique. They can nevertheless be valuable in special cases and an example of their application in this connection is discussed in the paper by Dr. Miller.

CONCLUDING REMARKS

We can see from the examples we have discussed that measurements of the isotope effect in diffusion in solids can be most informative. Because of its dependence on the correlation factor they can provide a useful confirmation in many cases of the mechanism of diffusion that is operating: they may be especially valuable in the future in helping to establish mechanisms in the more unfamiliar and less studied systems. In impurity diffusion, where usually the mechanism of diffusion is known, they provide valuable information on relative jump frequencies of solute and solvent species. Of especial interest is the values they can provide of the quantity ΔK, for this contains important and significant information on the dynamics of atom jump processes that is not readily available by other methods. Such information is only beginning to be made use of in furthering an understanding of diffusion in solids and the case of Na which we have discussed illustrates how decisive and searching a role it may play.

REFERENCES

1. A.D. LeClaire, "Correlation effects in diffusion in solids", Vol. 10, Chap. 5 of *Physical Chemistry—An Advanced Treatise* (Academic Press (to appear 1970)).
2. Sir J. Jeans, *Introduction to the Kinetic Theory of Gases* (C.U.P. 1946).
3. S. Chapman and T. G. Cowling, *Mathematical Theory of Non-Uniform Gases* (C.U.P. 1939).
4. E. A. Mason and B. Kronstadt, *J. Chemical Education*, **44**, 740 (1967).
5. I. Amdur and J. W. Beatty, *J. Chem. Phys.*, **42**, 3361 (1965).
6. E. A. Mason, B. K. Annis and M. Islam, *J. Chem. Phys.*, **42**, 3364 (1965).
7. K. Compaan and Y. Haven, *Trans. Faraday Soc.*, **52**, 786 (1956) and **54**, 1498 (1958).
8. J. R. Manning, *Phys. Rev.*, **136 A**, 1758 (1964).
9. A. Schoen, *Phys. Rev. Letters*, **1**, 138 (1958).
10. J. G. Mullen, *Phys. Rev.*, **121**, 1649 (1961).
11. A.D. LeClaire, *Phil. Mag.*, **14**, 1271 (1966).
12. G. H. Vineyard, *J. Phys. Chem. Solids*, **3**, 121 (1957).
13. J. N. Mundy, L. W. Barr and F. A. Smith, *Phil. Mag.*, **14**, 785 (1966).
14. L. W. Barr and A.D. LeClaire, *Proc. British Ceramic Soc.*, **1**, 109 (1964).
15. R. Feder and H. Charbnau, *Phys. Rev.*, **149**, 464 (1966).
16. C. P. Flynn, *Phys. Rev.*, **171**, 682 (1968).
17. M. Sakamato, *J. Phys. Soc. Japan*, **19**, 1862 (1964).
18. W. Eichenauer, W. Löser and W. Witte, *Z. Metallk.*, **56**, 287 (1965).
19. Y. Ebisuzaki, W. J. Kass and M. O'Keefe, *Phil. Mag.*, **15**, 1071 (1967).
20. L. W. Barr and N. Peterson, Argonne Nat. Lab. Progress Report ANL-7155 (1965).
21. S. Rothman and N. Peterson, *Physica Status Solidii*, **35**, 305 (1969).
22. N. Peterson, *Phys. Rev.*, **136 A**, A 568 (1964).
23. C. M. Walter and N. Peterson, *Phys. Rev.*, **178**, 922 (1969).
24. N. Peterson and S. Rothman, *Phys. Rev.*, **163**, 645 (1967).
25. Th. Heumann and R. Imm, *J. Phys. Chem. Solids*, **29**, 1613 (1968).
26. E. M. Pell, *Phys. Rev.*, **119**, 1014 (1960).
27. G. M. McCracken and H. M. Love, *Phys. Rev. Letters*, **5**, 201 (1960).
28. A. J. Bosman, Thesis, University of Amsterdam (1960).
29. R. C. Frank, D. E. Swets and R. W. Lee, *J. Chem. Phys.*, **35**, 1451 (1961).
30. W. M. Jones, *J. Amer. Chem. Soc.*, **75**, 3093 (1953).
31. L. W. Barr and J. N. Mundy, (1964) unpublished.
32. W. Jost and A. Widman, *Z. Physik. Chemie*, **45B**, 285 (1940).
33. Th. Heumann and D. Primas, *Zeit. f. Naturforsch.*, **21a**, 260 (1966).
34. W. Eichenauer, W. Löser and W. Witte, *Zeit. f. Metallk.*, **56**, 287 (1965).
35. R. C. Frank, W. L. Lee and R. L. Williams, *J. Appl. Phys.*, **29**, 898 (1958).
36. R. J. Friauf, *Phys. Rev.*, **105**, 843 (1957).
37. L. W. Barr, N. Peterson and A.D. LeClaire (to be published).

Centrifugal fields as a tool in the study of diffusion in solids

L. W. BARR

Solid State Division, A.E.R.E.,
Harwell, Didcot, Berks

ABSTRACT

In addition to his more obvious contributions to the study of diffusion, Graham made an indirect contribution through the continuation of his diffusion studies by his research assistants. One of these, Roberts-Austen, made the first quantitative measurements of diffusion in a solid, gold in lead, and thus founded the subject of solid state diffusion.

Roberts-Austen's results are still of great interest because of the very high diffusion coefficients which he obtained and recently much work has been done on fast noble metal diffusion in lead and similar metals. While the hypothesis has been advanced that such diffusion is by interstitials, direct proof is still lacking.

Recently gold and silver in sodium and potassium have similarly been shown to be very fast diffusers even at room temperature. Because of this it is possible to set up equilibrium distributions of these solutes in large samples in reasonable times. This feature has been used to determine the effective mass of gold in sodium and potassium in a centrifugal field. The size of the effective mass confirms that in these solvents gold is in interstitial solution. These results and applications of the technique to other systems are discussed.

1   INTRODUCTION

"... my long connection with Graham's researches made it almost a duty to attempt to extend his work on ... diffusion to metals". With these words Roberts-Austen, formerly Graham's assistant, gives his reason for that interest in diffusion which resulted in the first quantitative measurement of diffusion in a solid, the diffusion of gold in lead[1]. In this remarkable paper Roberts-Austen reports not only the gold-lead measurement but also the use of a constant temperature furnace, with continuous temperature recording; a preliminary measurement of the diffusion of gold in silver, which measurement led him to propose a correlation between melting temperature and diffusion rates; and the first measurement of the distribution of a solute in a solid under the action of an external field, namely the distribution of gold in gold-lead and gold-bismuth alloys in a thermal gradient.

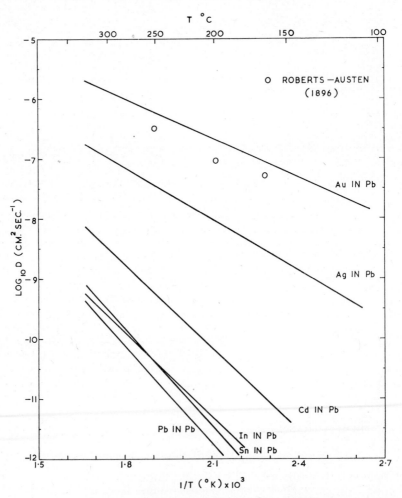

Figure 1   Self and impurity diffusion rates in lead

At the time Roberts-Austen's results on gold diffusion in lead excited interest because of the large values of the diffusion coefficients he found. As more information on the rates of self- and impurity-diffusion became available, notably the value for self-diffusion in lead[2], the very high diffusion rates of gold and silver in lead became increasingly striking. As a consequence it was proposed that these metals dissolve and diffuse interstitially in lead.

More recently there has been renewed interest in the diffusion of group 1b metals in lead[3,4], tin[5], indium[6], thallium[7], sodium[8,9] and potassium[10]. In these studies the concept of relative size of solute and solvent has been fruitful in suggesting possible systems where high diffusion rates might be found. While the available evidence favours interstitial solution and diffusion as an explanation for the observed fast diffusion in these systems, direct proof is lacking. X-ray measurements, which might distinguish between interstitial and substitutional solutions, seem not be to practical because of low solubilities.

It is useful at this point to compare self and impurity-diffusion in lead, as a typical example of the heavy metals, and in sodium, as an example of the alkali metals. This is done in Figs. 1 and 2.

In Fig. 1 the diffusion rates of Au, Ag, Cd[11], In and Sn[12] in lead are shown as a function of temperature. The very rapid diffusion of the noble metal solutes is clearly seen, as is the strikingly close agreement between Roberts-Austen's results, obtained by chemical techniques, and the latest radiotracer measurements made seventy years later. The decreasing diffusion rate with increasing valence in the sequence Ag, Cd and In is in marked contrast with the behaviour of the same solutes in noble metal solvents, e.g. Cu and Ag. In these solvents, of course, a satisfactory explanation has been given for the impurity diffusion rates in terms of electrostatic binding between the substitutional solute and vacancies[13]. In and Sn diffuse at rates very comparable with the rate of lead self-diffusion and while the reliability of the In and Sn results is doubtful (they are early chemical diffusion measurements) there is a possibility that Sn is faster than In diffusion.

In Fig. 2 are shown the diffusion rates of the same solutes in sodium. There is obviously a close resemblance to the lead system. Sn, however, certainly diffuses more rapidly than In, thus reversing the trend noted above. The diffusion of Au in potassium is also shown.

Central to the problem of accounting for these results is to determine, unambiguously, the mechanism of solution and diffusion of the noble metal

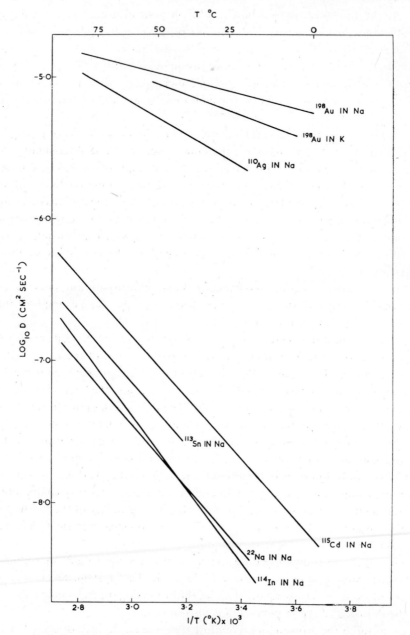

Figure 2 Self and impurity diffusion rates in sodium and gold diffusion in potassium

solutes. Because of their rapid diffusion equilibrium, or steady state, distributions of these solutes can be set up in reasonable times under the action of external fields (e.g. thermal and gravitational). In the particular case of gold in sodium and potassium, where the diffusion rate at room temperature is about 7.10^{-6} cm²sec^{-1}, equilibrium distributions can be set up in about a day in specimens about one cm long.

Because it is possible to subject light metals, like the alkali metals to extremely high centrifugal fields ($\sim 10^5$ g) at room temperature, the equations describing the equilibrium distribution of solute in a solution subjected to such fields have been examined and it is found[14] that these provide a basis for identifying the nature of the solution.

In this paper an account will be given of the kind of information centrifugal field experiments can yield, results obtained on the gold in sodium and potassium systems will be reported and a brief review will be given of other systems where useful information might be obtained by use of this technique.

2 THEORY

2.1 General Comment

Because centrifugal field techniques have been extensively used to determine molecular weights in solutions there is an extensive literature on the subject. Much of the content of this can be applied directly to centrifugation in solids. Among useful books on the subject can be mentioned the works by Svedberg and Pedersen[15], Fujita[16] and Schachman[17].

2.2 The Equilibrium Distribution

The equilibrium distribution of solute in a centrifugal field can be derived thermodynamically and is given, for a binary solution, by the equation[15]

$$\frac{1}{N}\frac{dN}{dx} = \frac{(M - V(x) \cdot \varrho(x))\,\omega^2 x}{RT\,(1 + \partial \ln f(x)/\partial \ln N)} \tag{1}$$

where N is the mole fraction of solute at a distance x from the axis of rotation and ω is the angular velocity. V, ϱ and f are the partial molar volume of the solute, the density of the solution and the activity coefficient respectively, all evaluated at x. M is the molecular weight of the solute, R and T are the gas constant and temperature.

For a very dilute ideal solution and if the compressibility of the solution can be neglected, V and ϱ are constant and $f = 1$. Equation (1) is then

$$\frac{1}{N}\frac{dN}{dx} = \frac{(M - V\varrho)\,\omega^2 x}{RT}. \tag{2}$$

Since in most chemical applications the quantity being determined is the molecular weight, this equation is more often encountered in the form

$$\frac{1}{N}\frac{dN}{dx} = \frac{M\,(1 - \bar{V}\varrho)\,\omega^2 x}{RT}$$

where $\bar{V}$ is the *partial specific volume* of the solute.

If v' is written for the ratio of the partial molar volumes of the solute and the solvent then $V\varrho = v'M_s$ where M_s is the molecular weight of the solvent. Equation (2) can now be written in the form

$$\frac{d\ln N}{dx^2} = \frac{(M - v'M_s)\,\omega^2}{2RT}. \tag{3}$$

It is useful to derive equation (3) by a kinetic argument to give insight into the physical content of the equation. Consider two adjacent lattice planes of unit area spaced a distance a apart on which the concentrations of diffusing atoms are C_1 and C_2, as shown in Fig. 3a. The jump frequency of the diffus-

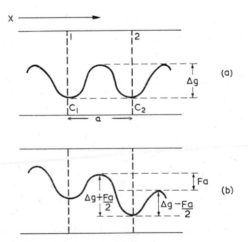

Figure 3 Schematic representation of the potential of an atom (a) without an applied field, (b) with a centrifugal field F applied in the x direction

ing atoms is written as $v \exp(-\Delta g/kT)$ where v is the vibrational frequency of the atoms and Δg is the Gibbs' free energy of activation. If a centrifugal field is applied, normal to the planes, the potential barrier is lowered for jumps in the direction of the field and raised for jumps against the field by $\frac{1}{2} Fa$, as shown in Fig. 3b. F is the centrifugal force/atom and is given by

$$F = (m - v\varrho)\,\omega^2 x$$

where m is the mass of the diffusing atom and v the volume change on *adding* the atom to the solution. $(m - v\varrho)$ is clearly the effective mass of the atom in solution, $v\varrho$ being the buoyancy correction.

The net flux of atoms is

$$J = \frac{1}{2} v C_1 a \exp\left(\frac{-\Delta g + \frac{1}{2} Fa}{kT}\right) - \frac{1}{2} v C_2 a \exp\left(\frac{-\Delta g - \frac{1}{2} Fa}{kT}\right).$$

Since $\frac{1}{2} Fa \ll kT$ the exponential can be expanded to give

$$J = \frac{1}{2} v a^2 \exp\left(\frac{-\Delta g}{kT}\right)\left[-\frac{\partial C}{\partial x} + \frac{FC}{kT}\right].$$

If F is zero this reduces to Fick's law with the well-known random walk expression of the diffusion coefficient.

In equilibrium with the centrifugal field applied the net flux vanishes and

$$\frac{1}{C}\frac{dC}{dx} = \frac{(m - v\varrho)\,\omega^2 x}{kT} \tag{4}$$

which is equivalent to equation (2) after multiplying the right-hand side above and below by Avogadro's number.

If the substitution $\varrho = m_s/v_s$ is made in equation (4) where v_s is the volume and m_s the mass of a solvent atom, then again equation (3) is recovered by multiplying the right-hand side above and below by Avogadro's number, remembering that $v' = v/v_s$.

With typical values of $\omega = 40.10^3$ rev min^{-1}, $x \sim 9$ cm and $T = 300°K$, equation (3) predicts about a two-fold change in concentration across a one cm specimen if $(M - v'M_s)$ is $\sim 10^2$. This equilibrium distribution should be readily measurable in a light solvent such as sodium.

If there is no relaxation of the lattice V, v and v', for a solute atom on an interstitial site, will all be zero. Similarly, if the solute goes into substitutional solution $v' = 1$. If the solute goes into substitutional solution with an associated vacancy $v' = 2$. This latter situation was considered necessary if the high diffusion rate of gold in sodium was to be explained in terms of a vacancy mechanism[9].

The right-hand side of equation (3) can take up a set of values corresponding to $v' = 0$, 1 or 2 so allowing the nature of the solution to be found from a measurement of the left-hand side. Clearly the larger the value of M_s the more readily can v' be resolved. From this point of view the gold-lead system is almost ideal. A measure of the relaxation round the solute can be obtained from the departure of v' from these integral values. If the values of M and M_s do not permit of sufficiently high resolution it may not be possible, in some cases, to distinguish unambiguously between mechanisms on the basis of measurements of v' alone.

2.3 Time to Equilibrium

A rough estimate for the time, t, to set up an equilibrium distribution in a centrifugal field can be found by setting the mean diffusion length equal to the specimen length L.

$$\sqrt{Dt} = L. \tag{5}$$

A more accurate method for finding t is given by Van Holde and Baldwin[18]. In Table 1 the times to equilibrium obtained using their method are given for two values of L for some values of the diffusion coefficient. The table was prepared using $M = 150$ and a mean rotor radius of 8.5 cm. The calculated times are, in fact, relatively insensitive to the exact values used for these quantities. The values in brackets were calculated using equation (5). The table shows that the dominant factor determining t is the specimen length. While the time to equilibrium calculated using Van Holde and Baldwin's method is always less than that calculated from equation (5), the difference is scarcely significant and it is probably better to use the higher value of t.

When the centrifugal field is switched off the gradient of solute will, of course, relax. Clearly the equilibrium distribution set up by the field must be measured before significant relaxation has taken place. A rough estimate for the time available for the measurement can be found by using equation (5) with L replaced by the section thickness used in finding the distribution of solute.

3 EXPERIMENTAL

The techniques evolved at Harwell for the study of diffusion in the alkali metals have been fully described elsewhere[19,20]. For the centrifugal field experiments saturated solutions of high specific activity gold, [198]Au, in sodium and potassium were prepared by diffusion at room temperature[8,9]. Dilute solutions ($\sim 10\%$) were made from this stock material by melting with suitable amounts of pure metal. Some experiments have been done with more concentrated ($\sim 30\%$) solutions.

The gold-alkali metal solution was cast into thin walled polyethylene tubes, Fig. 4. After cutting to a suitable length (~ 1 cm) under water-free paraffin oil (prepared by treating it with freshly extruded sodium) the specimen was

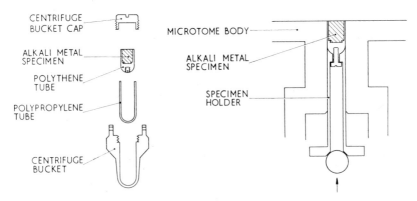

Figure 4   Experimental arrangement for centrifuging and sectioning alkali metal solutions

placed in a polypropylene tube contained in a vacuum tight centrifuge bucket. A thin film of oil was left on the specimen surface to minimize oxidation during centrifugation.

The centrifugal fields were produced in a thermostated preparative ultra-centrifuge (M.S.E. Superspeed 65) run at 40.10^3 rev min^{-1} at temperatures in the range 17–21 °C. At these temperatures the diffusion coefficients for gold in potassium and sodium are about 5 and 7.10^{-6} cm^{2}sec^{-1}, respectively. Reference to Table 1 shows that the equilibration time would be 27–36 hours. Normally specimens were centrifuged for three days to ensure equilibrium.

After centrifuging the specimens were quickly cut into sections ~ 0.5 to 1 mm thick using the microtome arrangement shown in Fig. 4. The re-entrant

Table 1 Time in hours to attain equilibrium

Diffusion coefficient cm²sec⁻¹	Specimen length	
	1 cm	0.1 cm
7.10^{-6}	26 (40)	0.3 (0.4)
10^{-6}	180 (270)	2 (2.7)
10^{-7}	–	20 (27)
10^{-8}	–	190 (280)

well in the base of the specimen tube prevented the metal lifting out of the tube during the final cuts. The total time from stopping the centrifuge until completion of sectioning was usually about ten minutes. Application of equation (5) shows that no appreciable relaxation of the equilibrium distribution occurs in this time.

The gold content and thickness of each section was found, in the usual way, by counting and weighing.

4 RESULTS

4.1 Gold in Potassium

Because, as indicated earlier, the greater discrimination between different types of solution is obtained for the larger values of M_s, the case of gold-potassium solution will be considered first. In Fig. 5 is shown an example of the equilibrium gold distribution as a function of the square of the section distance from the axis of rotation of the centrifuge. The dashed lines were calculated using equation (3) for values of $v' = 0, 1$ and 2 corresponding to interstitial, substitutional and impurity-vacancy pair substitutional solution, each with no relaxation of the lattice. From comparison of the dashed lines with the experimental points it is inferred that gold is in interstitial solution in potassium. From a number of experiments the effective mass of gold in potassium has been found to be 190 ± 4 corresponding to a v' of 0.2 ± 0.1.

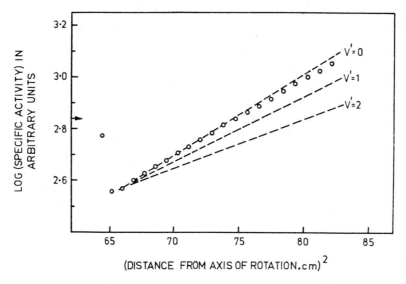

Figure 5 The distribution of Au in K after centrifuging at 40.10^3 r.p.m. for three days at $17\,^{\circ}$C. The initial point is high because of slight oxidation of the specimen surface causing gold trapping. The arrow marks the original, uniform, gold concentration

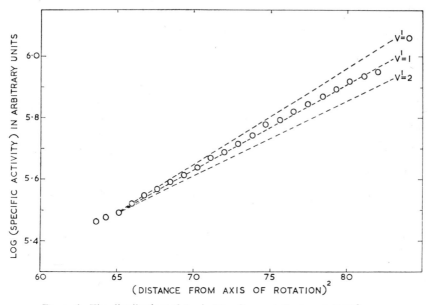

Figure 6 The distribution of Au in Na after centrifuging at 40.10^3 r.p.m. for three days at $18\,^{\circ}$C

A close examination of the experimental points reveals a barely perceptable curvature which suggests that v' increases slightly at greater distances from the axis of rotation. This point will be discussed further below.

4.2   Gold in Sodium

An example of the equilibrium distribution obtained by centrifuging a gold-sodium solution is shown in Fig. 6. There is a marked decrease in the resolution due to the use of a lighter solvent. The curvature of the experimental points is more marked in the case of sodium. The average value of the effective mass of gold in sodium is, from several experiments, 180 ± 5 corresponding to $v' = 0.8 \pm 0.2$.

5   DISCUSSION

5.1   Gold-potassium and Gold-sodium Solutions

It must be emphasized that centrifugal field experiments in solids measure the *partial molar volume of the solute*. No information is obtained about the mechanism of diffusion.

The results for gold-potassium solutions indicate that gold is wholly in interstitial solution. This inference makes it difficult to avoid the conclusion that gold also diffuses interstitially. The slight departure of the experimental points from the $v' = 0$ line indicates that gold slightly dilates the potassium lattice.

The results for gold-sodium solutions could be interpreted as a purely interstitial solution with marked dilation of the lattice; as a partially substitutional solution; or as a substitutional solution with a contraction of the lattice. Purely effective mass measurements cannot, in this case, distinguish between these alternatives. However, the close resemblances in diffusion behaviour of gold in sodium and potassium leave little doubt that the solution is entirely interstitial.

It is possible on the basis of a rough size argument to see how the experimental values of v' can be accounted for on the basis of wholly interstitial solutions. In Table 2 are listed the atomic and ionic radii of some of the elements of interest and the radius of the octahedral sites in sodium and potassium. If the point of view is adopted that ionic rather than atomic sizes control the interstitial solubility of the noble metals[6], then the gold ion is almost exactly the size of the octahedral site in potassium. Only a very slight dilation of the lattice would be required to accommodate the gold ion. In sodium, on the other hand, a substantial dilation of the lattice would be necessary thus accounting for the larger value of v'. Similar size arguments apply to tetrahedral sites.

Table 2 Atomic and ionic radii of some elements in angstroms. Atomic radii from Slater[21], ionic from Pauling[22]

Element	Ionic radius	Atomic radius
Au	1.37	1.35
Ag	1.26	1.60
Cu	0.96	1.35
Cd	0.97	1.55
In	0.81	1.55
Sn	0.71	1.45
Na	0.95	1.80
K	1.33	2.20

	Radius of octahedral site
Na	1.19
K	1.34

This point of view is supported by the measured solubilities and heats of solution of gold in sodium[9] and potassium[10]. These solubilities are shown as a function of temperature in Fig. 7. The solubility of gold is approximately ten times greater in potassium than in sodium with a correspondingly lower heat of solution.

Table 2 also suggests that silver should dissolve interstitially in both sodium and potassium. Experimentally it is found that silver diffuses rapidly in both sodium (see Fig. 2) and potassium, indicating interstitial solution. Preliminary measurements suggest that the solubilities of silver in sodium and potassium are very similar though lower in magnitude than gold in sodium. Copper on the other hand seems not to be soluble in sodium at room temperature which suggests that when the ion size is very much less than the size of the interstitial site, solution does not occur. The *atomic size* of copper does not, of course, favour substitutional solution.

Turning now to the curvature in the experimental gold distributions, it must be remembered that there is a pressure gradient along the specimen associated with the centrifugal field. The pressure at the end of a one cm sample is about 200 atmospheres for the fields used in this work. It seems possible that this pressure causes compression of the alkali metal lattice leading in turn to an effective increase in v'. The fact that the change in v' is larger in the case of sodium is in qualitative agreement with this hypothesis.

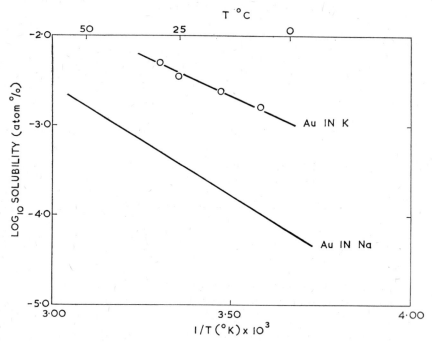

Figure 7   The solubility of Au in Na and K as a function of temperature

5.2   Cadmium, Indium and Tin Solutions in Sodium

From Table 2 it is clear that the ionic radii of these metals, being similar to that of copper, do not favour interstitial solution. Their atom radii, on the other hand, are beginning to be in the favourable region for substitutional solution. There seems little doubt that the mechanism of In and Sn diffusion is the same as that of sodium self-diffusion. While the mechanism for this has not been established unambiguously, the weight of evidence favours a vacancy mechanism[19]. It therefore seems likely that In and Sn dissolve substitutionally and diffuse by a vacancy mechanism* This conclusion is supported by the observation that Tl, which is known to dissolve substitutionally, diffuses at a rate intermediate between Sn and Cd[23].

Clearly it would be desirable to have measurements for v' for Cd.

5.3   Applications to other Systems

It was remarked earlier that if a two-fold concentration change is required across a specimen then, for centrifugal fields presently available, a specimen length of ~ 1 cm is required. Table 1 shows that in this case the solute diffusion coefficient must be $\gtrsim 4.10^{-6}$ cm^{2}sec^{-1} if equilibrium is to be attained in reasonable times. In Table 3 is a short list of solutes and solvents satisfying this condition. Clearly much useful information could be obtained from measurements in these systems but such measurements must await the construction of a high temperature centrifuge capable of rotating at $\sim 10^3$ rev sec^{-1} at the temperatures listed.

Table 3   Solutes having diffusion coefficients
$\gtrsim 4 \times 10^{-6}$ cm^{2}sec^{-1} at the temperatures listed

Solute	Solvent	Temp. °C	Ref.
Ag	Na	50	–
Cu	Pb	260	4
Au	Pb	320	4
Au	In	150	6
Cu	Bi$_2$Te$_3$ ($\perp$ to c-axis)	100	24
Ag	Bi$_2$Te$_3$ ($\perp$ to c-axis)	300	25
Ag	beta-alumina	250	26

Table 1 shows that if shorter specimens could be used, solutes with much lower diffusion coefficients could be studied in reasonable times. Clearly greater precision in measuring the solute distribution would then be needed. In this connection Fig. 8 is of interest. This figure shows the variation of

---

* If these solutes do dissolve substitutionally and diffuse by a vacancy mechanism the observation that Sn diffuses faster than In is readily explained on the basis of an impurity-vacancy binding energy which increases with the valence of the impurity. The case of In and Sn diffusion in sodium then strongly resembles similar impurity diffusion in the noble metals[13].

isotopic ratio of two sodium isotopes with what is essentially the square of the distance into a diffusion sample[9]. The main point of interest is that though the concentration of ^{22}Na is varying with respect to ^{24}Na by less than 10%, precise measurements are possible. This happens because the *ratio of isotopic*

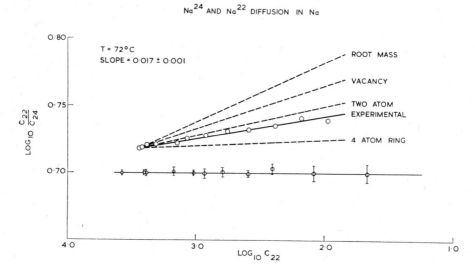

Figure 8   The variation of the isotopic ratio of ^{22}Na and ^{24}Na with the square of the penetration ($\propto \log_{10} C_{22}$) into a crystal of Na

concentrations can be found very accurately by purely radioactive counting techniques. Errors due to weighing and variable counting geomtry from one sample to another, are eliminated. It seems likely that application of this technique to the determination of the solute distribution would greatly improve the precision with which it can be found. In such an application, of course, one of the isotopes would be the solute, the other some suitable isotope of the solvent uniformly distributed throughout the samples. For example, in the case of gold in sodium the solvent isotope could be ^{24}Na and half-life discrimination used to find the $^{198}Au/^{24}Na$ ratio.

Calculations show that using this method satisfactory measurements should be possible with specimen lengths ~ 1 mm. Reference to Table 1 indicates experiments would then be possible for solutes having diffusion rates $\sim 10^{-7}$ cm^{2}sec^{-1}. For example, measurements on Cd in sodium would be possible at $\sim 40°$C and for Rb and K in sodium and Na in potassium at room temperature.

6 CONCLUDING REMARKS

The purpose of this paper was to show the kind of information obtainable on the nature of solid solutions from measurements of the equilibrium distribution of solute in a centrifugal field. In the course of doing this it was shown that gold dissolves, and presumably diffuses, interstitially in potassium and sodium.

In discussing these systems a detailed knowledge of the solubilities was useful. It seems likely that this is generally true for fast diffusing systems where interstitial solutions are suspected. These systems offer the opportunity of studying separately both equilibrium (partial molar volumes and heats of solution) and kinetic properties (motion energies and activation volumes).

The ability to measure partial molar volumes (activation volumes of formation) is of particular interest since a close connection has been postulated[19,27] between the dilation round a diffusing species and ΔK, the fraction possessed by the diffusing atom, of the total translational kinetic energy associated with the decomposition of the saddle point configuration. The product $f \Delta K$ can be found from isotope effect measurements and for interstitials f is one. If, as is the case for gold in sodium the dilation round the interstitial is large a small ΔK might be expected, particularly if $M < M_s$.*

At the beginning of this paper emphasis was laid on the strong resemblance between diffusion rates in lead and sodium. An argument based on analogy strongly suggests that gold dissolves interstitially in lead. But arguments from analogy, though useful, are not conclusive. If the experimental techniques described in this paper can be applied to the gold-lead system, an unambiguous determination of the nature of the solution should be possible.

ACKNOWLEDGEMENTS

I should like to thank Mr. F.A.Smith for his invaluable collaboration in the experiments described in this paper and to thank Mr. A.D. LeClaire for many useful discussions. I also wish to thank Dr. A.P.J.Phillips of the M.R.C. Unit, Harwell for permission to use the M.S.E. Superspeed 65 and for much advice on techniques of centrifugation.

---

* This observation follows immediately from LeClaire's rationalization[27] of the empirical relationship between ΔK and the activation volume[19]. For if the kinetic energy of the atoms involved in the diffusion process is proportional to the product of their mass and a characteristic length, then $\Delta K \sim (1 + nv'M_s/3M)^{-1}$. Here n is a weighed average of the number of atoms involved in the dilation around the solute, all the atoms assumed to have mass M_s.

REFERENCES

1. W.C.Roberts-Austen, *Phil. Trans. Roy. Soc.*, **187**, 383 (1896).
2. G.Hevesy and A.Obrutsheva, *Nature*, **115**, 674 (1925).
3. G.V.Kidson, *Phil. Mag.*, **13**, 247 (1966).
4. B.F.Dyson, T.Anthony and D.Turnbull, *J. Appl. Phys.*, **37**, 2370 (1966).
5. B.F.Dyson, *J. Appl. Phys.*, **37**, 2375 (1966).
6. T.Anthony and D.Turnbull, *Phys. Rev.*, **151**, 495 (1966).
7. T.Anthony, B.F.Dyson and D.Turnbull, *J. Appl. Phys.*, **39**, 1391 (1968).
8. L.W.Barr, J.N.Mundy and F.A.Smith, *Phil. Mag.*, **14**, 1299 (1966).
9. L.W.Barr, J.N.Mundy and F.A.Smith, *Phil. Mag.*, **20**, 389 (1969).
10. F.A.Smith and L.W.Barr, *Phil. Mag.*, **21**, 633 (1969).
11. J.W.Miller, *Phys. Rev.* (in press) (1969).
12. W.Seith and T.Heumann, *Diffusion in Metallen*. Springer-Verlag (1955).
13. A.D. LeClaire, *Phil. Mag.*, **7**, 141 (1962).
14. L.W.Barr and A.D. LeClaire, *Phil. Mag.*, **20**, 1289 (1969).
15. T.Svedberg and K.O.Pedersen, *The Ultracentrifuge*. Oxford University Press. 1940.
16. H.Fujita, *Mathematical Theory of Sedimentation Analysis*. Academic Press 1962.
17. H.K.Schachman, *Ultracentrifugation in Biochemistry*. Academic Press. 1959.
18. K.E. Van Holde and R.L.Baldwin, *J. Phys. Chem.*, **62**, 734 (1958).
19. J.N.Mundy, L.W.Barr and F.A.Smith, *Phil. Mag.*, **14**, 785 (1966).
20. L.W.Barr, J.N.Mundy and F.A.Smith, *Phil. Mag.*, **16**, 1139 (1967).
21. J.C.Slater, *J. Chem. Phys.*, **41**, 3199 (1964).
22. *Handbook of Chemistry and Physics*, Chemical Rubber Publishing Co. (1968).
23. L.W.Barr and F.A.Smith (to be published).
24. R.O.Carlson, *J. Phys. Chem. Solids*, **13**, 65 (1960).
25. J.D.Keys and H.M.Dutton, *J. Phys. Chem. Solids*, **24**, 563 (1963).
26. Yung-Fang Yu Yao and J.T.Kummer, *J. Inorg. Nucl. Chem.*, **29**, 2453 (1967).
27. A.D. LeClaire, *Phil. Mag.*, **14**, 1271 (1966).

Self-diffusion mechanism and isotope effect in body-centred-cubic sodium

J. WORSTER, R. C. BROWN and N. H. MARCH

Department of Physics, The University, Sheffield

and

R. C. PERRIN and R. BULLOUGH

Theoretical Physics Division, Atomic Energy Research Establishment, Harwell

ABSTRACT

The rate theory of Vineyard is used to compare the jump rates ω_1 and ω_2 for two isotopes of masses M_1 and M_2 diffusing in body-centred-cubic sodium. Assuming the vacancy mechanism of diffusion, we have calculated the isotopic mass dependence of the frequency of the localized mode in which the diffusing isotope is constrained at the saddlepoint, and coupled to its near neighbours by the pair potential of Paskin and Rahman.

It is shown in this way that the Vineyard theory, with the vacancy mechanism of diffusion, cannot explain the observed isotope effect in sodium. Therefore, studies of the energetics of vacancy and interstitial mechanisms have been carried out. Though, with both pair potentials we have used in the calculations, the vacancy mechanism is favoured energetically, it is argued on essentially geometrical grounds that the interstitial migration energy is very small in the body-centred-cubic structure and it seems possible that reasonable changes in the pair potential could favour the interstitial mechanism.

With this split-interstitial mechanism three atoms are involved at the saddlepoint. If these three atoms are assumed strongly coupled to one another, but to interact only weakly with the rest of the lattice, the measured isotope effect follows.

1 INTRODUCTION

Recent experimental work on diffusion of isotopes has focussed attention on the isotopic mass dependence of the jump rates (see, for example, Barr and LeClaire[1]). We have made a detailed study of the theory of the dynamics of isotopic jumps in a variety of crystals elsewhere (Brown, Worster, March, Perrin and Bullough, referred to below as I[2]). Our primary purpose here is to report on the consequences of this theory for diffusion in body-centred-cubic Na.

The basis of the theory we use may be summarized as follows:

1) If D_1 and D_2 are diffusion coefficients for self-diffusion by the vacancy mechanism of two isotopes of masses M_1 and M_2, then

$$\frac{D_2}{D_1} - 1 = f\left[\frac{\omega_2}{\omega_1} - 1\right]. \tag{1.1}$$

Here f is the usual correlation factor taking account of the fact that when an atom exchanges with the vacancy, its next jump is most probably back into the vacancy and ω_1 and ω_2 are the jump frequencies of the two isotopes.

2) The rate theory as applied by Vineyard[3] is used for the jump rate which turns out to be directly proportional to the imaginary frequency, iv say, of the localized normal mode in which the jumping atom is constrained at the saddlepoint. Then, as shown in I, we may write

$$\frac{D_2}{D_1} - 1 = g\left[\sqrt{\frac{M_1}{M_2}} - 1\right] \tag{1.2}$$

where g can be calculated in terms of v and, if the mass difference δM between the isotopes is small, is given by

$$g = -\frac{\delta v^2}{\delta M}\frac{M}{v^2} \tag{1.3}$$

where M is the mass of the diffusing atom. g depends upon the sharing of kinetic energy between the diffusing atom and its neighbours. If there is no sharing of energy, $g = 1$ and g is lowered as more kinetic energy is shared.

Thus the problem is to calculate the frequency of the local mode. Experimentally, for Na^{22} and Na^{24} diffusing in Na^{23} we find from (1.1) and (1.2) that, in a vacancy mechanism, for which the correlation factor $f = 0.7272$, $g \simeq 0.5$.

2 NORMAL MODE IN BODY-CENTRED CUBIC SODIUM

When the diffusing atom is at its saddlepoint in the vacancy mechanism of diffusion, its near neighbour atoms lie at the vertices of two equilateral triangles, orthogonal to the jump direction (see Fig. 1).

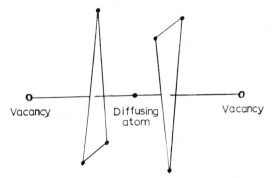

Figure 1   Saddlepoint configuration for vacancy mechanism of diffusion in Na

Our assumption then is that the coupling of the diffusing atom to these six near-neighbour atoms dominates the frequency of the localized normal mode. If this coupling is neglected, then this leaves simply an Einstein model, with $g = 1$ as remarked above.

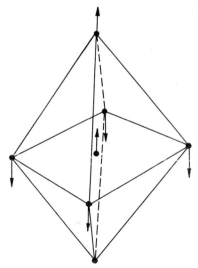

Figure 2   Normal mode of regular octahedral molecule such as SF_6

The treatment of the normal mode is analogous to that of a classical problem in the vibrations of molecules. Thus Herzberg[4] discusses the normal modes of a regular octahedral molecule like SF_6. In particular he describes a mode in which the central atom and the six peripheral atoms all vibrate in the direction of one of the fourfold axes of symmetry as shown in Fig. 2.

There are three such degenerate modes and a linear combination of these gives a mode in which the central atoms vibrate along one of the threefold axes.

The configuration existing at the saddlepoint in sodium is a regular octahedron compressed along one of its threefold axes, which still persists as a threefold axis of symmetry, and so one of the normal modes of this arrangement will be similar to that described above.

2.1   Force Constants of Normal Mode

The normal mode is drawn schematically in Fig. 3.

There will be six force constants relating the three independent displacements x, y, z.

If the diffusing atom has mass M_1 and the other atoms M_2, three equations of motion may be written for the system:

$$M_1\ddot{x} = ax + cy + dz$$
$$6M_2\ddot{y} = cx - by + ez \qquad\qquad (2.1)$$
$$6M_2\ddot{z} = dx + ey - fz$$

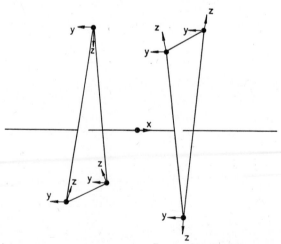

Figure 3   Normal mode of vibration at saddlepoint

where $a, b, f > 0$ and the other constants are undetermined. In a harmonic approximation we may write

$$\ddot{x} = \nu^2 x$$

$$\ddot{y} = \nu^2 y \qquad (2.2)$$

$$\ddot{z} = \nu^2 z$$

where ν is the frequency of the normal mode. The three equations above then have a non-trivial solution if the determinant of the coefficients vanishes

$$\begin{vmatrix} a - M_1\nu^2 & c & \dot{a} \\ c & -(b + 6M_2\nu^2) & e \\ d & e & -(b + 6M_2\nu^2) \end{vmatrix} = 0. \qquad (2.3)$$

This is a cubic equation in ν^2 which can be solved by the method of Tartaglin and Cardan, ν^2 being given by the real root of the equation. Hence g can be calculated from equation (1.3).

2.2 Limits of g

In the limit of no coupling between the atoms, $c = d = 0$ and the real root of (2.2) is simply $\nu^2 = a/M_1$ which trivially gives a value $g = 1$.

The lower limit of g is obtained by varying the force constants and computing g for extreme cases. It is found that g never falls below $\frac{1}{2}$ in the present model.

2.3 Force Constants for Sodium

The force constants appropriate for the saddlepoint configuration in the case of the vacancy mechanism were obtained by first determining the relaxed configuration of atoms surrounding the diffusing atom. This was done by considering a lattice of 432 discrete atoms which were allowed to interact via the pair potential LRO 2 shown in Fig. 4. One atom was removed from the central region and a neighbouring atom was moved to the saddlepoint position. This atom was held fixed while the surrounding atoms were allowed to relax to their equilibrium positions using the procedure described by Bullough and Perrin[6]*.

---

* See § 3 for some further discussion relating to this method.

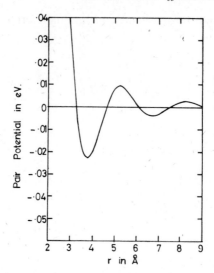

Figure 4 Pair potential LRO 2 of Paskin and Rahman[5]. Potential LRO 1 is of the same general form, being deeper at the first minimum and with larger amplitude oscillations

The relaxations of the six normal mode atoms were:—

0.0187   lattice parameters in the direction of diffusion

0.0185   lattice parameters perpendicular to the diffusion direction

The force constants obtained were:—

$$a = 0.39$$

$$b = 5.23$$

$$c = -0.025$$

$$d = 0.42$$

$$e = 0.75$$

$$f = 6.70$$

which are in arbitrary units since only the ratios enter the calculation of g.

The constants coupling the force on one atom with the displacement of another have a very small effect, and the calculated value of g is unity within one part in 10^2.

It might be questioned whether the potential we have employed in the calculation is sufficiently accurate for our purposes. It is clear that it may well not be a fully quantitative potential, but the fact that we obtain but small departures of *g* from unity suggests strongly that the conclusion that *g* given by the rate theory is near to unity is not sensitively dependent on the choice of potential. A more detailed argument supporting this conclusion is described in the Appendix.

If the rate theory is correct, then it is clear that another mechanism than vacancy diffusion must be sought in body-centred-cubic Na. We therefore turn to discuss the results of calculations on the energetics of vacancy versus interstitial mechanisms for diffusion in Na.

3 ENERGIES OF FORMATION AND MIGRATION FOR VACANCIES AND INTERSTITIALS IN SODIUM

Using the two potentials LRO 1 and LRO 2 discussed by Paskin and Rahman[5] (cf.) calculations have been carried out for the formation and migration energies of a vacancy and an interstitial in body-centred-cubic Na.

We should say at the outset that, for fully quantitative work, the best approach for calculating the vacancy formation energy is via electron theory, without direct recourse to pair potentials. This approach has been employed by Stott, Baranovsky and March[7] for face-centred-cubic metals, but in the more open body-centred-cubic structure relaxation effects play a more important role. Though the results obtained below by direct use of pair potentials are less basic, we shall see that the most important qualitative conclusions (which involve the migration energies rather than the formation energies) appear not to be critically dependent on the details of the calculation.

Using the methods of Bullough and Perrin, the relaxations round vacancies and interstitials have been found and the appropriate energies calculated. For example, to obtain the interstitial formation energy, an atom was inserted in the central region of the lattice described in § 2.3. Each atom is considered in turn and the force on it due to neighbouring atoms is calculated. This force is allowed to act for a certain time interval τ and the atom's new position and velocity at time τ determined. This procedure is applied to each atom in turn and the total kinetic energy is determined after the time interval τ. With a suitable choice of τ, this builds up to a maximum after a few steps. At this point the kinetic energy is put equal to zero and the procedure is repeated. This takes us towards the equilibrium situation and the process is continued until equilibrium is attained to within a prescribed numerical accuracy. The principal results are summarized in Table 1 for potentials LRO 1 and LRO 2. The results for the interstitial are for the $\langle 110 \rangle$ split interstitial (see Fig. 5) which is the most stable configuration.

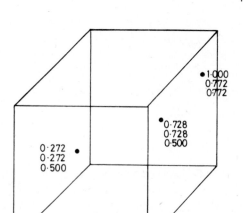

Figure 5   Two positions of the split interstitial configuration before and after migration. Atomic coordinates are in units of the lattice parameter

Table 1   Formation and migration energies of vacancies and interstitials in Na, calculated from Paskin-Rahman potentials (All energies are given in electron volts)

	LRO 1	LRO 2
Vacancy formation energy	0.26	0.15
Vacancy migration energy	0.30	0.19
Interstitial formation energy	0.71	0.56
Interstitial migration energy	0.01	0.01

Though, in each case, the vacancy mechanism is favoured energetically according to these results, the energy differences involved are not major and suggest to us that it is feasible that reasonable changes in pair potential could reduce the sum of formation plus migration energies for the interstitial below that for the vacancy.

We want to stress again that the present procedure is not fully quantitative for calculating the formation energies as it does not give a complete account of volume changes. It should be more reliable for the migration energies and fortunately our main conclusions from the Table depend on these energies.

One important conclusion from the results in the Table is that the migration energy for the interstitial is very small. Closer study reveals that this is largely a consequence of the geometry of the body-centred-cubic structure and should hold therefore for any reasonable potential. The vacancy migration energies, on the other hand, are about the same as the vacancy formation energies, and a large factor greater than the interstitial migration energies. It is this circumstance, coupled with the isotope effect discussed earlier, which leads us to tentatively propose an alternative mechanism involving the split interstitial.

4   INTERSTITIAL MECHANISM AND ISOTOPE EFFECT

We are at present attempting to apply the rate theory in a fully quantitative way to the isotope effect for interstitial diffusion, but no numerical results are available as yet. Nevertheless it seems to us that the isotope effect can then be understood in general terms, as we shall now argue.

As we remarked earlier, the stable interstitial configuration is a split interstitial, where two atoms share a single lattice site. The migration of this defect then takes place by one of the atoms displacing a near-neighbour atom and forming with it a split interstitial about its lattice site, whilst the other atom comes to rest on the lattice site which the first atom has vacated.

At the saddlepoint, three atoms are then crucially involved in the motion (see Fig. 6). If there were essentially infinite coupling between these three atoms and no coupling between them and the rest of the lattice, the frequency of the normal mode would be given by

$$\nu^2 = \frac{\text{constant}}{\sum\limits_{i=1}^{3} M_i}. \tag{4.1}$$

Therefore it would follow from equation (1.3) that g would be approximately one third for the roughly equal masses involved in practice. A similar result follows by an argument given earlier by Barr and Mundy[8].

We have examined the correlation factor for this type of motion and a fuller discussion will be given in I. However the conclusion is that the factor f

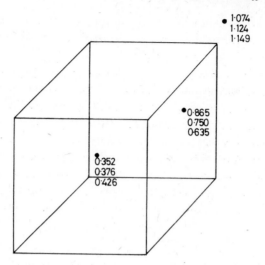

Figure 6   Saddlepoint for split interstitial mechanism. Atomic coordinates
are in units of the lattice parameter

in equation (1.1) is larger than in the vacancy case and giving up the vacancy mechanism leads to a value of g such that $0.5 > g > 0.36$. A quantitative calculation is in progress, to see whether the force constants from LRO 1 and LRO 2 lead to a situation near to the limiting case discussed above.

5 CONCLUSION

The rate theory applied to Na leads irrevocably to disagreement between the calculated and measured isotope effect, when the theory is based on the vacancy mechanism.

Thus, an alternative interpretation has been put forward in which the self diffusion in Na is by interstitial motion. It seems possible to understand the isotope effect in general terms from an interstitial diffusion mechanism.

Should subsequent experimental work show that the vacancy mechanism of diffusion is operative, the applicability of the rate theory of diffusion would be seriously in doubt.

ACKNOWLEDGEMENTS

Financial support is acknowledged from an SRC postgraduate studentship (R.C.B.) and from an EMR contract with AERE, Harwell (J.W.). One of us (N.H.M.) wishes to acknowledge the stimulus derived from visits to AERE, Harwell as a Consultant during the course of this work. Particular thanks are due to Drs Barr, LeClaire and Lidiard for many valuable discussions.

APPENDIX: VALUE OF g FOR VACANCY DIFFUSION

It will be argued here, on general grounds, that the normal mode involved for vacancy diffusion in Na would be expected to give a large value of g.

Computer calculations show quite clearly that the value of g depends on the relative values of the force constants a, b and f, which relate the force and displacement of one coordinate and the constants c, d and e which relate the force in one coordinate to the displacement in another. A high value of g is obtained when a, b and f are larger than the others.

A simple model of the system taken to its harmonic limit shows that for any force law the sign of the coupling constant c depends upon the geometry of the octahedron of atoms. An elongated octahedron would give a negative c, a compressed octahedron would give a positive c. The critical shape, corresponding to vanishing c, is that in which the length of the diffusion axis and the perpendicular distance of the six atoms from it are in the ratio 1.28:1. The ratio in Na is 1.40:1 and c is expected to be small, as the calculations show. The same model can be used to estimate d. This force constant vanishes in the limit of extreme elongation or compression of the octahedron. The value it should have in Na is about one third of its maximum possible value.

REFERENCES

1. L. W. Barr and A. D. LeClaire,, *Proceedings of British Ceramic Society*, **1**, 109 (1964).
2. R. C. Brown, J. Worster, N. H. March, R. C. Perrin, and R. Bullough, 1969 (to be published).
3. G. H. Vineyard, *J. Phys. Chem. Solids*, **3**, 121 (1957).
4. G. Herzberg, *Infrared and Raman Spectra of Polyatomic Molecules* (Van Nostrand: New York, 1945).
5. A. Paskin and A. Rahman, *Phys. Rev. Letters*, **16**, 300 (1964).
6. R. Bullough and R. C. Perrin, *Proc. Roy. Soc.*, **A 305**, 541 (1968).
7. M. J. Stott, S. Baranovsky and N. H. March, 1969, *Proc. Roy. Soc.*, **A 316,** 201 (1970).
8. L. W. Barr and J. N. Mundy, *Diffusion in Body-Centered Cubic Metals* (American Society of Metals, Ohio, 1965), p. 171.

Interstitial solute–vacancy pairs in lead alloys*

J. WESLEY MILLER

Argonne National Laboratory
Argonne, Illinois

ABSTRACT

Dyson *et al.*[1] have explained the extremely rapid diffusion of noble metals in polyvalent metals by proposing that these solutes dissolve both interstitially and substitutionally by the dissociative mechanism[2], but diffuse primarily by the interstitial mechanism. To study the role of solute valence in such solutions, the diffusivity of cadmium in lead and the enhancement of lead self-diffusivity by cadmium additions were determined. The small linear enhancement effect disallows the interpretation that cadmium diffuses by the simple vacancy mechanism. It is suggested that cadmium also dissolves dissociatively in lead, but that interstitial cadmium ions bind strongly to vacancies. A detailed treatment of solute and self-diffusivity in a dissociative alloy shows that the kinetics of interstitial-vacancy pair diffusion agree with the data, and predicts a very small isotope effect for cadmium diffusion. This isotope effect has been measured, and confirms the prediction. If such interstitial-vacancy pairs are also present in noble metal lead alloys, a linear enhancement of the self-diffusion may result. This effect has been measured for additions of silver and gold, and is interpreted as further evidence for interstitial solute vacancy pairs in lead alloys.

* This work was started under a contract supported by the U.S. Office of Naval Research and Advanced Research Projects Agency at Harvard University and is continuing at Argonne National Laboratory under the auspices of the United States Atomic Energy Commission.

1   INTRODUCTION

Lazarus[3] developed a theoretical model to explain the observed differences in solute activation energies for vacancy diffusion. The valence dependence of solute activation energies was attributed to nearest-neighbour impurity-vacancy interactions, which, in the absence of significant impurity size effects, are electrostatic in nature. It was postulated that screening of these interactions by the conduction electrons is sufficiently strong in most metals that only nearest-neighbour interactions are significant. More rigorous calculations[4] of the electronic potential surrounding the effective point charge of an impurity ion confirm this idea. The model predicts that impurities of greater valence than the solvent will interact attractively with vacancies, thereby increasing the total vacancy content of the alloy crystal above that of a pure solvent crystal. The solute activation energy is thus reduced by the amount of the impurity-vacancy binding energy. Consequently, as is generally observed, solutes of greater valence than the solvent diffuse at a rate faster than the self-diffusivity while those of lesser valence diffuse more slowly.

Lidiard[5] has developed an ingenious kinetic model for vacancy diffusion in a dilute *fcc* alloy, in which he also makes the reasonable assumption that impurity interactions are limited to the nearest-neighbour distance. The fractions of impurity-vacancy pairs and free vacancies are calculated, and jump frequencies are assigned, on this basis. The model explains the observed linear enhancement of self-diffusion by solute additions, which may be written

$$D_1 = D_1^0 (1 + b_{11} X_2), \tag{1}$$

where D_1^0 is the self-diffusivity in the pure solvent, D_1 the self-diffusivity in an alloy containing solute atom fraction X_2, and b_{11} is the linear enhancement factor. More recently, Howard and Manning[6] have refined the calculation of the enhancement factor b_{11} by including solvent correlation effects, which are distinctly different for those solvent atoms that are nearest-neighbour to an impurity-vacancy pair. The three parameters for diffusion in a dilute alloy are: the ratio of solute- to self-diffusivity in pure solvent D_2^0/D_1^0, the linear enhancement factor b_{11} and the correlation factor for solute diffusion f_2. These parameters may all be measured experimentally in many alloy systems and are uniquely determined by the atomic jump frequency ratios w_4/w_0, w_3/w_1 and w_2/w_1, as defined by Howard and Manning.[6] Experimental measurement of all three of these parameters, at a given temperature, then permits calculation of the jump frequency ratios. If it is only possible to measure D_2^0/D_1^0 and b_{11}, the locus of all possible sets of the jump frequency ratios is defined, and one may be able to calculate limits for the correlation factor f_2.

It has been shown more recently[7] that the geometry of diffusion by the simple vacancy mechanism imposes a physical limitation on the allowed values of b_{11}. Values of b_{11} less than

$$(b_{11})_{min} = -18 + 1.9448 \, (D_2^0/D_1^0) \qquad (2)$$

are incompatible with diffusion by the simple vacancy mechanism. This minimum enhancement is approached as the jump frequency ratios $w_3/w_1 \to 0$, $w_4/w_0 \to 0$, and $w_2/w_1 \to \infty$, the extreme condition for tightly bound impurity-vacancy pairs. Since there are physical limitations on impurity-vacancy binding, it is reasonable to expect that vacancy diffusion in real alloys does not approach such an extreme. In accord with this expectation, Fig. 1 shows that the observed enhancement factors b_{11} for substitutional

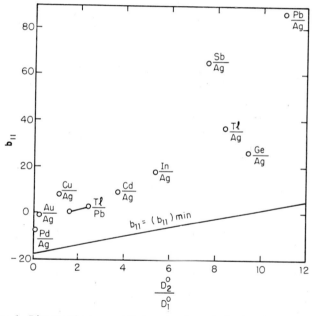

Figure 1   Linear enhancement factors b_{11} for substitutional impurities in fcc metals as a function of the solute to self-diffusivity ratio D_2^0/D_1^0. The data are to be compared to the solid line $b_{11} = (b_{11})_{min}$, the minimum linear enhancement factor consistent with impurity diffusion by the vacancy mechanism

alloy systems are indeed much greater than the minimum $(b_{11})_{min}$. The successful fit of most experimental results for substitutional fcc alloy systems to this model suggests that the method may be extended to explain other diffusion behaviour in metals.

2 EXPERIMENTAL STUDIES OF DIFFUSION IN LEAD

The diffusion behavior of the noble metals in lead, as reported by Seith and Keil[8] in early work, and more recently by Dyson, Anthony and Turnbull[1], is unexplained by these models for vacancy diffusion. These impurities diffuse at rates several orders of magnitude greater than the lead self-diffusivity, whereas most substitutional impurity diffusivities differ from the self-diffusivity by less than an order of magnitude at the melting point. If dissolved substitutionally and interacting with vacancies in accordance with the screened electrostatic interaction model, noble metal ions should be repelled by vacancies and diffuse at a rate slower than lead self-diffusivity. The magnitudes of the pre-exponential factors and activation energies of the diffusion constants suggest that these solutes diffuse interstitially.

Seith and Keil, and Dyson, Anthony and Turnbull, recognizing the possibility that these solutes might diffuse by the interstitial mechanism, and realizing that fast diffusion of solutes by the vacancy mechanism necessitates a large enhancement of the self-diffusion, investigated the self-diffusivity of lead in gold-lead[1,8] and copper-lead[1] alloys. Since no enhancement of lead self-diffusivity was observed, it was proposed that these solutes diffuse primarily by the interstitial mechanism. In analogy with the findings of Frank and Turnbull[2] for the solute copper in germanium, it was suggested that the noble metal solutes dissolve dissociatively, i.e. in both the substitutional and interstitial sites, in lead, but that the effective diffusion constant is dominated by the interstitial contribution in these dissociative alloys.

Anthony also determined that the noble metal solutes diffuse in much the same manner in other polyvalent metals[10,11,12] and proposed the same mechanism for dissolution and diffusion in these alloy systems. To test this proposal for one of these systems, the solute gold in indium, Anthony[9] performed a centrifugal migration experiment. In a centrifugal field, the solute atoms of a dilute dissociative alloy behave as a dilute gas of substitutional atoms, having an effective mass $m_{solute} - m_{solvent}$, and of interstitial atoms, having an effective mass m_{solute}. Observation of the concentration gradient established when equilibrium is attained permits calculation of the fraction dissolved in interstitial sites. The experimental result indicated that gold dissolves entirely in the interstices of indium.

The results of these investigations prompted a diffusion study of the cadmium-lead system[7] in an effort to understand the valence dependence of this tendency towards interstitial dissolution. Considered as a substitutional solute diffusing by the vacancy mechanism, the diffusivity of cadmium in

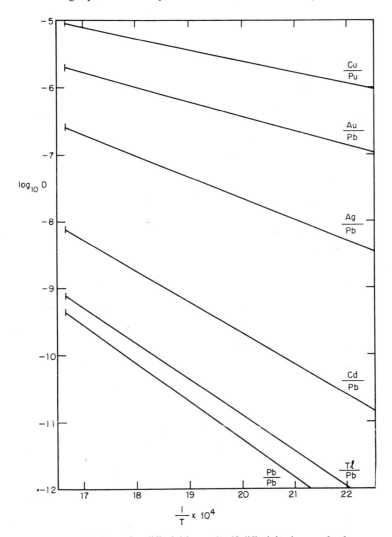

Figure 2 Impurity diffusivities and self-diffusivity in pure lead

lead is very fast indeed, in direct contradiction with the predictions of the screened electrostatic interaction model, but not nearly so fast as the noble metal solutes, as may be seen in Fig. 2. The linear enhancement factors for self-diffusion b_{11}, determined at several temperatures, are distincly less than $(b_{11})_{min}$, as shown in Fig. 3, and are empirically observed to obey the rela-

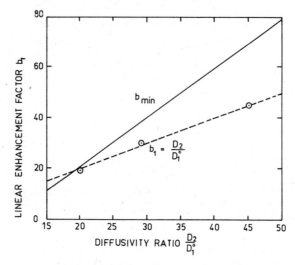

Figure 3 Comparison of the observed linear enhancements and the minimum enhancement consistent with the simple vacancy mechanism of solute diffusion

tion $b_{11} \sim D_2^0/D_1^0$. The data are thus incompatible with the vacancy mechanism of diffusion. To explain these data by means of dissociative dissolution and diffusion, one must also account for the observed positive linear enhancement of self-diffusion, which, assuming that self-diffusion proceeds by the vacancy mechanism, implies an attractive solute-vacancy interaction. Since attractive interstitial cadmium-vacancy interactions are in accord with the screened electrostatic interaction model, it was proposed[7] that cadmium dissolves dissociatively in lead, and that the divalent cadmium interstitial ions are strongly attracted by vacancies.

To further investigate this proposal, it was decided to apply the Lidiard method to the problem of solute and self-diffusion in a dissociative alloy. Before discussing these results, we shall review briefly other recent applications of the Lidiard method to diffusion problems.

3   RECENT APPLICATIONS OF THE LIDIARD METHOD

Although the Lidiard model successfully explained the enhancement of self-diffusion by solute additions, the physical cause of an equally striking effect, the enhancement of solute diffusivity by small solute additions, has remained somewhat of a mystery. It may be shown that solute pairs, rather than isolated single solute atoms, are required to explain a linear dependence of solute diffusivity on solute concentration. The magnitude of the effect suggests that an attractive binding energy is also responsible, although solute-solute interactions should in general be repulsive according to the screened electrostatic interaction model. These ideas have been resolved by the proposal that the nearest-neighbour interaction of two solute ions with a single vacancy is responsible for the effect[13]. If such doubly associated vacancy defects are present in dilute substitutional alloy systems, there will result a pronounced linear enhancement of the solute diffusion, and a quadratic enhancement of the self-diffusion.

Since nearest-neighbour solute-solute interactions occur in only one of the four unique configurations of the double associated vacancy defect, a simplified form of the result may be derived by neglecting such interactions. The binding energy of a second solute atom is taken to be $-b$, which may of course differ from the binding energy of a single solute atom to a vacancy. The fractions of free vacancies, impurity-vacancy pairs, and doubly associated vacancies are calculated in the same manner which Lidiard[5] employed. Assuming that the impurity-vacancy exchange frequency is unaltered in these configurations and that solute correlation effects are approximately the same as for the impurity-vacancy pair configuration, a very simple expression for the solute linear enhancement factor b_{21} results[13]:

$$b_{21} = 11 \left(e^{-b/kT} - 1 \right). \tag{3}$$

The solute diffusivity D_2 in a dilute alloy containing solute atom fraction X_2 is given by

$$D_2 = D_2^0 (1 + b_{21} X_2), \tag{4}$$

in direct analogy with Eqn. (1). The importance of the result (3) lies in the very simple dependence of the effect on the binding energy $-b$. This binding energy of a second solute atom to a vacancy may be calculated directly from experimental results of this effect, which are available for the systems Pb in Ag[14], In in Ag[15], Cd in Ag[15] and Tl in Pb[16]. The binding energies calculated were all of reasonable magnitude, agreed with the predictions of the screened electrostatic interaction model, and were nearly equal to values of the binding energy for the respective single solute atoms to a vacancy. In addition, the temperature dependence of the observed solute enhancement effects agreed remarkably well with the predicted dependence of Eqn. (3).

The self-diffusivity in a crystal containing both the singly and doubly associated vacancy defects has also been calculated.[13] The self-diffusivity D_1 may be expressed

$$D_1 = D_1^0 (1 + b_{11}X_2 + b_{12}X_2^2) \tag{5}$$

where b_{12} is the quadratic enhancement factor for self-diffusion. Solvent correlation effects were not included explicitly; it was instead assumed that the various correlation factors for each type of initial solvent jump $f_i = f_0$, the correlation factor for self-diffusion in a pure fcc crystal. The resultant linear enhancement factor b_{11} is identical to Lidiard's result, and the quadratic solvent enhancement factor b_{12} is similar in form, but a function of the additional solvent jump frequencies of the doubly associated vacancy. The calculation demonstrates the manner in which higher order association of impurity atoms with vacancies results in higher order enhancements of the self-diffusion. Higher order enhancement effects are evident in many experimental studies of diffusion in alloys, frequently at less than 5 atomic percent solute addition, and are always of the same sign as the linear enhancement. This is to be expected if the interaction of additional solute atoms with a vacancy is very similar to that of the first.

The Lidiard method has also been applied to calculate the contribution of vacancies and divacancies to self-diffusion. To do this correctly, one must include the contribution of associative and dissociative jumps, of jump frequencies k_1 and k_2 respectively, as well as those jumps which retain the association, of jump frequency k_0. Since the calculation of correlation effects for this general case has not been completed, these effects will be represented by an effective correlation factor for divacancy diffusion f_d. If we write the self-diffusion constant D_1 as the addition of vacancy and divacancy contributions, D_{1v} and D_{2v} respectively, the following expressions are obtained for these contributions:

$$D_{1v} = 2s^2 w_0 f_0 \, e^{-g_v/kT} (1 - 18 \, e^{-g_v/kT}) \tag{6}$$

$$D_{2v} = 8s^2 k_0 f_d \, e^{-(2g_v+b)/kT} \left[1 + \frac{7k_2}{2k_0} \right]. \tag{7}$$

Here s is the atomic jump distance, g_v the standard Gibbs free energy of vacancy formation and $-b$ the divacancy binding energy. It may, however, be more useful to consider this modification of monovacancy self-diffusivity D_1^0 as an enhancement of this self-diffusivity due to divacancies. The resultant enhancement effect depends linearly on the monovacancy concentration X_v,

$$D_1 = D_1^0 (1 + b_{2v} X_v). \tag{8}$$

The linear enhancement factor b_{2v} may then be written

$$b_{2v} = -18 + \frac{4k_1}{w_0} \left[\frac{k_0}{k_2} + \frac{7}{2} \right]. \tag{9}$$

When statistically significant measurements of the contribution of divacancies to the self-diffusivity and to the isotope effect for self-diffusion are accomplished, it may be possible to deduce the atomic jump frequency ratios and the divacancy binding energy in the same manner that solute diffusion parameters are utilized with the Lidiard[5] and Howard and Manning[6] results.

4 DIFFUSION IN A DILUTE DISSOCIATIVE FCC ALLOY

Peterson and Rothman[17] determined that the activation energies for diffusion of many impurities in aluminium were very nearly equal to that for self-diffusion, and varied only slightly with impurity valence. This suggested that substitutional impurity-vacancy binding energies are small in aluminium. They pointed out that this was in accord with the calculations of March and Murray[4], which show that the electronic screening of impurity charge is almost complete at nearest-neighbour distance in many polyvalent metals. It is thus expected that substitutional impurity-vacancy binding energies in lead will also be very small. The data for diffusion of thallium in lead[16] are evidence in support of this expectation. For this reason, substitutional solute-vacancy interactions were neglected in the model for diffusion in a disso-ciative alloy[18], and only nearest-neighbour interstitial solute-vacancy inter-actions were considered to be significant.

The postulation of close interstitial-vacancy pairs as an impurity defect in equilibrium with both the substitutional and interstitial fractions of solute in a dissociative solution provides a logical intermediate step in the kinetics of dissociation. That an interstitial solute of small valence should interact attractively with a nearest-neighbour vacancy is a consequence of the screened electrostatic interaction model. The equilibrium fractions of free interstitial solute atoms, associated interstitial solute-vacancy pairs and free vacancies are calculated, and the solute and solvent jump frequencies are assigned as depicted in Fig. 4.

If the close pairs are tightly bound, dissociative interstitial jumps (k_1) are energetically disfavoured. The fraction of free interstitials may then be redu-

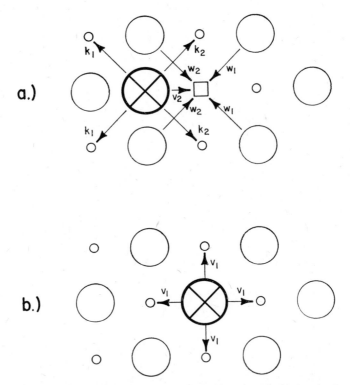

Figure 4 An interstitial solute-vacancy pair (a) and a substitutional solute atom (b) are shown in the (100) plane to illustrate the assignment of atomic jump frequencies. The Cd^{2+} solute ions, crossed circles, and Pb^{4+} solvent ions, open circles, are shown approximately to scale

ced to the extent that the interstitial contribution to the solute diffusivity is very small compared to the contribution of close pairs. Since interstitial jump frequencies are generally much larger than vacancy jump frequencies, an associated solute atom may randomize its position with respect to its vacancy in less than the time required for a vacancy jump to occur. It may then be reasoned that the effective diffusion of both solute and solvent depend upon the solvent-vacancy exchange frequency. The kinetics of diffusion by the tightly bound close pair mechanism are thus very similar to those for solute vacancy diffusion in the limit of tight binding. The limit of tight binding may be more physically realizable in this case, however, because of the smaller impurity-vacancy interaction distance and the characteristically large interstitial jump frequencies.

The solute and solvent diffusivities have been calculated in the manner used by Lidiard[5], and solute correlation effects explicitly included. Interstitial jump frequencies are assumed to be much larger than vacancy jump frequencies, the limit of tight binding is calculated, and the free interstitial contribution accordingly neglected. It is shown[18] that the correlation factor for solute diffusion becomes very small while that for solvent diffusion approaches that for self-diffusion in a pure fcc crystal. In this limit, the linear enhancement factor for solvent diffusion reduces to

$$b_{11} \cong f_0 \, \frac{D_2^0}{D_1^0}. \tag{10}$$

Since this result agrees well with the observed linear enhancement of lead self-diffusion by additions of cadmium, it is proposed[19] that cadmium diffuses in lead primarily by this tightly bound interstitial solute-vacancy pair mechanism.

If this close pair mechanism is responsible for cadmium diffusion in lead, the correlation factor for solute diffusion must be very small. The isotope effect for cadmium diffusion in lead must therefore be very small. In addition, if these close pair defects are an equilibrium defect in dissociative alloy solutions, and are bound in accordance with the screened electrostatic interaction model, lead self-diffusion will be enhanced by additions of other solutes which dissolve dissociatively, even though the solute diffusion is dominated by the free interstitial contribution.

5   RECENT EXPERIMENTAL STUDIES OF DISSOCIATIVE LEAD ALLOYS

The isotope effect for diffusion of cadmium in lead was performed[19] as a critical test of the proposed interstitial-vacancy pair mechanism of solute diffusion. Two radio-isotopes of cadmium, Cd^{109} and Cd^{115m}, were simultaneously diffused into a single crystal of lead. After sectioning, it was ascertained that there had been no surface hold-up effects and that the penetration profile was Gaussian for both isotopes for all slices of sufficient activity for an isotope effect determination. The relative activities of Cd^{109} and Cd^{115m} in each slice were determined by counting the 22 KeV X-ray resulting from the decay of Cd^{109} and the 1.63 MeV beta radiation of Cd^{115m}, using energy discrimination and a beta filter.

The isotope effect for the diffusion of cadmium in lead was determined to be $(D_{109}/D_{115m} - 1) = 0.0032 \pm 0.001$ at 248 °C, from which it was calculated that $f_2 \Delta K = 0.12 \pm 0.03$ for a mechanism involving single atomic jumps. The possibility of vacancy, interstitial, simple interchange and ring mechanisms were examined and rejected because of inconsistency with the data for D_2^0/D_1^0, b_{11} and $f_2 \Delta K$. Since these data are all remarkably consistent with diffusion by the proposed close pair mechanism, it was concluded[19] that cadmium dissolves dissociatively in lead, and diffuses primarily by the interstitial-vacancy pair mechanism.

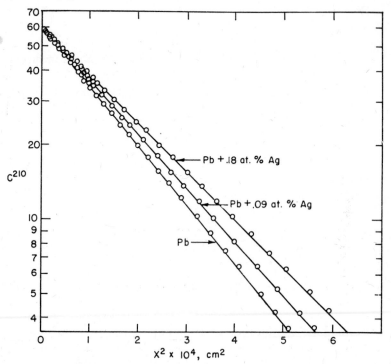

Figure 5   Penetration profiles for Pb^{210} self-diffusion in pure lead, lead +0.09 atomic percent Ag and lead +0.18 atomic percent Ag, at 300.3 °C. The Pb^{210} specific activity is plotted logarithmically in arbitrary units

On the basis of the screened electrostatic interaction model, it is anticipated that interstitial gold and silver ions in lead solution should also interact attractively with vacancies, although not so strongly as interstitial cadmium ions. A weaker interaction for these solutes is also consistent with the proposed interstitial mechanism of solute diffusion[1], since the resulting increase in dissociation enhances the fraction of interstitial solute. It was supposed, nevertheless, that the interstitial-vacancy interaction might be sufficiently large that a resultant linear enhancement of the self-diffusivity could be measured.

The observed enhancement effects[20] were in fact somewhat larger than expected. For example, the lead diffusivity in a Pb + 0.089 atomic percent Au alloy is approximately five times the lead diffusivity in pure lead, at 215.2°C. The penetration plots for lead diffusion in pure lead and silver-lead alloys at 300.3°C are shown in Fig. 5, and the final data for the enhancement effect by silver additions is shown in Fig. 6. The linear enhancement factor determined for additions of silver to lead[20] is b_{11} (Ag/Pb, 300.3°C) =

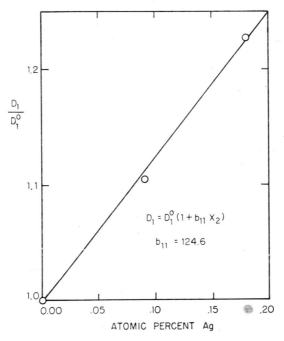

$$D_1 = D_1^0 (1 + b_{11} X_2)$$

$$b_{11} = 124.6$$

Figure 6   Linear enhancement effect of silver solute additions on lead self-diffusivity, at 300.3°C

136.8 $\pm$ 5.1, and those determined for additions of gold to lead[20] are b_{11} (Au/Pb, 199.4 °C) = 5726 $\pm$ 298 and b_{11} (Au/Pb, 215.2 °C) = 4312$\pm$93.

Although these enhancement factors appear very large, the minimum enhancement factors consistent with the vacancy mechanism of solute diffusion $(b_{11})_{min}$, Eqn. (2), are much larger. These data thus confirm the proposal[1] that silver and gold diffuse primarily by the interstitial mechanism in lead.

In accordance with the screened interaction model and the calculations of March and Murray, the interaction of the substitutional fraction of gold or silver solute with vacancies in lead should be weakly repulsive, while the interstitial-vacancy interaction for these solutes should be more strongly attractive. It is thus proposed[20] that the observed enhancements of lead self-diffusivity by gold and silver additions result from an attractive interstitial solute-vacancy interaction.

Measurements of the isotope effect for diffusion of silver in lead are presently being made at Argonne National Laboratory. S. Rothman has developed a new technique for determining the relative activities of two radio-isotopes emitting similar radiation which cannot be resolved completely by energy discrimination, and whose half-lives are too long to permit an accurate half-life separation. This technique is now being used in this experiment for the isotopes Ag^{105} and Ag^{110}. It is hoped that the results will lead to greater understanding of the mechanism of dissolution and diffusion of noble metal solutes in lead.

ACKNOWLEDGEMENTS

I am deeply indebted to Professor Turnbull for advice and counsel concerning the experimentation performed at Harvard University. I would also like to thank Dr. John Mundy for comments on the manuscript and Dr. T. R. Anthony for permission to quote from his doctoral thesis.

REFERENCES

1. B. F. Dyson, T. R. Anthony and D. Turnbull, *J. Appl. Phys.*, **37**, 2370 (1966).
2. F. C. Frank and D. Turnbull, *Phys. Rev.* **104**, 617 (1956).
3. D. Lazarus, *Phys. Rev.* **93**, 973 (1954).
4. N. H. March and A. M. Murray, *Proc. Roy. Soc.* **A 261**, 119 (1961).
5. A. B. Lidiard, *Phil. Mag.* **5**, 1171 (1960).
6. R. E. Howard and J. R. Manning, *Phys. Rev.* **154**, 561 (1967).
7. J. W. Miller, *Phys. Rev.* **181**, 1095 (1969).
8. W. Seith and A. Keil, *Z. Physik. Chem.* **B 22**, 350 (1933).
9. T. R. Anthony and D. Turnbull, *Phys. Rev.* **151**, 495 (1966).

10. T. R. Anthony, B. F. Dyson and D. Turnbull, *J. Appl. Phys.* **39**, 1391 (1968).
11. B. F. Dyson, T. R. Anthony and D. Turnbull, *J. Appl. Phys.* **38**, 3408 (1967).
12. T. R. Anthony, Ph.D. Thesis, Harvard University, 1967.
13. J. W. Miller, (to be published).
14. R. E. Hoffman, D. Turnbull and E. W. Hart, *Acta Met.* **3**, 417 (1955); *Acta Met.* **5**, 74 (1957).
15. A. H. Schoen, Ph.D. Thesis, University of Illinois, 1958.
16. H. A. Resing and N. H. Nachtrieb, *J. Phys. Chem. Solids* **21**, 40 (1961).
17. N. L. Peterson and S. J. Rothman, Argonne National Laboratory Annual Progress Report, Metallurgy Div., ANL-7299 (1965).
18. J. W. Miller, *Phys. Rev.* (to be published).
19. J. W. Miller and W. A. Edelstein, *Phys. Rev.* (to be published).
20. J. W. Miller, (to be published).

3.5

Diffusion under mechanical stress

A. F. BROWN

Department of Physics, The City University, London

ABSTRACT

In vacancy diffusion, the diffusion coefficient D is proportional to the concentration c of vacancies. If then c is increased above its equilibrium value, for example if excess vacancies are created by dislocation movements during plastic deformation, then D should be increased.

It is now generally accepted that the lifetimes of vacancies at diffusion temperatures are too short for enhancement to be observable by conventional methods of measurement, and that the large enhancements of D by plastic deformation which were claimed in the early 1960's were due to experimental error. However, where diffusion can be observed at very small values of $(Dt)^{1/2}$, for example by the growth of age-hardening precipitates, enhancement should be observable and is a cause of overageing in fatigue and of some forms of work-softening. But the claim that sonic vibration can markedly increase the rate of age-hardening at temperatures lower than those used in conventional diffusion measurements now appears to have been based on faulty experimental technique.

HISTORICAL INTRODUCTION

In vacancy diffusion, the diffusion coefficient D is proportional to the concentration c of vacancies. If then c is increased above its equilibrium value, for example if excess vacancies are created by dislocation movements during plastic deformation, then D should be increased.

In 1952 Seitz[1] had shown that movement of dislocations would lead to production of vacancies, and in proposing experiments to test his theory he said: "It would be interesting to know if the rate of diffusion of a foreign atom or a radioactive tracer can be increased by continuous plastic flow." If

219

at that time vacancy kinetics had been as well understood as it is now, it is probable that the experiments would never have been tried, at least not by direct methods. As it turned out, an answer was already in print in the much quoted paper of Buffington and Cohen[2]. In this work α-iron was deformed in compression during the diffusion anneal and the self-diffusion coefficient under strain, D_s, measured and compared with the self-diffusion coefficient in unstrained α-iron, D_u. The authors found a linear relation

$$D_s/D_u - 1 = 180,000\dot{\varepsilon} \tag{1}$$

where $\dot{\varepsilon}$ is the strain rate in s^{-1} and 180,000 was the constant relevant to a anneal temperature of 890 °C ($0.64T_m$). At their highest strain rate ($\dot{\varepsilon} = 7.5 \times 10^{-5}\,s^{-1}$ or about 20 % per hour) the enhancement ($D_s/D_u - 1$) was about 15. These results were withdrawn in 1958 and replaced by another set [3] not essentially different but giving enhancements of about 24 times at 750 °C. It is, however, the earlier results which have become enshrined in the literature and have been widely applied. They have, for example, been built into a theory of creep.[4]

The work of Cohen's group will be discussed later along with all the multitude of subsequent results by other authors, but meantime it is appropriate to record that they repeated their experiments in 1963[5] at strain rates up to a hundred times higher, with consequent increases of enhancement up to factors of the order of 2500. This remains almost the only work on b.c.c. metals and, in consequence, tended to be left out of the controversy which took place in the 1960's, as it was not then certain that diffusion in b.c.c. metals was primarily by a vacancy mechanism. Thus experimental refutation[6] of their results had to wait till 1967.

Indirect observation of acceleration of diffusion-controlled reactions by mechanical deformation had been observed before 1952 and explanations were now forthcoming in terms of Seitz's theory. Thus, for example, Cottrell[7] explained the Portevin-LeChatelier effect of jerky flow and Seitz himself explained the Gyulai-Hartly effect, where plastic deformation increases electrical conductivity in ionic crystals. The present paper will rather concentrate on the effect of deformation, particularly cyclic deformation, on age-hardening where the results are important in the theory of fatigue and are still under discussion. These go back at least to 1930 when Mahoux[8] claimed that high frequency vibrations could speed up the penetration of certain elements into steel. Since then, there have been many claims that age-hardening of alloys is speeded up by ultrasonic irradiation of quite small intensity:

the more extravagant claims of the 1950's would have had us believe that the time to a given state of ageing was reduced by factors of up to 1000 by insonation, which would mean that diffusivities were increased by these large factors. Such claims, as well as the modest claims of the 1960's, have recently been reviewed by Haynes and Shyne[9] in presenting their own results of which a typical example is shown in Fig. 1. This shows that the precipitation

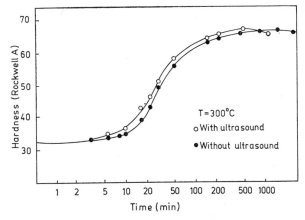

Figure 1 The influence of 20 kHz acoustic energy on the age hardening kinetics of a Cu–Be alloy annealed at 300 °C. Maximum acoustic stress $\sim 3.5 \times 10^8$ dyn cm^{-2}. After Haynes and Shyne[9]

hardening of a beryllium–copper alloy is accelerated by a factor of 1.5–2 in the early minutes of insonation. Explained in terms of theories based on strain-enhancement of diffusion[10], sonic energy would have set dislocations in motion and the consequent production of vacancies gives the required acceleration.

Another apparent manifestation of diffusion enhancement by sonic strain is being currently debated: this is the effect of sonic energy on metal working processes, and has again been the subject of recent reviews[11]. What is observed is illustrated in Fig. 2[12] which is the work of a group in Vienna and shows the stress-strain of a zinc single crystal with and without superposed ultrasonic energy. There are two effects: firstly, when the vibrator is switched on, there is an immediate fall in the yield stress which is simply due to a superposition of the static stress and the peaks of the applied stress; the same effect is obtained by tapping the apparatus. Its apparent triviality does not mean it has no technological importance.[11] The second effect is that the rate of work-hardening is reduced; recovery, it appears, takes place faster under insonation since the defects producing work-hardening can diffuse faster to sinks. As with age-hardening, the same effect is produced by raising the temperature.

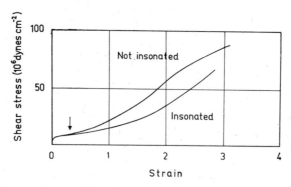

Figure 2   Room temperature stress-strain curves for a zinc single crystal with and without insonation. The arrow indicates when the ultrasonic generator was switched on. Maximum sonic intensity estimated[10] at $\sim 10^7$ dyn cm^{-2}. After Oelschlagel[12]

   The above summary of some manifestations of the problem mentions some of the more extreme claims (enhancements of D, by static and sonic energy, of the order of thousands) and also some moderate claims (doubling of D). These experiments were some of many dozens performed in the 1960's to see if mechanical deformation enhanced diffusion. Some groups in many parts of the world consistently obtained large enhancements, not often by factors of a thousand, but sometimes by factors of 100.[12] Other groups (equally widely spread) obtained small enhancements, while other groups obtained no effect at all. Where enhancement was observed the linear relation between $(D_s/D_u - 1)$ and $\dot{\varepsilon}$ was also usually observed. This paper makes no attempt to review all this work: as far as direct measurement techniques are concerned such a task has recently been completed by Barry[13]. Instead, a simple theory is developed and used to point some morals and indicate possiblities for the future. An attempt is also made to justify the immense amount of work done to disprove the early claims. The theory concentrates on the vacancy effect, merely making reference to papers where the effect of dislocation short-circuiting is discussed, whether in undirectional[14] or in vibrational[10] deformation.

THEORY

The diffusion coefficient D is given in terms of the atomic fraction c of vacancies and the frequency v with which they change places with diffusing atoms by

$$D = \alpha a^2 v c, \tag{2}$$

where a is the lattice spacing and α is a geometrical factor. In the equilibrium case, without applied strain, c is the equilibrium thermal vacancy concentration and D corresponds to D_u as defined above. If now c^* is the atomic fraction of excess vacancies produced by plastic deformation, then the diffusivity under strain D_s is given by

$$D_s = \alpha a^2 v (c + c^*). \tag{3}$$

The concentration of excess vacancies c^* is found by equating the rate at which they accumulate (dc^*/dt) to the difference between the production rate and the destruction rate:

$$\frac{dc}{dt} = \frac{d}{dt}(P\varepsilon) - \frac{c^*}{\tau}$$

where ε is the strain, P the atomic fraction of vacancies produced by unit strain and τ the mean lifetime of a vacancy.

In equilibrium $(dc/dt) = 0$ and

$$c^* = P\dot{\varepsilon}\tau. \tag{4}$$

From (2), (3), (4) the strain enhancement of diffusivity as defined above is

$$D_s/D_u - 1 = (P\tau/c)\,\dot{\varepsilon} \tag{5}$$

which has the same form as the empirical relationship (1). Assuming that P varies little with temperature the parameter $(P\tau/c)$ has the temperature dependence of D^{-1}. Now v, the number of times a vacancy makes a jump per second, can be related to τ by

$$v = n_0/\tau, \tag{6}$$

where n_0 is the mean number of jumps a vacancy makes before being captured by a trap. Hence, putting (4) and (6) into (5) and using (2) to eliminate c gives

$$D_s = D_u + \alpha a^2 n_0 P\dot{\varepsilon}, \tag{7}$$

and if, as in most experiments, the strain rate is constant for the time t of the diffusion anneal:

$$D_s t = D_u t + \alpha a^2 n_0 P\varepsilon. \tag{8}$$

If the strain rate is not constant but, for example, increases by a series of small jumps, then there is a numerical factor of order unity in the last term of (8).

The interesting cases are those where $D_s \gg D_u$. These include the many cases in the 1950's and early '60's where large enhancements of diffusivity were claimed in diffusion-under-strain experiments at temperatures where D_u is large enough to measure. Also included are cases where diffusion-controlled processes such as ageing have been observed, particularly under cyclic stresses, at temperatures much lower than normal. Particularly interesting are cases where overageing has been held responsible for room temperature fatigue.

For these cases the diffusion distance

$$(D_s t)^{1/2} = (\alpha a^2 n_0 P\varepsilon)^{1/2} \tag{9}$$

and is apparently independent of temperature as well as of strain rate and depends only on total strain. Taking the latter conclusion first, it is still quite valid (apart from a constant of order unity) if the strain rate is not constant as was assumed in deriving equation (8); in particular a series of jumps was a perfectly adequate way of deforming the specimen and there was no need to devise elaborate devices to prevent this. Again, there was nothing to be gained by using high speed or even explosive deformation to give large values of $\dot\varepsilon$; there was in fact much to be lost by doing so, since time-to-temperature corrections became unmanageable, and specimen cracking led to distortion of diffusion profiles. As to the conclusion that strain-enhanced vacancy diffusion can be produced at any temperature, however low, this is not quite correct, since this simple derivation conceals the obvious requirement that that τ, the vacancy lifetime, must be small compared with t, the time of diffusion anneal. The conclusion does, however, lead us to question the assumption that diffusive processes cannot contribute to fatigue at very low temperatures.

To put in numerical values; for f.c.c. metals, if a is taken as the unit cube edge, then $a^2 \sim 10^{-15}$ cm^2 and $\alpha = \frac{1}{12}$. $\varepsilon \sim 1$ since while most experiments which were carried out reached, in one way or another, a total strain of about unity. Thus the only adjustable factors are P and n_0. Now the maximum credible value of P, about 10^{-3}, was obtained from our own experiments on diffusion under stress by assuming that all the mechanical work used to deform the specimen went to form vacancies: any higher value contradicts conservation of energy while most quoted values, based on electrical resistivity measurements, are an order of magnitude lower. As to n_0, if it is assumed that the main traps for vacancies are dislocations, then $n_0 \sim 10^9$ corresponds to a dislocation density f of about 10^6 cm^{-2} and to a vacancy lifetime of about 10^{-1} s at $T = \frac{2}{3} T_m$. On the basis of quenching experiments, this is, if anything, a rather high value for τ (and consequently for n_0) in a well annealed metal. Thus $n_0 \sim 10^9$ but no higher, can be accepted as relevant to those experiments where the stress diffusion runs were carried out at diffusion temperatures. Higher values of the dislocation density and correspondingly lower values of n_0 are relevant to those experiments where the deformation took place at temperatures lower than those of normal diffusion (i.e. the fatigue and insonation experiments) and the diffusion was observed by growth of precipitates or by annealing.

Thus, for well annealed metals at temperatures of normal diffusion the maximum diffusion distance

$$(D_s t)^{1/2} \sim 1000 \text{ Å},$$

while at lower temperatures, where the dislocation density might be several orders of magnitude higher, the estimated $(D_s t)^{1/2}$ would be still less.

DISCUSSION

If this analysis is accepted, then three questions arise:

1 Why were large values of $(D_s/D_u - 1)$ or, what is the same thing, values of $(D_s t)^{1/2}$ of the order of tens of microns found by so many experimenters in the early 1960's?

2 What would have been the consequences of uncritical acceptance of their results?

3 Can strain-enhanced vacancy diffusion be observed at all?

1 Why were Large Apparent Enhancements Found?

The main reason for experimental error was the need to make $D_u t$ small at the same time as $\alpha a^2 n_0 P \dot{\varepsilon}$ (Eqn. 8) was made large. Thus small values of $(Dt)^{1/2}$ had to be measured under conditions of large strain. Bad resolving power invariably leads to overestimate of diffusion coefficients and these overestimated coefficients were attributed to strain-enhancement. Equation (8) shows, inserting 1000 Å as the numerical value of the second term on the right hand side, that even a 10% enhancement required measurement of $(Dt)^{1/2}$ of only 1 μm. This is beyond the resolving power of conventional and most unconventional techniques of diffusion measurement operating under best conditions. It was even more impossible when the surface was rumpled as a result of efforts to obtain large strains.

At the same time, it is unlucky that the sources of error in so many techniques contrived to give linear plots of (concentration) against (depth)^2, that is to simulate the true diffusion curve. Apparent diffusion plots simulating true diffusion were obtained by autoradiography (when the poor resolution was due to finite range of the radioactive emissions) and also by turning off surface sections from bars when the circular cross-section had been distorted by twisting (loss of resolution due to mixing of concentrations from different depths). In a recent repeat of early work Ruoff[6] shows that simulated diffusion profiles are even obtained by a process of mechanical mixing where surface activity is pushed deep into the metal by normal slip processes. A particularly good demonstration of this 'cussedness of nature' effect has recently been given by Borg and Lai[15] who plotted erf (concentration) against depth for various specimens of iron roughened by repeated $\alpha \leftrightarrow \gamma$ phase transitions. However rough the specimen, a linear plot was obtained but with slope which decreased with increasing roughness. If these slopes had been uncritically interpreted as proportional to $(Dt)^{-1}$ then the apparent values of D would have increased with increasing roughness.

It was also rather unlucky that a simulated dependence of $(D_s/D_u - 1)$ on $\dot{\varepsilon}$ was obtained as expected from equation (5). To see why this happened, note that the apparent diffusion depth $(D_a t)^{1/2}$ is given by

$$D_a t = D_u t + C,$$

where D_u is the true diffusivity at the temperature of the anneal and, as we have seen, independent of strain. C is a constant depending on the method of measurement: it might, for example, depend on the range of β-particles or it might be a measure of surface roughness. Dividing by $D_u t$ gives

$$(D_a/D_u - 1) = (C/D_u)\, t^{-1}. \tag{10}$$

Now most sets of experimental results were so arranged that all specimens reached approximately the same strain but in different times t. Thus $t^{-1} \propto \dot{\varepsilon}$ so that if D_a, the apparent diffusivity, is interpreted as D_s, the diffusivity under strain, equation (10) has the same form as equations (5) and (1) and the parameter in brackets on the right has the correct temperature dependence.

To disprove the claim of large enhancement involved proving a negative, which is always difficult. Some of the most convincing demonstrations of the non-effect borrowed a technique from biology and annealed a control (unstrained) sample along with the deformed sample.[9,13]

2   The Theoretical Consequences of Strain-enhancement of Vacancy Diffusion

The larger apparent enhancements of diffusivity at normal diffusion temperatures required that the quantity $(\alpha a^2 n_0 P \varepsilon)^{1/2}$ should be of the order of tens of microns, that is some 100 times larger than calculated above. Only n_0 was available for adjustment, so some ingenious ideas were put forward for reasons why vacancies should make up to 10^4 more jumps before destruction at high temperatures than at low. Vacancy lifetimes of hundreds of seconds at temperatures as high as $0.85T_m$ had to be explained. It was, for example, noticed that when steel ball models of crystals were shaken violently to simulate temperature, missing atoms (vacancies) lost their identity and dissociated into partial vacancies or regions of low density. In real crystals the occurrence of such an effect would certainly have meant that all the partial vacancies would have had to be collected together before a complete vacancy could be captured, but such dissociation would have involved a different diffusion mechanism at high temperatures. Other explanations involved an elaborate model of vacancies and divacancies, the net effect of which was to introduce additional parameters which could be manipulated to give agreement between theory and results. But the commonest idea was that dislocations ceased to be effective traps for vacancies at high temperatures. This requires that the binding energy between a vacancy and a dislocation should be much smaller than is now accepted. Moreover, the long vacancy lifetimes, thus explained, mean unacceptably high vacancy supersaturations which should have become visible as porosities.

If the strain diffusion experiments have any value, it is in showing that n_0 does not increase with temperature, that vacancy lifetimes extrapolated from annealing data $(T = \frac{1}{3}T_m)$ are valid and that consequently the higher values of dislocation-vacancy binding energy, proposed by the theoreticians, are to be preferred.

3   Have Strain-produced Vacancies any Observable Effect on Diffusion?

It is clear that strain-enhanced vacancy diffusion cannot be observed in conventional diffusion experiments where $(Dt)^{1/2}$ is required to be at least 1 μm.

There is, however, clear evidence that diffusive processes take place during fatigue at temperatures well below normal and in those regions subjected to most strain. Thus Forsyth[16] and many others have shown that age-hardening precipitates grow in the persistent slip bands which are a feature of fatigue. Again, Broom, Molineux and Whittaker[17] reviewing the different behaviours of high strength alloys under fatigue conditions point out that in an alloy such as DTD 683 where the static strength derives from unstable precipitates, the poor fatigue properties are due to overageing under fatigue stresses. On the other hand, in a medium strength alloy, such as aluminium–7 per cent magnesium, where static strength derives from Cottrell atmospheres, enhanced diffusion produced by fatigue strains may have a self-healing effect by helping to reform the atmospheres.

To apply equation (9) to fatigue-enhanced diffusion, assume that within a persistent slip band at fatigue temperatures the dislocation density is about 10^9 cm^{-2} which corresponds to $n_0 \sim 10^6$. Within such a slip band the strain ε might easily reach 10. Thus:

$$(D_s t)^{1/2} \sim 100 \text{ Å}$$

which is probably enough to account for overageing by fatigue.

Enhancement of diffusion rates by sonic energy has received a great deal of uncritical study. Although sonic energy is much more expensive than thermal energy there was the hope that some special characteristics might be found in the sonic-aged alloys. Again, there was the negative aspect if sonic stresses, for example in aircraft, could lead to overageing in the structure. In recent work, however, Haynes and Shyne[9] irradiated a Cu–Be alloy with 20 kHz acoustic energy and took special care to keep the specimen temperature constant. They found (Fig. 1) that there was only a small ($\times$ 1.5) increase in the hardening rate and that the large increases claimed by earlier workers could be attributed to failure to take account of sonic heating. The small increase they did find was entirely caused by acoustic enhancement of grain boundary precipitation. This is reasonable, since acoustic energy in the kHz frequency range (which was universally used in all previous experiments) cannot be absorbed by, and produce to-and-fro motion of, dislocations. It can, however, be absorbed at grain boundaries.

For the same reasons as discussed above, a direct measurement of the effect of sonic energy on diffusion using conventional sectioning techniques is not practicable. Thus the negative results of Walker *et al.*[18] who measured diffusivity in zinc by tracer techniques, both with and without 60 kHz insonation, are correct but hardly surprising, since in their case $(D_u t)^{1/2} \sim 40$ μm.

As to the effect of vibrational energy on metalworking processes, what is hoped for is some sort of work-softening which will, to quote a recent popular article,[19] "ease the metal-worker's load". In the case of fatigue the gross slip bands are themselves evidence that such a mechanism takes place, for a slip band is a region of intense work-hardening, and work-hardening is relieved by recovery processes which involve diffusion. If then a slip band is also a region where diffusive processes go specially fast it will soften rapidly and become a likely region for further slip whether in the same direction as before or in the contrary direction. That the excess vacancies are there to assist diffusion and can themselves diffuse together to form voids was first shown by Forsyth.[16] Whether sonic energy applied to the whole metal specimen can produce a similar effect is much more doubtful, for it now appears[11] that the only effect of superposing vibrational stresses on static ones is to produce deformation at the peaks. If rises of temperature are eliminated, then any work softening is eliminated too.

ACKNOWLEDGEMENT

The writing of this paper has been considerably eased by the work of Mr. B. E. Barry[13], whose thesis includes a very full critical review of all attempts at direct observation of stress-enhanced diffusion.

REFERENCES

1. F. Seitz, *Adv. in Physics*, **1**, 43 (1952).
2. F. S. Buffington and M. Cohen, *Trans. Met. Soc. AIME*, **194**, 859 (1952).
3. N. Ujiiye, B. L. Averbach, M. Cohen and V. Griffiths, *Acta Met.*, **6**, 68 (1958).
4. A. J. Kennedy, *Processes of Creep and Fracture in Metals*, Oliver and Boyd, 1962, p. 265.
5. K. Hirano, M. Cohen, B. L. Averbach and N. Ujiiye, *Trans. Met. Soc. AIME*, **227**, 950 (1963).
6. A. L. Ruoff, *J. Appl. Phys.*, **38**, 3999 (1967).
7. A. H. Cottrell, *Phil. Mag.*, **44**, 829 (1953).
8. G. Mahoux, *C. R. Acad. Sci. Paris*, **191**, 1328 (1930).
9. G. A. Haynes and J. C. Shyne, *Metal Sci. J.*, **2**, 81 (1968).
10. A. F. Brown, *Applied Mat. Res.*, **5**, 67 (1966).
11. C. E. Winsper and D. H. Sansome, *J. Inst. Metals*, **96**, 353 (1968).

12. D. Oelschlägel, *Z. Metallk.*, **53**, 367 (1962).

13. B. E. Barry, M.Sc. Thesis, University of Edinburgh, 1969.

14. A. L. Ruoff and R. W. Balluffi, *J. Appl. Phys.*, **34**, 1634, 1848, 2862 (1963).

15. R. J. Borg and D. Y. F. Lai, *J. Appl. Phys.*, **39**, 2738 (1968).

16. P. J. E. Forsyth, *Phil. Mag.*, **2**, 437 (1957).

17. T. Broom, J. H. Molineux and V. N. Whittaker, *J. Inst. Met.*, **84**, 357 (1955–6).

18. B. F. Walker, V. A. Johnson, W. C. Hahn, Jr. and J. D. Wood, *Trans. Met. Soc. AIME*, **242**, 1233 (1968).

19. F. Wheeler, "Vibration eases the metalworker's load", *New Scientist*, **40**, 302 (1968).

3.6

The heat of transport in solids

R. V. HESKETH

Berkeley Nuclear Laboratories, Berkeley, Glos.

ABSTRACT

Simple models of the scattering of phonons and electrons at lattice defects show that the heat of transport, $Q^{*'}$, is identical with the vibrational entropy term of the Gibbs Free Energy of the defect:

$$Q^{*'} \equiv -T\,\Delta S_v.$$

A general principle governs the direction of drift of defects in a temperature gradient their unrestricted motion always carries heat down the temperature gradient.

The models are compared with experiments in thermal and electric gradients. Isotopes, vacancies, substitutional solutes, interstitial solutes and dissolved gases are considered. Metals and ionic crystals are considered. Credible results are obtained.

> For instance, I found it quite impossible to refute a theory which described a single object as both hot and cold, though I knew that a thing couldn't possibly be hot and cold at the same time.
>
> LUCIAN, *Menippus goes to Hell*

1 INTRODUCTION

The heat of transport of a defect in a solid is an intensive variable, it is the quantity of heat possessed by a defect. The definition includes the restriction "in an isothermal situation" and this restriction is important. This quantity of heat, usually called $Q^{*'}$, is a quantity of energy, and its gradient gives rise to a force which is perfectly real and Newtonian:

$$F = -\nabla Q^{*'} = -\frac{dQ^{*'}}{dT}\nabla T = -T\frac{dQ^{*'}}{dT}\cdot\frac{\nabla T}{T}. \qquad (1)$$

231

Thus an experiment which depends for its success on the existence of a force pushing the defect along the temperature gradient would yield a null result if $Q^{*\prime}$ were independent of temperature. We are now, unavoidably, in head-on collision with the literature (see reference (1) and the references there cited), which unanimously writes

$$F = -Q^* \frac{\nabla T}{T},\tag{2}$$

and asserts the existence of a force

$$F_0 = -Q_0^* \frac{\nabla T}{T},\tag{3}$$

where Q_0^* is independent of temperature (reference 1, equations (7) and (15)). The prime is omitted from Q^* in (2) and (3) to avoid what is, for the moment, a side-track, the distinction between "heat of transport" and "energy of transport". This distinction will be taken up later.

The heat of transport is an intensive variable. It is an energy which is an intrinsic property of the defect. It must therefore appear in the Gibbs free energy of the defect; we cannot leave it out. Thus if we write the Gibbs free energy of one defect as

$$\Delta G = \Delta H - T(\Delta S_v + \Delta S_c),\tag{4}$$

where ΔH is the enthalpy of formation of the defect and where subscripts v and c denote the vibrational and configurational parts of the entropy ΔS, this expression must somewhere contain the heat of transport $Q^{*\prime}$. We shall see that

$$Q^{*\prime} \equiv -T \Delta S_v.\tag{5}$$

Both quantities arise from local perturbations of the atomic and electronic densities and the elastic stiffness around the defect; the change in density and stiffness leads to a change in vibrational frequency, and this gives the entropy change ΔS_v directly. A local change of frequency also leads to the scattering of phonon and electron waves, and such scattering is the origin of the heat of transport. Since $Q^{*\prime}$ and $-T \Delta S_v$ are identical we expect any two calculations of the two quantities to yield similar answers. When two parallel calculations[2,3] yield quite different answers, we are forced to conclude that one must be wrong. Equation (1) has already excluded large terms from the generally accepted expressions[1] for $Q^{*\prime}$. Criticism of remaining terms is made below.

This paper presents an elementary calculation of the momentum transfer, from elastic waves, to a defect in a solid. Just as photons show both wave and particle properties, phonons too show a duality, and I have found it useful to initially think of them as particles. Introductory equations for particle scattering are therefore given. For very simple defects such as a vacancy which does not perturb the elasticity of the solid, or a macroscopic void (which does), the calculation may be compared with formal arguments which associate $\frac{1}{2}kT$ with each degree of freedom of the defect. The calculation is seen to be good to a few per cent. There are vacancies in real solids which approximate to very simple defects, and for these the experimental data both verify the calculation of the transferred momentum and demonstrate the validity of identity (5).

Phonon scattering gives rise to the segregation of isotopes, and the behaviour[4] of Li^6 is perfectly regular, it provides no evidence for a composite mechanism. The sense and magnitude of motion are those expected of a simple vacancy. Phonon scattering is also seen on its own in insulators, and the data[5] on radioactive Sr^{++} in NaCl are considered. The experimental Arrhenius plot does not however immediately yield $Q^{*\prime}$. The necessary thermodynamics to extract $Q^{*\prime}$ from the measured slope are therefore given at the beginning of the paper.

These thermodynamics are also important to such defects as carbon in solution in iron[6], and in fact to all defects which, because they are conserved, are present in arbitrary concentration. (Vacancies are not conserved, in the steady state we have as many as thermal equilibrium requires.)

In metals, the conduction electrons also transfer momentum to a defect, and in either a thermal or an electric gradient they exert a force. It is here calculated empirically in terms of the measured electrical resistivity of the defect.

The two calculations of momentum transfer, from phonon and from electron scattering, are then compared with some of the experimental data. The agreement ranges over the entire adjectival spectrum. Where it is both necessary and possible I would attribute disagreement to two causes, first the binding together of two different defects, and second, the failure of the electron and phonon systems to come into thermal equilibrium over a distance small compared to that between two defects in the solid. Each of these phenomena give rise to an additional force on the defect, a force omitted from the present calculation. The first is unfamiliar in the present context, but it is necessary to invoke it to account for the change of sign of the pressure[7] which occurs across an iron diaphragm, during the thermo-osmosis of hydrogen, as the temperature is lowered below 530°C. This change of sign is not explicable in simple terms. The second is familiar as the phonon drag term of thermo-electricity. It is conspicuous when the defect has dissimilar scattering volumes for phonons and for electrons, and when the defect concentration is high. "High", in this context, means a few parts per million.

Two examples of electro-migration are considered, silver[8] and, because the two isotopes[4] feel an identical force but have different mobilities, Li^6 in Li^7.

2 PARTIAL THERMAL EQUILIBRIUM

A specimen in which a steady temperature gradient is maintained is not in total thermal equilibrium, for total thermal equilibrium requires uniform temperature. However, if restrictions are placed on two relaxation times of the system, the specimen may be said to be in "partial equilibrium" and the customary thermodynamic functions may be used[9].

We consider as a total system, a large heat source at temperature T_I, a small specimen, and a large heat sink at temperature T_{II}. The adjectives "large" and "small" ensure that the relaxation time of the specimen, Δt, is much less than the relaxation time of the whole system. We now choose, within the specimen, two adjacent regions, each so small that its relaxation time is much less than that of the specimen, Δt. For each of these regions, for times $\sim \Delta t$, we have a definable entropy[9]. Let the two regions have mean temperature T_1 and T_2. Consider the transfer of a small quantity of heat, dQ, from region 1 to region 2:

$$dQ = -dQ_1 = +dQ_2. \tag{6}$$

For an irreversible process

$$\frac{dQ_1}{dt} \leqslant T_1 \frac{dS_1}{dt}, \tag{7a}$$

$$\frac{dQ_2}{dt} \leqslant T_2 \frac{dS_2}{dt}, \tag{7b}$$

and

$$dt \ll \Delta t. \tag{8}$$

Thus

$$dQ_1 + dQ_2 = 0 \leqslant T_1\, dS_1 + T_2\, dS_2. \tag{9}$$

If dQ is an infinitesimal amount of heat, and if partial equilibrium has been attained, then the transfer is reversible, and (9) becomes an equality. (That sentence contains, and perhaps submerges, the crux of the argument. Let us pass on.) The foregoing is well known. It is set out to make clear the distinction between partial and total equilibrium; in partial equilibrium T_1 differs from T_2.

We now expand the terms on the right hand side of (9) in standard thermodynamic functions (reference 9, equation (24.7)):

$$0 = dW_1 - v\, dN_1\, dP_1 - \mu_1\, dN_1 + dW_2 - v\, dN_2\, dP_2 - \mu_2\, dN_2. \quad (10)$$

In this equation μ is the chemical potential of the defects, P is an external pressure, v is the activation volume of each defect, and the number of defects transferred from region 1 to region 2 during the transfer of the heat dQ is

$$dN = -dN_1 = +dN_2. \quad (11)$$

The heat function is

$$dW_1 = dN_1\, \Delta H, \quad (12)$$

where ΔH is the enthalpy of one defect. Thus, in differential form, (10) becomes

$$0 = v\, \nabla P + \nabla \mu. \quad (13)$$

If no external force acts on the defects, ∇P is zero, and *the chemical potential is constant throughout the specimen*. This conclusion may seem obvious. It is emphasised because it differs from that found in the literature, (in particular the literature[10] on interstitial solutes.) The chemical potential is constant even though different parts of the specimens are at different temperatures. We see that there is a real difference between total and partial thermal equilibrium; in total equilibrium we maximise the entropy, ΔS, in partial equilibrium we maximise the entropy term $T\Delta S$. (The specimen is not a closed system.) Equation (9) defines this maximum, for we could write it

$$\Delta (T\Delta S) \to 0. \quad (14)$$

It reduces to total equilibrium when $T_1 = T_2$. (Paragraph 25 of reference 9 has misprints and omissions. These acknowledged, it leads to (13) above.)

For defects free from external restraint we may write the constant chemical potential

$$\mu_0 = \Delta H - T\Delta S_v + kT \ln(c). \quad (15)$$

ΔS_v is the vibrational entropy, and the defect is present in concentration c. We restrict discussion to ideal solutions. We have no *a priori* knowledge that ΔS_v is a constant, and we must therefore assume

$$\Delta S_v = a + bT + c'T^2 + \cdots . \tag{16}$$

The slope of an Arrhenius plot of (15) is then

$$-kT\frac{\nabla c}{c}\frac{T}{\nabla T} = \mu_0 - \Delta H - bT^2 - 2c'T^3 - \cdots . \tag{17}$$

When, in the more general case, a force of external origin acts on the defects, integration of (13) gives additional terms in the slope of the Arrhenius plot:

$$-kT\frac{\nabla c}{c}\frac{T}{\nabla T} = \mu_0 - \Delta H - bT^2 - 2c'T^3 - (vP - v\,dP)$$

$$+ T\frac{d(vP - v\,dP)}{dT}. \tag{18}$$

Equations (17) and (18) represent physical situations. If the specimen is free from restraint, and if, *also*, there is a saturated source and sink of defects at both the hot and cold end of the specimen, (17) applies. This case is not of great interest since it only occurs experimentally for defects which are not conserved, i.e. configurational defects such as vacancies and interstitials, and it is not customary to make an Arrhenius plot of these, (a fierce experiment!). Equation (18) applies to thermo-osmosis, in which a differential gas pressure is applied across a thin specimen, and also in general to defects which are conserved. Instead of a gas pressure, we may have the differential pressure caused by the absence of a source of defects at one side of the specimen, and the continued presence of a sink at the other side, i.e. the side to which defects drift and precipitate out. We must regard the source as being absent from any specimen which contains a fixed and arbitrary amount of solute. The force caused by the withdrawal of the source is given by (1):

$$v\,\nabla P = T\frac{dQ^{*'}}{dT}\cdot\frac{\nabla T}{T} = -T\frac{d(T\Delta S_v)}{dT}\cdot\frac{\nabla T}{T}. \tag{19}$$

The importance of this equation is stressed by De Groot[11] (paragraph 9). It is also the Clausius–Clapeyron equation (reference 9, paragraph 78). Substitution of (19) reduces (18) to

$$-kT\frac{\nabla c}{c}\frac{T}{\nabla T} = \mu_0 - \Delta H + T\,\Delta S_v - 2T\frac{d}{dT}(T\Delta S_v). \tag{20}$$

This is not a very simple result at which to arrive. I have tried to show that it depends on the absence of a source of defects at the side of the specimen from which they are drifting, and that otherwise, (17) applies. Again we have to acknowledge that we differ from the literature. This would omit the terms μ_0, $-\Delta H$ and $-2T\,d\,(T\,\Delta S_v)/dT$ from (20). Experiments on interstitial solutes have provided "one of the notable problems of the subject", and since the neglected terms are usually much greater than the remaining term this is not surprising. Values of μ_0 are seldom available from experiment.

In experiments which measure the drift velocity of atoms past inert markers in the crystal lattice, the coefficient a of (16) is not lost, as it is from (17). The flux of atoms is[11]

$$J = \frac{D}{kT} \times \text{Force},\qquad(21)$$

where D is a diffusion coefficient. Evaluating the force from (1), (5) and (16):

$$J = \frac{D}{kT}\,(aT + 2bT^2 + 3c'T^3 + \cdots)\,\frac{\nabla T}{T}\qquad(22)$$

$$= \frac{D}{kT}\,(-Q^{*\prime} + bT^2 + 2c'T^3 + \cdots)\,\frac{\nabla T}{T}.\qquad(23)$$

Thus a is retained and the terms in higher powers of T are more strongly represented than one might expect.

We now calculate the coefficients a and b, associating a with phonon scattering and b with electron scattering.

3 PARTICLE SCATTERING

Consider a stream of particles, each of mass m and of velocity $\vec{v}$, incident upon an obstacle which completely destroys the momentum of each particle. If the obstacle is a cube of volume Ω, we may let it present a cross-sectional area $\Omega^{2/3}$ to the stream of particles. Let the number of particles per unit volume, in the stream, be ϱ. The force $\vec{F}$ on the obstacle is the rate at which momentum is destroyed:

$$\vec{F} = (\varrho\Omega^{2/3}v_0)\,(m\vec{v}).\qquad(24)$$

v_0 is the particle speed, the scalar value of the velocity. The first bracket is the number of particles striking the obstacle in unit time, the second bracket is the momentum lost by each particle. Equation (24) may be written

$$\vec{F} = 2\Omega^{2/3} E\vec{i}, \tag{25}$$

where E is the energy density in the particle stream and $\vec{i}$ is the unit vector parallel to $\vec{v}$. If the obstacle is bombarded from all three directions, and if the energy density of this cloud of particles is a variable, there is a net force on the obstacle:

$$\Delta F = -2\Omega^{2/3}\vec{i}\,\Delta E/3 = -\tfrac{2}{3}\Omega\,\nabla E, \tag{26}$$

or alternatively,

$$\Delta F = -\frac{2}{3}\frac{\Omega}{\omega}\nabla(\omega E), \tag{27}$$

where ω is a characteristic volume. In crystalline solids it is usefully identified with the volume of the Wigner–Seitz cell. Then Ω/ω is a pure number, the number of Wigner–Seitz cells in the scattering volume characterising a defect, and ωE is the energy per cell of the particles (or waves) which are being scattered. The factor of one third appears in (26) and (27) because only one of the three components of momentum is relevant, that along the energy gradient.

Equation (27) thus provides a basic formula by which to calculate the forces from phonon and electron scattering. We have to determine the extent to which the postulate of (24) is satisfied, that there is a momentum loss $m\vec{v}$. Thereafter we have to calculate the scattering volume Ω. The energy density per Wigner–Seitz cell is, for electrons γT^2, where γT is the electronic specific heat. Since the experimental temperatures greatly exceed the Debye temperature, we may write for phonons $\omega E = 3kT$.

4 PHONON SCATTERING

Initially, we treat the solid as an isotropic elastic medium, and the defect as a spherical volume, in which there is a mass, or density, defect and also an elastic modulus defect. This volume scatters elastic waves[12].

The amplitude of the scattered wave, at an angle θ to the direction of incidence contains the factor

$$\left(\frac{\delta G}{G} + \frac{\delta D}{D}\cos\theta\right), \tag{28}$$

where δG and δD are the local changes in elastic modulus and density within the scattering volume. Now the wave amplitude is a measure of the momentum. Thus

$$\left(\frac{\delta G}{G} + \frac{\delta D}{D}\cos\theta\right)\cos\theta \qquad (29)$$

is a measure of the component of the momentum in the direction of incidence. This component corresponds to the longitudinal momentum loss $m\vec{v}$ of the particle considered above. We sum over all θ to obtain

$$\frac{1}{3}\frac{\delta D}{D}. \qquad (30)$$

We see that no longitudinal momentum is transferred by scattering at the elastic modulus defect. As is well known[12], the modulus defect behaves as a simple source; it scatters energy isotropically. The mass defect however is a dipole source[12]. It therefore receives longitudinal momentum. I emphasise these properties because they are directly responsible for a general principle:

The scattering forces are always such as to transport mass *down*
the energy gradient. (31)

(We note *en passant* the equanimity of the literature in reporting mass transport up an energy gradient. I believe that in such cases one must look for neglected forces, or energy gradients; water does not freely flow uphill, and similar statistical probabilities operate in diffusion.)

The *fraction* of the scattered momentum which is received as longitudinal momentum is obtained by dividing (30) by an average of (28); we come to this in a moment. (30) must also be multiplied by three to sum the three possible modes of polarisation[13].

The cross-section for elastic scattering at this defect, summing over all θ, and counting all three modes of polarisation, is [13]

$$\sigma(q) = \frac{q^4}{4\pi}\omega^2\left[\left(\frac{\delta D}{D}\right)^2 + \left(\frac{\delta G}{G}\right)^2\right], \qquad (32)$$

where q is the wave vector of the phonon. (The $\delta G/G$ term has previously been prefixed[13] with a factor of either $\frac{3}{2}$ or 2. A formal reason is given below for writing unity.) This equation is correct for an elastic solid, and also for long wave-length phonons. For short wave-length phonons, of wave-length comparable to the radius of the defect, it must be reduced in magnitude[14]; it varies less strongly than q^4. We assume the simple Debye phonon spectrum:

$$D(q) = 3q^2 \, dq/q_D^3, \tag{33}$$

for $q < q_D$, where

$$q_D = 6\pi^2/\omega. \tag{34}$$

Integrating (32) for all values of q, and using Morse's correction[14] for short wave-length phonons, we obtain a total cross section

$$\sigma = 1.04\omega^{2/3} \left[\left(\frac{\delta D}{D} \right)^2 + \left(\frac{\delta G}{G} \right)^2 \right]. \tag{35}$$

The accuracy of the hand calculation which gives the factor 1.04 is seen in the formal reasoning given below; this indicates a value 0.9982. For a spherical obstacle the cross sectional area is that of a circle, and the corresponding scattering volume is

$$\Omega = \frac{4}{3\sqrt{\pi}} \sigma^{3/2}. \tag{36}$$

The average of (28), by which (30) must be divided, is, for all modes of polarisation, the square root of the square bracket in (35). (This bracket represents the square of the amplitude.) Thus (36) gives, with (30), the scattering volume to be used in calculating the longitudinal momentum transfer from all phonons:

$$\Omega = \omega \frac{4 \times 1.06}{3\sqrt{\pi}} \frac{\delta D}{D} \left[\left(\frac{\delta D}{D} \right)^2 + \left(\frac{\delta G}{G} \right)^2 \right], \tag{37}$$

and the force exerted by the phonons is

$$\Delta F_p = -\frac{2}{3} \cdot \frac{4 \times 1.06}{3\sqrt{\pi}} \frac{\delta D}{D} \left[\left(\frac{\delta D}{D} \right)^2 + \left(\frac{\delta G}{G} \right)^2 \right] \nabla (3kT). \tag{38}$$

Identifying this force with equation (1), we have the part of the heat of transport which is due to phonon scattering:

$$Q_p^{*\prime} = \frac{3}{2} kT \cdot \frac{\delta D}{D} \left[\left(\frac{\delta D}{D} \right)^2 + \left(\frac{\delta G}{G} \right)^2 \right]. \tag{39}$$

We may consider this result by defining two defects with very simple properties. The first is a vacancy in a monatomic solid. For a vacancy $\delta D/D = -1$. We next suppose that the interatomic forces are unperturbed at the vacant site; that the modulus defect is zero. Then $Q_p^{*\prime} = -\frac{3}{2}kT$. Now potential energy is stored in the elasticity of the solid, but since $\delta G = 0$, it is unaffected by the creation of the vacancy. In taking out one atom we remove no potential energy, only kinetic energy. For an atom with three degrees of freedom this kinetic energy is $\frac{3}{2}kT$. If, secondly, we suppose a vacancy for which $\delta G/G = -1$, we have $Q_p^{*\prime} = -3kT$. Since $\delta G/G = -1$ no potential energy is stored at the vacancy, and in creating such a vacant site we remove the heat associated with the three degrees of freedom of the potential energy and that associated with the three degrees of freedom of the kinetic energy.

Thus the removal of heat stored in the lattice vibrations is synonymous with the transfer of phonon momentum to the defect. Nor have we any good reason to doubt[3] the reality of momentum transfer, it is evident in Brownian motion; we see phonons in a liquid transferring momentum to a small particle.

In a crystalline solid, momentum and quasi-momentum exist. The latter is determined only to within the addition of a constant vector[9], and the physical equivalence of the two is well known[9]. The distinction between the two is irrelevant to the present purpose; experimental measurements are on a macroscopic scale, they refer to the time averaged motion of each atom, and to the space averaged position of all the atoms i.e. we measure displacements with respect to the crystal lattice. In a frame of reference firmly attached to the crystal lattice, momentum and quasi-momentum are identical[9].

If we turn this problem over we find the question "What about Umklapping?". The experimental data refer to temperatures well above the Debye temperature, and the collision of a phonon with a defect has a high probablity of generating or annihilating a second phonon. The question is answered by stating the sum rules developed for the Mössbauer problem[15]: *in a large number of such events the average of the energy or momentum transferred is simply the classical value*. In diffusion we have a large number of events, and Umklapping therefore leaves equation (39) intact.

In the two preceding paragraphs, statistical aspects of the problem have been stated. The quantity we aim to calculate, the heat of transport, is itself a statistical quantity. I find it necessary to quote[9]

"Temperature, like entropy, is clearly a quantity of purely statistical character, having a meaning only for macroscopic bodies". (40)

Readers unfamiliar with the literature may consider this emphasis unnecessary, but I must make clear the statistical nature of the force acting on the defects; all the quantities on the right of equation (1) are of a statistical character, and so too is the resultant force.

The magnitude of the heat of transport can be reduced or increased by relaxation round a defect. Consider a vacancy which, instead of producing a density change $\delta D/D = -1$ over one Wigner–Seitz cell, produces a density change of $\delta D/D = -1/n$ over n Wigner–Seitz cells. Summing over all n cells by (39):

$$Q_p^{*\prime} = -\frac{3}{2} kT \left[\frac{1}{n^2} + \left(\frac{\delta G}{G} \right)^2 \right], \tag{41}$$

instead of

$$Q_p^{*\prime} = -\frac{3}{2} kT \left[1 + \left(\frac{\delta G}{G} \right)^2 \right]. \tag{42}$$

This relaxation may well be important in the alkali metals, especially if the modulus defect is small. When the surrounding atoms relax into the vacancy n is less than one.

Ideally, the equations of this section refer to small density and modulus defects, but for, say, gold in copper, $\delta D/D \simeq 2$, while for an interstitial atom in the noble metals $\delta G/G$ is probably of the same order. Takeno[16] has calculated a large resonant cross section for very heavy defects. Presumably the low frequency resonance of a heavy defect again reduces the magnitude of $Q^{*\prime}$ below that calculated by (39).

If we knew, numerically, the thermal resistivity of a defect, we could, as an alternative to (39), calculate empirically the scattering volume of the defect. For phonons the formal method is simple and preferable. For electrons it is difficult. The empirical method, using experimental values of the electrical resistivity of the defect, and building in the phase factor from a knowledge of the comparative electronic specific heats of the defect and the matrix, is a useful alternative.

5   ELECTRON SCATTERING

The assumptions of the empirical method are simple, but they require the residual electrical resistivity of a defect to be known. They also require the scattering cross-section thus calculated to be a good description of the defect's properties at the high temperatures encountered in thermo-migration and electromigration experiments. The calculation gives the momentum transfer from the electrons acting collectively. The exchange of momentum between electrons of equal mass does not affect the sum of the momenta which they may transfer to a defect. Therefore, in a free electron calculation, electron-electron collisions do not appear. Equally, since we require the collective momentum transfer, and since this is a vector quantity, the Fermi velocity of each electron, when summed for all electrons, disappears from the calculation. If Matthiesen's rule is obeyed, the presence of other species of defects may be ignored, and the calculation becomes classical.

Consider an electron of charge $-e$ and effective mass m^*, in a potential gradient $\nabla(V)$. It experiences an acceleration $\nabla(V)\,e/m^*$, and its mean velocity, $\bar{v}$, is given by

$$(\bar{v})^2 = \frac{s}{2}\,\frac{e}{m^*}\,\nabla(V), \tag{43}$$

where s is the distance it travels between collisions in which its momentum is destroyed. This destruction of momentum is that assumed in (24). Let each defect have a cross-section $N\omega^{2/3}$ for such scattering. If we specify a defect concentration of one atomic per cent, then

$$s = 100\omega^{1/3}/N. \tag{44}$$

If there are n free electrons per atom, the current flowing along a pipe of cross-sectional area $\omega^{2/3}$ is $-ne\bar{v}$, and Ohm's Law reads

$$\nabla(V) = Rn\,e\bar{v}/\omega^{2/3}, \tag{45}$$

where R is the residual electrical resistivity of one atomic per cent of defects. From (43), (44) and (45)

$$N = \frac{R^2 n^2\,e^3}{\omega m^*}\,\frac{100}{2\nabla(V)}. \tag{46}$$

Making a numerical reduction, in the correct units,

$$N = 22.5 R^2 n^2 m/(m^*\omega), \tag{47}$$

where R is in units of micro-ohm centimetres per atomic per cent of defects, ω is in cubic Angstroms, and m^*/m is the ratio of the mass of a conduction electron to that of a free electron. If we assume isotropic scattering, and raise the circular cross-section to a spherical volume, the scattering volume of the defect for electrons scattering, is

$$\Omega_e = \frac{4}{3\sqrt{\pi}}\,\omega(N)^{3/2},\tag{48}$$

and

$$\Omega_e/\omega = \frac{4\times106}{3\sqrt{\pi}}\,R^3 n^3 \left(\frac{m}{m^*\omega}\right)^{3/2}.\tag{49}$$

The phase shift of the scattering must be built into this empirical calculation. It's sign may be obtained without ambiguity. In equation (39) we see that the phase shift makes the heat either positive or negative, according to the sign of $\delta D/D$. In the present calculation we may use this result in reverse; if the electronic specific heat of the defect is greater than that of the host, the electronic, defect, specific heat will be positive, and so will the phase factor.

The magnitude of the phase factor is less tractable. If, in (39), we could write $\delta G = 0$ and retain only the δD term, we could write

$$Q_p^{*\prime} = \pm\frac{3}{2}\,kT\left|\left(\frac{\delta D}{D}\right)^3\right|,\tag{50}$$

where the plus or minus sign was to be selected according to the sign of the $\delta D/D$. An empirical calculation of $(\delta D/D)^2$ would then be sufficient; the empirical phase factor would be unity, either plus or minus. Therefore, if we could now, for electron scattering, set the elasticity term equal to zero, the magnitude of the phase factor to be used in (49) would again be unity. To what extent may this be done? Each defect is electrically neutral, it consists of a charge to be screened and an equal and opposite screening charge. Formally, these charges are distributed over the entire volume of the solid, and without placing any restriction of their distribution, the electrical neutrality of the solid guarantees their equality. A conduction electron is scattered by attraction to or repulsion from parts of the defect; the collision consists of successive sub-collisions with parts of screening charge. If each sub-collision could be regarded as an encounter between two, free, electrons (and not between the conduction electron and two or more electrons acting co-operatively) the "elasticity" of the sub-collision would be given by the Coulomb force between two electrons; it would be unchanged by the presence of the defect, and we should have $\delta G = 0$. Therefore, to the extent that the momentum transfer occurs by successive, independent, electron–electron collisions within the screening charge, the phase factor may be set equal to unity.

Substitution of equation (49) into (27) and comparison with (1) gives the part of the heat of transport due to electron scattering:

$$|Q_e^{*\prime}| = \frac{2}{3} \cdot \frac{4 \times 106}{3\sqrt{\pi}} R^3 n^3 \left(\frac{m}{m^*\omega}\right)^{3/2} (\gamma T^2)$$

$$= 53 R^3 n^3 \left(\frac{m}{m^*\omega}\right)^{3/2} (\gamma T^2), \qquad (51)$$

where γT^2 is the electronic heat content per atom of the host matrix, where the sign of $Q_e^{*\prime}$ is determined by the sign of the electronic defect specific heat, as stated above, and where the phase factor is assumed to be unity. Since the phase factor may be less than one (51) gives a maximum value of $Q_e^{*\prime}$.

The sign of m^*/m is of considerable interest. In (51) it occurs to the three halves power, and a negative value would give an imaginary value of $Q_e^{*\prime}$. We note an ambiguity in the sign of $\bar{v}$ given by (43), and that this sign appears in (45). Its choice is not arbitrary; it must correctly describe the momentum carried by the conduction electron which is to be scattered at the defect. Suppose that this conduction electron is more properly described as a positive hole. We may adopt either of two conventions[17] in choosing the signs of the mass and charge. If we take the more general convention of regarding electrons and holes as both being of positive mass but of respectively negative and positive charge, the momenta of the two particles are of *opposite* sign; the positive of the product of mass and velocity in the case of electrons, but the negative of that product in the case of positive holes (reference 17, equations 106 and 107). The same physical conclusion is of course reached from the alternative convention, and if, in (43), we change *either* the sign of e or m^* to describe momentum transfer from positive holes, we must simultaneously change the sign of $\bar{v}$ when we take the square root. This leaves (45) unaltered, and (51) gives real values of $Q_e^{*\prime}$ for both electron and positive hole conduction by the host matrix. Moreover, we cannot conclude[3] that the sign of $Q_e^{*\prime}$ is determined by the sign of the carrier, for both positive and negative values can occur in one host matrix by introducing into it two defects of entirely different properties. We may readily see the principal governing the sign of $Q_e^{*\prime}$, by making comparison with the phonon case, in which the direction of the force is such as to drive *down* the phonon energy gradient a defect containing an excess of heat (a positive defect specific heat) and to drive *up* the energy gradient a defect with a negative defect specific heat. I feel reasonably sure that a similar principle occurs in electron (or hole) scattering; defects of positive defect electronic specific heat are pushed down the electron energy gradient. Defects having a negative defect electronic specific hear are pushed in an opposite direction. Two species of defects, moving in opposite direction, may coexist in the one solid. If two such defects are bound together their combined motion carries heat down the temperature gradient.

From the experimental data[7,18,19] the presence of a second, bound, species of defect must frequently be assumed; one is otherwise left with the unconvincing spectacle of heat flowing from a colder to a hotter place.

6 PHONON DRAG

In equation (26) we introduced a factor of $\frac{1}{3}$, which represented the isotropic distribution of momentum between the three Cartesian directions. The isotropy is a condition of local thermal equilibrium, and appears in (39) and (51). If there is thermal equilibrium, the total heat of transport is simply

$$Q^{*\prime} = Q_p^{*\prime} + Q_e^{*\prime}. \tag{52}$$

There are no additional terms, because there is no system, other than phonons and conduction electrons, which contains heat. There are no cross terms because, obviously, the phonon system does not pertub the thermal equilibrium of the electron system.

However, equilibrium is not always achieved in real experiments. Consider the collision with a lattice defect of a phonon travelling down the energy gradient. After the collision the phonon energy may be unaltered (if a second phonon is neither created nor destroyed) but longitudinal momentum will have become transverse momentum. Now consider many such scattered phonons, incident on a second defect, downstream from the first. This stream of phonons is deficient in longitudinal momentum, but has excess transverse momentum. We could represent this by writing a factor less than one third into (26), (39) and (51). Let us suppose instead that before they strike the second defect these scattered phonons undergo many collisions, either with phonons of different wave vector or with electrons. The equilibrium between longitudinal and transverse momentum will then be restored, and when the stream of phonons strikes the second defect it carries no information about its collision with the first defect; it is again in thermal equilibrium. Electron–phonon collisions are more important than phonon-phonon collisions in restoring thermal equilibrium, and (52) is valid if

The mean free path of electron-phonon collisions is much smaller than the separation of lattice defects. (53)

This criterion is the opposite of that for the validity of Matthiesen's rule[20].

At 1000 °C in copper the electron-phonon mean free path is about thirty interatomic distances, and thus thermal equilibrium only exists if the defect concentration is much less than thirty parts per million. This is a stringent limit. At greater defect concentrations there exists, over distances large compared to the inter-defect spacing, a steady exchange of momentum between the electron and phonon systems. This is familiar at low temperatures as the phonon drag term of thermo-electricity. It is important in thermomigration because of the small distance over which thermal equilibrium must be restored. For defects in small concentration, such as lattice vacancies, the disequilibrium may be insignificant, but for a strongly scattering defect, such as cobalt in copper, present in appreciable concentration, the disequilibrium may be severe. I shall in this paper use (52), and simply note that in cases of disequilibrium it overestimates the magnitude of the heat of transport.

7 ISOTOPE EFFECTS

a Thermo-migration

An isotope is the simplest of all defects, and we therefore consider it first. It does not scatter electrons nor does it have an elastic modulus defect. By virtue of its mass defect, is scatters phonons and diffuses at a rate different from the host atoms. The phonon scattering is the more important in thermomigration, while only the diffusion rate is relevant to electro-migration. The data[4] on Li^6 and Li^7 illustrate both these effects.

We specify host atoms, isotope atoms, and lattice vacancies, by the subscripts a, b and v. Each Wigner–Seitz cell is occupied by one of these species, so that for the fractional concentrations c:

$$c_a + c_b + c_v = 1. \tag{54}$$

When the isotopic separation in the specimen has attained a steady value, the velocity of a and b atoms must be identical in any frame of reference. If we measure reduced fluxes[10,11], denoted by J', with respect to the crystal lattice, then

$$J_a'/c_a = J_b'/c_b \neq 0. \tag{55}$$

Recalling that equation (1) is correct and that equation (2) is not, the fluxes in a temperature gradient are

$$J_a' = -\frac{\nu_a c_a c_v}{kT}\left[kT\left(\frac{\nabla c_a}{c_a}\right)_T + T\frac{dQ_a^{*'}}{dT\cdot}\frac{\nabla T}{T}\right], \tag{56}$$

and similarly for J_b'. Lithium may reasonably be assumed to have similar properties to sodium, so that the jump frequencies[21,22] are related by

$$\nu_a = (1 - 0.0266)\,\nu_b. \tag{57}$$

In phonon scattering the terms involving the difference $(\nu_a - \nu_b)$ are negligibly small. With $\nu_a = \nu_b$, (55) and (56) become

$$kT\left[\frac{\nabla c_a}{c_a} - \frac{\nabla c_b}{c_b}\right]_T = \left(T\frac{dQ_b^{*'}}{dT} - T\frac{dQ_a^{*'}}{dT}\right)\frac{\nabla T}{T}. \tag{58}$$

Now vacancies are not conserved, and infinite sources and sinks for them exist at the boundaries of the specimen. Therefore no external force acts upon them, and

$$(\nabla c_v)_T = 0. \tag{59}$$

Thus (54) gives

$$(\nabla c_a + \nabla c_b)_T = 0, \tag{60a}$$

$$c_a + c_b = 1 - c_v \simeq 1, \tag{60b}$$

and the left hand side of (58) becomes

$$-kT\left[\frac{\nabla c_b}{c_b}\cdot\frac{1}{1-c_b}\right]_T. \tag{61}$$

For an isotope, the defect enthalpy is zero, and the defect entropy is constant, whence (see equation (102) below)

$$\left[\frac{\nabla c_b}{c_b}\right]_T = \frac{\nabla c_b}{c_b}. \tag{62}$$

Putting $c_b = 0.074$ for the Li^6 isotope, and removing the x variable, we have

$$-kT\frac{\Delta c_b}{c_b}\cdot 1.08 = \left(T\frac{dQ_b^{*'}}{dT} - T\frac{dQ_a^{*'}}{dT}\right)\frac{\Delta T}{T}. \tag{63}$$

We might at first sight expect to set $T \, dQ_a^*/dT$ equal to zero, since the host lattice consists of a atoms on normal, undisturbed, lattice sites. However, this would ignore the transport of host atoms through the lattice by, say, a vacancy mechanism. Let us assume a simple vacancy mechanism of diffusion. *If* there is no relaxation round the vacancy (see (41) and (42)), the phonon part of the scattering is

$$T \frac{d}{dT} (Q_{a,p}^{*'}) = -\frac{3}{2} kT \left[1 + \left(\frac{\delta G}{G}\right)^2 \right]. \qquad (64)$$

Diffusion of b atoms occurs when these exchange places with a vacancy. To what extent should the isotope and vacancy be regarded as a single defect rather than two independent defects? I don't know, but an upper bound (to the magnitude) is provided by assuming them to share one site:

$$T \frac{d}{dT} (Q_{b,p}^{*'}) \geqslant -\frac{3}{2} kT \cdot \frac{8}{7} \left[\left(\frac{8}{7}\right)^2 + \left(\frac{\delta G}{G}\right)^2 \right]. \qquad (65)$$

The electronic scattering is unchanged by the presence of the isotope at the vacancy, and so disappears in the subtraction in (63). Also, if

$$\delta G/G \ll 1.85, \qquad (66)$$

and if we therefore negelect it,

$$T \frac{d}{dT} (Q_{b,p}^{*'} - Q_{a,p}^{*'}) \geqslant -0.735 kT \qquad (67)$$

for the Li^6 isotope in natural lithium. Thus the steady state isotopic separation specified by (63) is

$$\frac{\Delta c_b}{c_b} \leqslant 0.68 \frac{\Delta T}{T}. \qquad (68)$$

We see that for $\Delta T = 40°K$ at a temperature of $400°K$, the isotopic separation will not exceed 7%. The data[4] of Thernquist and Lodding give an enrichment of 0.7% over this temperature interval, and, while a steady state had not been achieved[23], it is clear that the magnitude given by the upper bound of (65) is not achieved. A lower bound is obtained by assuming, in place of (65), that the isotope and vacancy are completely independent, that their entropies are additive. The minimum isotopic separation is 0.04%. The observed value is the median value, and the entropy of the isotope–

vacancy therefore exceeds that of the two components. Now entropy is a negative term in the Gibbs Free Energy, and the increase may be described as a binding entropy; the data indicate that Li^6 is bound to the vacancy with a binding entropy of $0.07k$. The only adjustable parameter is n of equation (41), and experiment therefore indicates the reasonable value $n = 3$ for a vacancy divided in a saddle point configuration.

For the present purposes the important thing is that the light isotope is enriched at the hot end of the specimen, and that this is in no way incompatible with a simple vacancy mechanism of diffusion.

b   Electromigration

In electromigration mass is transported down a gradient of electron energy. The phenomenon is similar to that of the previous paragraph, in which mass is transported down a gradient of phonon energy, and it may be treated by the same equations. The energy gradient is

$$\nabla \left(n(-e) V \right), \tag{69}$$

where n is the number of free electrons (or holes) per atom of the matrix, $-e$ is the electronic charge and V is the conventional potential. For the reasons stated in section 5 we do not distinguish between electronic conductors and hole conductors. From (21) and (27) we may write the flux of defects due to the voltage gradient as

$$J = \pm \frac{D}{kT} \cdot \frac{2}{3} \frac{\Omega_e}{\omega} \nabla \left(neV \right), \tag{70}$$

where Ω_e/ω is given by (49) and the positive or negative sign is to be chosen according to whether the phase factor is plus or minus one. For a vacant lattice site, whose electronic specific heat is always less than that of the surrounding metal, the phase factor is negative (compare with (37) which, for similar reasons, also gives a negative phase factor). Thus the *vacancy* flux is

$$J = -D \left(\frac{2}{3} \frac{\Omega_e}{\omega} \frac{ne}{kT} \right) \nabla(V) = -DX \nabla(V), \tag{71}$$

where X is a positive quantity. The equation corresponding to (56), for the *atom* flux, is therefore

$$J_a' = -v_a c_a c_v \left[\left(\frac{\nabla c_a}{c_a} \right)_T - X \nabla(V) \right]. \tag{72}$$

Let us consider steady state conditions, defined by (55). We have

$$v_a \left[\frac{\nabla c_a}{c_a} - X \nabla(V) \right] = v_b \left[\frac{\nabla c_b}{c_b} - X \nabla(V) \right], \tag{73}$$

the suffix indicating "at constant temperature" being redundant. Using (57) and (60) to simplify this, and for the specific case of Li[6] in natural lithium:

$$\frac{\nabla c_b}{c_b} v_b \frac{1 - 0.0266 c_b}{1 - c_b} = 0.0266 X v_b \nabla(V), \tag{74}$$

or

$$\frac{\nabla c_b}{c_b} = \frac{0.0266}{1.08} X \nabla(V). \tag{75}$$

Now the total atom flux simplifies to

$$J_a' + J_b' = v_a c_a c_v (1 + c_b) X \nabla(V) = D X \nabla(V), \tag{76}$$

whence

$$\frac{\nabla c_b}{c_b} = \frac{0.0266}{1.03} \frac{J_a' + J_b'}{D}, \tag{77}$$

where D, is the measured self diffusion coefficient of lithium. Equation (77) indicates that a steady isotopic concentration gradient is established, the Li[6] being enriched at the end of the specimen towards which there is a bulk transport of atoms. This is the anode[4]. We may evaluate $(J_a' + J_b')$ for sample 30 of Thernquist and Lodding[24], and we may also evaluate[25] D. For this specimen then, at 136°C, we expect

$$\frac{\nabla c_b}{c_b} = 1.8\% \text{ per cm.} \tag{78}$$

The isotopic concentration gradient[4] has been measured at the same current density (Lodding, private communication) and the directly comparable experimental result is

$$\frac{\nabla c_b}{c_b} = 3\% \text{ per cm.} \tag{79}$$

Equation (77) leaves little room for manoeuvre. It contains measured quantities and one adjustable parameter, the difference in jump frequencies. This seems to be greater in lithium than in sodium; the quantity[21,22] $f\Delta K$ is 0.6 rather than 0.36, closer to the simple value of 0.73 for the b.c.c. lattice. The isotopic diffusion data[25] are compatible with this; the low diffusivity of Li[7] in Li[6] is explained by the negative binding entropy between the Li[7] and a vacancy. A free energy barrier has to be overcome before the two may come together. The pre-exponential factor for Li[7] diffusion is therefore reduced.

8   VACANCY ELECTROMIGRATION

The necessary equations have been given in the previous section. The flux of all atoms is the sum of the isotopic fluxes:

$$J = DX\, \nabla(V), \tag{80}$$

where X is given by (71) and the empirical calculation leading to (49). The free electron model is at its closest to reality in the noble metals, and I shall therefore compare the calculation with the electromigration data on silver[26]. The only sensitive parameter is the residual electrical resistivity of the vacancy. This is temperature dependent; at 78°K, 1000°K and 1230°K, experiment[26,27,28] gives $R = 1.7$, 3.37 and about 5, micro-ohm cm per atomic per cent of vacancies. The diffusion coefficient[29] at 1000°K is 1.2×10^{-10} cm^2 s^{-1}, whence, at this temperature, the expected flux of atoms towards the anode is

$$J = 3.7 \times 10^{-8} \text{ cm s}^{-1} \text{ per volt cm}^{-1}. \tag{81}$$

A marker, such as a surface scratch, may be coherent with the lattice, if it does not conduct electricity. (The parallel case of the coherence of a marker because of its failure to conduct heat is well discussed by Biersack and Diez[30].) In the frame of reference attached to the specimen ends, this marker then moves with a velocity $-J$. However, experimental artefacts cause the specimen shape to change, and this motion is superimposed on the marker movement at all points except the one at which the specimen is clamped. This combined motion gives the "isotropy factor", α, which is measureable. The marker velocity is then $-\alpha J$. Converting current density to voltage gradient by the total measured resistivity[31], and with the measured value[26] $\alpha = \frac{1}{3}$, the data give

$$J_{\text{expt}} = 3.63 \times 10^{-8} \text{ cm s}^{-1} \text{ per volt cm}^{-1}, \tag{82}$$

at 1000°K. I shall not claim that the agreement between (81) and (82) is a general measure of the accuracy of the empirical method of section 5, but at least the method gives a credible magnitude for metals with nearly free electrons, and there is no evidence that the empirical phase factor of ± 1 is wide of the mark.

The isotropy factor (see section 14) is not a parameter of the metal, but of the experimental apparatus. This is evident from the data of Kuz'menko[32], in which the marker velocities are some 60% greater, at all temperatures, than those reported by Ho and Huntington[26].

Equation (80) indicates that for vacancy motion the atom flux is always to the anode. The experimental data on other metals do not always support this conclusion, but two recent examples show that this prediction cannot yet be totally dismissed; in beta zirconium Dubler and Wever[33] find atom motion towards the anode, while Campbell and Huntington[19] find atom motion towards the cathode. It is my opinion that in the reactive metals one must allow the possibility that other defects, such as gas atoms, bound to the vacancies, may reverse the direction of vacancy flow.

Having established the credibility of the empirical calculation of the momentum transfer from electron scattering, I shall now use it, together with the gradient of the electronic heat content, $\nabla (\gamma T^2)$, to calculate fluxes and forces in thermomigration.

9 VACANCY THERMOMIGRATION: MONATOMIC SOLIDS

In a thermal gradient the flux of vacancies through the lattice is

$$J = - \frac{D}{kT} \cdot T \frac{dQ^{*\prime}}{dT} \cdot \frac{\nabla T}{T}. \tag{83}$$

A macroscopic marker, of thermal conductivity k_m, embedded in a lattice of thermal conductivity k_1, is observed to move, in a frame of reference defined by the specimen ends and not by the crystal lattice[30], with a velocity[30]

$$v = J \frac{k_1 - k_m}{k_1 + k_m/2}. \tag{84}$$

It is customary to have markers for which $k_1 \gg k_m$, in which case $v = J$. Since for all vacancies $T dQ^{*\prime}/dT$ is negative, vacancies should always flow, and the marker should always move, *up* the temperature gradient. If it does not, we are not observing the diffusion of simple vacancies. Alternatively, as in Fig. 6 of the paper by Jaffe and Shewmon[34], we may be misreading the data; this comment implies no criticism, for the authors' position is set out clearly. However, I believe the part of the line of positive slope, and not that of negative slope, contains the relevant information; the motion of a vacancy *down* a temperature gradient would require the flow of heat from a colder to a hotter place, and I do not believe such a flow can occur freely, as a result of diffusive motion.

We have some check on this conclusion from the identity

$$Q^{*\prime} \equiv -T \Delta S_v, \qquad (85)$$

since ΔS_v is obtainable from absolute, isothermal, dilatometric experiments[35]. For vacancies ΔS_v is positive. The magnitude measured in these isothermal experiments (summarised by McLellan[36]) is that given by the present calculation. In the noble metals the values of ΔS_v are close to the simple value $\frac{3}{2}k$ predicted by (42), but contain positive contributions from the shear modulus defect and from electron scattering[37], together with a negative contribution from the relaxation round the vacancy. Thus gold would appear to differ from silver and copper in having a larger relaxation associated with the vacancy.

Returning now to experiments in thermal gradients, let us consider the data[38] on cobalt. This illustrates, very clearly, the importance of electronic entropy. This is due to electron scattering, and we may make the general observation that a defect without electronic entropy would have zero electrical resistance. In the transition metals the vacancy electrical resistivity[39,40] is high, and the electronic entropy is both large and strongly temperature dependent. This leads to a marked curvature of the Arrhenius plot (see (17)). Detailed analysis will be given elsewhere. This same curvature of course occurs in a self diffusion plot, and must be considered in deducing a Soret coefficient. If we write

$$\Delta S = a + bT^n, \qquad (86)$$

the diffusion coefficient is

$$D = D_0 \exp -\frac{\Delta H - T \Delta S}{kT} = D_0 \exp \frac{a}{k} \cdot \exp \frac{bT^n}{k} \cdot \exp -\frac{\Delta H}{kT}, \qquad (87)$$

and the Soret coefficient, s_i, of Fig. 8 of reference 38 is

$$s_i = \frac{nQ^{*\prime}}{kT^2} \exp \frac{bT^n}{k}. \qquad (88)$$

Only one value, $n = 3$, fits the curvature of this Fig. 8, and only one value of b matches the absolute magnitude of s_i and the diffusion data. Taking the simple value $\frac{3}{2}k$ for the phonon entropy, and the thermal gradient data[38] for the electronic entropy, the total heat of transport of a vacancy in cobalt is

$$Q^{*\prime} = -\left(0.2\left(\frac{T}{T_0}\right) + 0.31\left(\frac{T}{T_0}\right)^4\right) \quad eV; \tag{89}$$

where $T_0 = 1600°K$. Similar calculations may be made for other metals. The thermal gradient data[41] on platinum give $n = 2$, and the vacancy resistivity data on silver[27,26,28] also give $n = 2$. In these two metals therefore, (49) is linear in temperature. In cobalt (49) is parabolic in temperature, but the absolute magnitude of the scattering volume cannot of course be obtained from a free electron model.

Experiments on noble metals in thermal gradients have so far given equivocal results, and I shall not discuss them. I must however refer to the data on beta-zirconium[19,33]. While the numerical values obtained from these two experiments differ by a factor two, both show atom transport towards the hot region of the specimen. Now unless we have a vacancy with a lattice specific heat or an electronic specific heat *greater* than that of an occupied lattice site, the vacancy always has a lower heat content than an occupied site. Thus the motion of a vacancy from a hot to cold region, and the converse motion of an atom, adds to the heat content in the hot region; heat is flowing up a temperature gradient. The specimen is not a closed system, but unless we provide some heat pump, *all* physical processes of heat diffusion, electron-phonon collisions, phonon-phonon collisions, (and equally) electron-defect collisions and phonon-defect collisions, lead to the flow of heat down the temperature gradient; the sense of the motion is defined by the second law. We do not observe rivers flowing uphill, though we may raise water in a bucket. For these two thermomigration experiments we must look for something corresponding to the bucket. As in the saga of Brer Rabbit and Brer Fox, a lighter bucket may go up provided a heavier bucket goes down. This is not a flippant analogy; I wish to emphasise the importance of the presence of a heavy defect moving down the temperature gradient, and the necessity that it is bound to the vacancy, and carries the vacancy down the gradient. This second defect must be heavy in order that there is a net flow of heat down the temperature gradient. I do not know what the defect is, but it seems likely that it is dissolved gas. The conflict between the results on electromigration[19,33] suggests that its electronic scattering power is of similar magnitude but opposite phase to that of the vacancy. The experiments on thermo-osmosis[7] also require the assumption of binding between dissolved gas and a vacancy moving up the temperature gradient. We therefore discuss them now.

10 THERMO-OSMOSIS

The fundamental equation is given by DeGroot[11] (paragraph 9, equation 23). De Groot's $Q*$ is not the isothermal heat of transport; and we must use (1) rather than (2) of the present paper. The pressure change across the diaphragm is

$$\Delta P = -T \frac{dQ*'}{dT} \cdot \frac{1}{\omega} \cdot \frac{\Delta T}{T}, \tag{90}$$

where ΔT is the temperature difference across the diaphragm. This equation does not require integration. (Equation (1) of Gonzalez and Oriani[7] is, incorrectly, an integration by Bearman. Nor does the integration reduce, upon differentiation, to the equation given by De Groot. Consequently, the numerical values derived are without value.)

We observe first that a higher gas pressure on the cold side of the diaphragm means that $T \, dQ*'/dT$ is positive. The pressure only reverses if the sign of $T \, dQ*'/dT$ reverses. Second, we note that for either hydrogen or deuterium dissolved in nickel, the quantity

$$\frac{\Delta P}{\Delta T} = -\frac{dQ*'}{dT} \cdot \frac{1}{\omega} \tag{91}$$

is experimentally found to have only a small temperature dependence. Because of the small temperature range, no distinction can be made between a constant value appropriate to phonon scattering, and a proportionality to T appropriate to electron scattering. On the other hand, when the two gases are dissolved in alpha iron, we find

$$\frac{dQ*'}{dT} \propto T^3. \tag{92}$$

This strong temperature dependence and the consequent reversal of the differential pressure below 530°C, are probably caused by the binding of the dissolved gas to a lattice vacancy. At high temperatures, at which the two are dissociated, the heavy defect, the interstitial gas, moves down the temperature gradient while the light defect, the vacancy, independently moves up the gradient. Association between the two defects increases as the temperature is lowered, and the associated gas and vacancy move in the direction of the net force. Experimentally this is up the temperature gradient. Thus the force on the vacancy exceeds that on the gas, and knowledge of the vacancy forces sets a limit to those on the gas. The further observation[7] that in electromigration through iron, hydrogen is transported to the cathode, indicates that the electron scattering force on the vacancy exceeds that on the gas. To get an idea of the binding energy E_B, we may associate it with the additional temperature dependence observed experimentally. If, for a temperature of 900°K,

$$\exp E_B/kT \propto T^2, \tag{93}$$

then,

$$E_B \sim 0.15 \text{ eV}. \tag{94}$$

A possible mechanism of migration of the associated gas and vacancy is the ring interchange of a gas and metal ion.

Phonon scattering from the mass defect alone would be 10^5 times smaller than that observed. Therefore, any phonon scattering must be principally due to the term

$$\frac{3}{2} kT \frac{\delta D}{D} \left(\frac{\delta G}{G}\right)^2. \tag{95}$$

For deuterium and hydrogen in alpha iron there is experimentally an indication of the isotopic difference (a factor of two) present in δD.

The assumption that all differential pressure was due to electron scattering would give a credible value for the resistivity of hydrogen in iron; $R = 14 \mu$ ohm cm per atomic per cent. Linde's rule would not be unduly stretched to accomodate it. The same cannot be said for the solution of hydrogen or deuterium in nickel, for which the observed pressures would require $R = 80 \mu$ ohm cm per atomic per cent. This is not comparable with the value 1.25μ ohm cm expected[39,40] for copper in nickel. It seems one must assume some phonon scattering. However, unlike alpha iron, there is no evidence in the observed values of $\Delta P/\Delta T$ of the isotopic difference given by (95).

11 VACANCY THERMOMIGRATION: BINARY ALLOYS

For substitutional binary alloys, the equations are an elaboration of those for isotopic separation. There are however two differences, and the first must be emphasised since it is not found in the literature. The total differential of

$\ln (c_b)$ is

$$\frac{\nabla c_b}{c_b} = \left[\frac{\nabla c_b}{c_b}\right]_T + \left(\frac{\delta c_b}{\delta T}\right)_x \frac{\nabla T}{c_b}. \tag{96}$$

and the second term is in general non-zero. The total Gibbs Free Energy of the alloy must be a minimum with respect to changes of concentration of the constituents i.e. with respect to their demixing. The total Gibbs Free Energy of N atoms is

$$G = N\,(\mu_a c_a + \mu_b c_b + \mu_v c_v), \tag{97}$$

where μ is the chemical potential of a species present in concentration c. As in (54),

$$c_a + c_b + c_v = 1 \approx c_a + c_b, \tag{98}$$

and

$$dc = dc_a = -dc_b. \tag{99}$$

Also

$$\mu_v = 0. \tag{100}$$

Differentiation of (97) with respect to concentration gives

$$\mu_a = \mu_b, \tag{101}$$

and the last term of (96) is given by

$$-kT\left(\frac{\delta c_b}{\delta T}\right)_x \cdot \frac{T}{c_b} = \varDelta H_a - \varDelta H_b + Q_a^{*\prime} - Q_b^{*\prime} - T\frac{d}{dT}\,(Q_a^{*\prime} - Q_b^{*\prime}). \tag{102}$$

Secondly, the diffusion rates of solvent and solute atoms are in general unequal, $\nu_a \neq \nu_b$. In this more general case, and with $c_b \ll 1$, (58) becomes

$$-kT\left[\frac{\nabla c_b}{c_b}\right]_T = \left(T\frac{dQ_b^{*\prime}}{dT} - \frac{\nu_a}{\nu_b}\,T\frac{dQ_a^{*\prime}}{dT}\right)\frac{\nabla T}{T}. \tag{103}$$

The experimental quantity is the total differential, obtained by summing (102) and (103), and the slope of the Arrhenius plot is

$$-kT\frac{\nabla c_b}{c_b}\frac{T}{\nabla T} = \varDelta H_a - \varDelta H_b + Q_a^{*\prime} - Q_b^{*\prime} - 2T\frac{d}{dT}\,(Q_a^{*\prime} - Q_b^{*\prime})$$

$$- \frac{\nu_a - \nu_b}{\nu_b}\,T\frac{dQ_a^{*\prime}}{dT}. \tag{104}$$

A simpler equation is obtained by abandoning the Arrhenius plot:

$$-\frac{d}{dT}(kT \ln c_b) = 2\frac{dQ_b^{*\prime}}{dT} - \frac{v_a + v_b}{v_b}\frac{dQ_a^{*\prime}}{dT}, \qquad (105)$$

but it is important to note that in the bracket on the left, T is a variable and not the mean temperature of the specimen. Equation (105) integrates to

$$-kT_0 \ln(c_b/c_{b,0}) = k\,\Delta T \ln(c_{b,0}) + 2\,\Delta Q_b^{*\prime} - \frac{v_a + v_b}{v_b}\Delta Q_a^{*\prime}. \qquad (106)$$

The mean temperature is T_0, and the mean concentration is $c_{b,0}$. We see that $\ln(c_b/c_{b,0})$ is linear in ΔT. More importantly, we see that for the small solute concentration used in radioactive tracer experiments[34,42] ($c_b \sim 10^{-8}$) the observed concentration gradient is principally due to $k\,\Delta T \ln(c_{b,0})$. Values of $Q_b^{*\prime}$ cannot be deduced when the solute concentration is unknown.

Consider the data[42] on antimony in dilute solution in silver. We may estimate the resistivity of lattice vacancies[26,28] to be $4.2\,\mu$ ohm cm per atomic per cent at $1145\,^\circ$K. The electronic specific heat of silver[43] and the dilatometric data[36] (for the phonon entropy) then give, at $1145\,^\circ$K, for a vacancy in a silver matrix,

$$\Delta Q_v^{*\prime} = -\Delta Q_a^{*\prime} = -1.5k\,\Delta T - 3\times4.3k\,\Delta T = -14.4k\,\Delta T. \qquad (107)$$

The diffusion of antimony in silver occurs by the Sb atom jumping into an Ag vacancy. We therefore have to evaluate $Q_b^{*\prime}$ for the Sb—vacancy defect. By assuming that the heavier Sb atom provides mass *at* the vacancy, one has for the phonon contribution

$$Q_{b,\text{phonon}}^{*\prime} = \frac{3}{2}kT\left(-\frac{92}{108}\right)^3 = -0.9kT. \qquad (108)$$

For a solute atom adjacent to a solvent vacancy, Matthiesen's rule is a good first approximation[20] for the combined electrical resistivity, and hence for the total electronic scattering. The vacancy has a negative electronic defect specific heat, and the solute atom, having an excess valence of three, has a positive electronic defect specific heat; the two defects scatter in antiphase, a phase factor of minus one must therefore be incorporated, and the resistivities of solute and vacancy must be subtracted. Taking the residual resistivity[40] of Sb at $1145\,^\circ$K to be $R = 7.5\,\mu$ ohm cm per atomic per cent, the electronic heat of transport of the solute-vacancy is

$$Q_{b,\text{electronic}}^{*\prime} = 2.2kT^2/T_0, \qquad (109)$$

the temperature dependence being less than that for the vacancy. From (108) and (109),

$$\Delta Q_b^{*\prime} = -0.9k\,\Delta T + 4.4k\,\Delta T = 3.5k\,\Delta T. \tag{110}$$

The absolute concentration of antimony, and the ratio of the solvent to the solute diffusion coefficient, are both known[42]. Numerically therefore, (106) becomes

$$\ln \frac{c_b}{c_{b,0}} = \frac{\Delta T}{T_0}\,(17.2 - 7 + 1.2 \times 14.4) = 27.5\,\frac{\Delta T}{T_0}. \tag{111}$$

The experimental data[42], show that $\ln (c_b/c_{b,0})$ is linear in ΔT, and that

$$\ln \left(\frac{c_b}{c_{b,0}}\right) = 13.3\,\frac{\Delta T}{T_0}. \tag{112}$$

Equation (111) would agree with (112) if its least certain component, the electronic entropy of (109), were to be trebled. This discrepancy of three is not unreasonable; the bold assumption of Matthiesen's rule, and of completely antiphase scattering, has underestimated the residual electrical resistivity of the solute-vacancy pair by only 30%. Thus the resistivity of the antimony at the vacancy does not exceed 9 μ ohm cm per atomic per cent, even for completely antiphase scattering. (The measured solute resistivity is that of the overwhelming fraction of solute which is not adjacent to a vacancy. It contains negligible information about the resistivity of the "saddle point" configuration of the diffusing solute.)

One observes that there is no such thing as a "reasonably small" concentration of solute; the smaller the concentration, the more important it is to know it, witness the figure of 17.2 in equation (111). This term is largely responsible for the difference of sign between the reported value $Q_b^{*\prime} = -1.26\,\mathrm{eV}$ and that found here, $Q_b^{*\prime} = +0.47\,\mathrm{eV}$. The positive, but not the negative, value is compatible with the rapid drift[52] of antimony towards the anode during electromigration.

12   INTERSTITIAL ALLOYS

In interstitial alloys the conservation equation, (54), is changed:

$$c_a + c_v = 1, \tag{113}$$

and c_b is arbitrary. If the lattice is not inert, the steady state is reached by diffusion of both solute and solvent, and is given by

$$-kT\frac{\nabla c_b}{c_b} \cdot \frac{T}{\nabla T} = \mu_{b,0} - \Delta H_b - Q_b^{*'} + 2T\frac{dQ_b^{*'}}{dT} - \frac{\nu_a c_\nu}{\nu_b} T\frac{dQ_a^{*'}}{dT}. \quad (114)$$

Like (104) it may be simplified:

$$-\frac{d}{dT}(kT \ln c_b) = 2\frac{dQ_b^{*'}}{dT} - \frac{\nu_a c_\nu}{\nu_b}\frac{dQ_a^{*'}}{dT}. \quad (115)$$

It differs from (105) because the conservation condition, expressed by (113), is different; this is the fundamental difference between interstitial and substitutional alloys.

Comparison with experiment is in general frustrated by a lack of knowledge of the amount of solute present.

The literature ignores all terms except $Q_b^{*'}$ in (114). These neglected terms are often an order of magnitude greater than the remaining term.

13 IONIC SOLIDS

I shall consider the steady concentration of divalent strontium dissolved in sodium chloride[5]. The system is interesting in that the Sr^{++} diffuses rapidly, despite negligible self diffusion in the NaCl lattice; the lattice has the inertness one associates with interstitial systems, but the defect itself is substitutional.

To match the data to equation (104) one must know, or calculate, ΔH_b. To calculate it, first imagine the creation of a Schottky vacancy pair in the NaCl lattice. The formation enthalpy of this pair is[44] 2.12 eV. Consideration of electrostatic effects may be avoided if we next carry out two operations simultaneously; we destroy the chlorine vacancy and at the same time replace a sodium ion by the divalent strontium ion. The destruction of the vacancy is exothermic; we regain its formation enthalpy. Replacement of an Na^+ ion by the Sr^{++} ion is also exothermic; we lose a small amount of enthalpy by straining the lattice to accomodate an ion of different size, but we gain a binding energy if we bring the Sr^{++} ion and the sodium vacancy on to neighbour Na^+ sites. This binding energy has been calculated[45]; it is 0.45 eV. The enthalpies may be calculated on two very simple assumptions, that each is the strain energy of a simple elastic jelly, and that this jelly may be dilated by removal or insertion of the filled shell of an ion; the L-shell of the Na$^+$ ion or the Cl$^-$ion. The enthalpy is then proportional to the volume dilatation. Sharing the enthalpy of the Schottky pair in the ratio of the two dilatations gives 0.03 eV for that of the chlorine vacancy. What dilatation is to be associated with the Sr^{++}ion? Its ionic radius exceeds that of the Na$^+$ which it replaces (1.12 Å as against 0.97 Å), but its N-shell is unfilled. I shall choose as the appropriate dilatation that of Cd^{++}, its neighbour in Group 2, whose N-shell is filled. This is speculative ground, but upon it I shall ignore the enthalpy associated with the substitution of an Na$^+$ ion by the Sr^{++} ion. The total enthalpy associated with the introduction of the Sr^{++} ion into the NaCl lattice is then

$$\Delta H_b = 2.12 - 0.45 - 0.03 = 1.64 \text{ eV}. \tag{116}$$

There is negligible dissociation of the Sr^{++} ion and the Na$^+$ vacancy; even at 1000°K they diffuse as one unit. Furthermore, at this temperature the associated defect diffuses about 100 times faster than the Na$^+$ ion, and about 1000 times faster than the Cl$^-$ ion. Thus we may put $\nu_a = 0$ in (104). In ionic solids there is no electronic contribution to $Q^{*\prime}$. The phonon contribution is linear in T, and therefore

$$2T \frac{dQ_b^{*\prime}}{dT} - Q_b^{*\prime} = Q_b^{*\prime}. \tag{117}$$

The maximum value of $Q_b^{*\prime}$ is obtained by considering the scattering from the Sr^{++} ion to be independent of that from the Na$^+$ vacancy. The unit of mass in the NaCl lattice is that of the two ions, so that this value of $Q_b^{*\prime}$, due to the mass defect, is

$$Q_b^{*\prime} = \frac{3}{2} kT \left[\left(\frac{87.6 - 23}{35.5 + 23} \right)^3 - \left(\frac{23}{35.5 + 23} \right)^3 \right]$$

$$= 1.92kT = 0.166 \text{ eV at } 1000°\text{K}. \tag{118}$$

The minimum value of $Q_b^{*\prime}$ is obtained by considering the Sr^{++} ion and the Na$^+$ vacancy to completely overlap, at one site, and is 0.047 eV. Thus, from (116), the slope of the Arrhenius plot is expected to lie between -1.59 eV and -1.47 eV. Allnatt and Chadwick's experimental value[5] is -1.56 ± 0.15 eV. It seems that assumptions leading to (116) are not outrageous. One notes that the heat of transport contributes not more than 10% of the measured Arrhenius slope.

The ionic solids AgCl and AgBr provide our firmest knowledge of the temperature dependence of the phonon contribution to the heat of transport[46]. Because of their low absolute temperature, the two hundred and fifty degrees from 150 °C to 400 °C provide a comparatively greater temperature range than any other data. Over this temperature range the heat of transport is closely proportional to the absolute temperature, as one expects from (39). Wirtz[47] has also observed that in PuO_2–UO_2 reactor fuel the heat of transport is proportional to the absolute temperature.

One may estimate the Arrhenius plot to be expected in other ionic solids, and Co^{++} in NaCl is typical. A binding energy of 0.32 eV may be estimated from the calculations of Bassani and Fumi[45]. The other two terms of the enthalpy remain the same. The heat of transport will not exceed 0.025 eV at 1000 °K, and so the expected slope is

$$- (2.12 - 0.32 - 0.03) + 0.025 = -1.79 \text{ eV}. \tag{119}$$

14 THE ISOTROPY FACTOR

The isotropy factor is an experimental artefact. It is a parameter of the apparatus and not a property of the specimen material. The departure of this factor from the value one is a measure of the imperfection of the technique. Evidence for this statement is provided by the unexplained differences[26] between the sets of data[26,32] on silver, and by the variations, within one specimen[19], which are not simply a function of temperature.

The isotropy factor is principally due to the *divergent* flow of heat in the specimen. Experimental specimens are frequently cylindrical. Ideally, the heat flow is along the axis of the cylinder, but in practice there is a radial heat loss. This is likely to be greatest at the hottest part of the specimen. In an electrically heated specimen the axial flux of heat is greatest in the cold part of the specimen. Thus, the ratio of radial to axial flow varies along the specimen, and the variation gives rise to a divergence. Creep may also contribute to the isotropy factor, but I shall neglect it here. Evaporation of the specimen may reduce the radial dimension, but since it produces no longitudinal deformation it is irrelevant to the isotropy factor.

Consider a macroscopic length of a cylindrical specimen, the shape of which is changing because of a divergent heat flow. In the absence of evaporation losses we may consider a conserved number of atoms. The volume is then constant, and the sum of the axial and radial strain rates is zero:

$$\dot{\varepsilon}_{xx} + 2\dot{\varepsilon}_{rr} = 0. \tag{120}$$

The velocity of separation of two markers, at each end of this length, has two components, one due to this shape change:

$$v_{xx} = \int \dot{\varepsilon}_{xx}\, dx, \qquad (121)$$

and one due to the motion of the crystal lattice at each marker in the centre of mass frame of reference [30]. This second component is quite distinct, and may be identified with the change in atom velocity between the two positions, v_a. If we regard one marker as fixed, the velocity of the second marker is

$$v_m = -v_a + v_{xx}, \qquad (122)$$

that is to say, the velocity of separation of the two markers exceeds $(-v_a)$ when the divergence of the heat flow causes the specimen diameter to shrink. In this case the isotropy factor, α, defined by

$$v_x = -v_m/\alpha \qquad (123)$$

is greater than one. Conversely, if the specimen diameter is increasing, α is less than one. The divergent heat flow shown in Fig. 1(a) is in the more probable of the two possible senses, and gives values of α which are less than one. However, there are no formal upper or lower limits to α; one could contrive, though it would be difficult, an experimental situation in which α exceeded

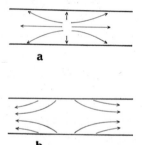

Figure 1 Two forms of divergent heat flow from the hot region of a specimen. That shown in (a) is for cooler surroundings and gives an isotropy factor less than one. That shown in (b) is for hotter surroundings, and would give an isotropy factor greater than one

one. It is easier to obtain, by a very divergent heat flow, values of α less than $\frac{1}{3}$, and such values are observed[1].

The decreasing transverse separation of surface markers[34] measures the radial atom flux, as well as the radial strain rate $\dot{\varepsilon}_{rr}$; we must again distinguish between motion of a crystal lattice in the centre of mass frame of reference [30], and the divergence of the flow. This decreasing transverse separation of markers is therefore not a simple measure of the isotropy factor.

15 THE WIRTZ MODEL

By considering diffusion at the atomistic level[48], the energy of transport has been expressed as a fraction, f, of the free energy of migration, E_m, of the diffusing species:

$$\Delta H + Q^{*\prime} = -f E_m. \tag{124}$$

The model divides the migration of an atom into three stages, each of which is associated with an energy barrier: (1) the free energy necessary to lift the atom out of its initial lattice site, E_1; (2) the free energy necessary to push it through the configuration intermediate between two lattice sites, E_2; and (3) the energy necessary to put it in the second site, E_3. This completes the migration process, and

$$E_m = E_1 + E_2 + E_3. \tag{125}$$

An atom migrating from site X to the adjacent site Y, has to surmount the barriers in the order $1 \to 2 \to 3$. An atom migrating in the reverse direction encounters barriers whose heights are still in the sequence, $1 \to 2 \to 3$, since once an atom has completely fallen into either site the two sites are indistinguishable.

Suppose that the site X is at a temperature T and that site Y is at $T + \Delta T$. Having surmounted the first barrier from site X the atom will have some intermediate temperature, say $T + \Delta T/3$. Let its temperature after surmounting the second barrier be $T + 2\Delta T/3$. The concentration of the diffusing species at the two lattice sites is now determined from the requirement of a zero net flux between the two sites, and *this flux must be zero at all points between X and Y.* Let the concentration be c_X, c_1, c_2 and c_Y, reading from X to Y, (c_1 and c_2 are the concentrations at $T + \Delta T/3$ and at $T + 2\Delta T/3$). The rate at which atoms leave X is

$$c_X \nu \exp\left(-E_1/kT\right), \tag{126}$$

where ν is the Debye frequency. The rate at which they arrive at X is

$$c_1 \nu \exp\left(-E_3/k\left(T + \Delta T/3\right)\right). \tag{127}$$

For a zero net flux (127) is equal to (126). Over the central barrier the flux $\overrightarrow{XY}$ is

$$c_1 \nu \exp\left(-E_2/k\,(T + \Delta T/3)\right), \tag{128}$$

and the flux $\overrightarrow{YX}$

$$c_2 \nu \exp\left(-E_2/k\,(T + 2\,\Delta T/3)\right). \tag{129}$$

Finally, the rate at which atoms arrive at Y is

$$c_2 \nu \exp\left(-E_3/k\,(T + 2\,\Delta T/3)\right), \tag{130}$$

and the rate at which they leave Y is

$$c_Y \nu \exp\left(-E_1/k\,(T + \Delta T)\right). \tag{131}$$

Since

$$\frac{c_X}{c_Y} = \frac{c_X}{c_1}\,\frac{c_1}{c_2}\,\frac{c_2}{c_Y}, \tag{132}$$

equation (126) to (131), taken in pairs, give

$$\frac{c_X}{c_Y} = \exp\left(\frac{E_1 + E_2 - E_3}{kT} \cdot \frac{\Delta T}{3T}\right) = 1 + \frac{E_1 + E_2 - E_3}{kT} \cdot \frac{\Delta T}{3T}. \tag{133}$$

Using the following equation, from which the $T\,dQ^*/dT$ term is incorrectly omitted,

$$\frac{\Delta c}{c} \cdot \frac{T}{\Delta T} = \frac{\Delta H + Q^{*\prime}}{kT}, \tag{134}$$

gives

$$\Delta H + Q^{*\prime} = -(E_1 + E_2 - E_3)/3. \tag{135}$$

A different result would be obtained by some other, equally arbitrary, assumption than that the temperature is $T + \Delta T/3$ beyond the first barrier from X, but the heights of all three energy barriers appear in (135), and without some further assumptions the equation gives no indication of the magnitude of the heat of transport.

Shockley[49] has indicated the fundamental difficulty of this approach, which is to build into the model the fact that the moving defect is always in thermal contact with the lattice. It cannot be assumed that at some arbitrarily chosen point on a potential hill between two lattice sites a defect receives a quantity of thermal energy and thereafter moves over the potential hill without further interaction with the lattice. The weakness of the Wirtz model is to postulate regions of thermal isolation from the lattice. The dictum expressed in (40) must be recalled.

16   DISCUSSION

a) The force arising from a gradient of a thermodynamic potential is perfectly real and Newtonian, for there is no physical quality in such a gradient by which to distinguish it from a Newtonian force. Indeed, if such a distinction could be drawn, one would question the validity of (13), and the validity of the thermodynamic function which gave rise to it. The quantity $-v\nabla P$ is a Newtonian force, a mechanical force, for v is an activation volume and P is a hydrostatic pressure. If $-\nabla\mu$ is not a Newtonian force then (13) and the thermodynamic functions (reference 9, equations 24.7, 24.8 and 24.9) contain quantities which are different in kind; the suggestion is absurd.

The reality of thermodynamic force is perhaps obscured by the desire to regard temperature and entropy, and hence thermal energy, as parameters which can describe microscopic systems, in particular a single, atomic, defect. These parameters refer only to macroscopic systems.

b) If one admits the theorem (or perhaps axiom) of microscopic reversibility, and thereby admits the use of standard thermodynamic functions, as at equation (9), the notation of the subject is greatly simplified. The quantity called by physicists the energy of transport and by chemists the heat of transport is the two terms

$$\varDelta H - T\varDelta S_v \tag{136}$$

of equation (15). The quantity called by physicists the heat of transport and by chemists the reduced heat of transport is the quantity

$$-T\varDelta S_v. \tag{137}$$

Equally, electromigration is described simply by the electronic entropy. Without electronic entropy a lattice defect would have no electrical resistance. Electronic entropy is responsible for electron scattering, and for the momentum transfer from electrons to defects which is the sole source of electromigration.

c) The parameter "effective charge" which is used in discussion of both thermomigration and electromigration is a snare and a delusion. It is a delusion because the charge on the defect is identically zero, the total screening charge must equal the charge which is screened. The solid as a whole is electrically neutral. The postulate[3] of an electrostatic force acting upon the defect is a violation of the Friedel sum rule[13]. The only effect of an electric field is to induce a dipole moment between the screened charge and the screening charge. This conclusion, that the electrostatic force is identically zero, is not new[50], nor is it accepted[51].

d) The concept of polarisation can be extended to ionic solids, for these also have an overall electrical neutrality; the charge in the cloud of cation defects is equal and opposite to that of the anion defects. In a temperature gradient a phonon wind blows through the lattice, exerting the greater force on the defect with the greater mass defect. The anion and cation defects therefore separate, until there is dipole between the two defect clouds, a dipole which balances the difference between the two phonon forces. Thereafter, the anion and cation defects are blown through the lattice with a common velocity. The separation of the two defect species in the thermal gradient generates the thermoelectric e.m.f.

e) The phase factor which in this paper is built into the electronic scattering is elsewhere[3] omitted. The same calculation[3] incorporates the electron mean free path into the part of the heat of transport due to momentum transfer from electrons. I believe this mean free path should disappear, as at equation (46) of the present paper.

f) We note again that there is no distinction between electron conductors and hole conductors, since the momentum, p, is respectively

$$p = mv \tag{138}$$

for electron conductors, but

$$p = -mv \tag{139}$$

for hole conductors[17].

g) Considerable confusion surrounds the transformation of fluxes and forces, despite the standard treatments[10,11]. To avoid this confusion I have worked throughout in "reduced" quantities, and in the frame of reference of the crystal lattice. (Some workers hold that vacancies will drift down a thermal gradient even though they are not subject to a driving force. This supposition is a violation of the Nernst-Einstein relation, see reference 1, equation 1.)

h) The conclusion reached in the present paper may be stated as a general principle:

Heat flows down a temperature gradient, even
when it is carried by defects in a crystal lattice. (140)

This conclusion parallels the known properties of phonon–phonon collisions and electron–phonon collisions; these always lead to the diffusive flow of heat down the temperature gradient. The principle is then simply a statement that the diffusive motion of defects in a crystal lattice is not driven by a heat pump. Their net diffusive motion does not pump heat from a colder to a hotter place.

The direction of motion of a defect in a temperature gradient is determined by its defect specific heat. A defect which increases the heat content of the lattice, i.e. which has a positive defect specific heat, diffuses down a temperature gradient. Conversely, a defect with a negative defect specific heat diffuses up the temperature gradient.

i) In establishing the defect specific heat, both the phonon and electron defect specific heats must be used. The second is often considerable at high temperatures, but it is not given by the intercept on an Arrhenius plot of diffusion or dilatometric data, instead it appears in the slope of such plots.

j) The electrical resistance of a defect, even though measured at high temperature, is principally that associated with the equilibrium configuration rather than that associated with the saddle point configuration of a defect in motion between two lattice sites. There are three reasons for this. The time spent by a defect in the saddle point configuration is a small fraction of that spent in the equilibrium configuration; it is of order $\exp(-E_m/kT)$, which has a maximum value $\sim 10^{-4}$ at the melting point. Thus if we look, as we must, at all of the defects in the solid, and if we measure the average conditions over a long time (about one second in resistance measurements), our measurements refer to 99.99% of the equilibrium configuration and 0.01% of the saddle point configuration. Secondly, the scattering volume calculable from (49) is often, and especially at high temperatures, several Wigner–Seitz cells. The disturbance associated with the saddle point occurs only at the centre of this scattering volume; it does not perturb the entire scattering volume. Thirdly, the plasma frequency, at which the screening charge relaxes, is 10^3 times greater than the lattice frequency, at which the ion, the screened charge, relaxes into a lattice site. Therefore, the screening charge is not appreciably disturbed by the diffusion of the ion.

k) The comparative magnitudes of the phonon and the electron forces on lattice defects do not reflect, and are unrelated to, the trasnport properties of the matrix. The diffusive heat current in an insulator is carried by phonon-phonon collisions. The heat current in a metal is carried by electron–phonon collisions. The properties calculated here depend on different events, phonon–defect collisions and electron–defect collisions. There is no reason to expect the properties of the defect to resemble the properties of the matrix, and we could demonstrate this by considering a variety of defects, with widely differing properties. The heat of transport of a defect is not necessarily dominated by electron effects, even in a good metal. The most striking example of this is the isotopic separation of lithium in a temperature gradient. This is entirely due to phonon effects.

ACKNOWLEDGEMENT

This paper is published by permission of the Central Electricity Generating Board.

PROOF NOTE

Equation (41) considers vacancy splitting, but does not include the negative dilation of the lattice caused by the inward relaxation of the atoms surrounding the vacancy. This dilation is given by the activation volume of vacancy formation. When it is included, one has $n = 6$ in lithium. It seems likely that a cube of the low temperature f.c.c. phase forms at the vacancy, and that the vacancy is split into six parts, each at approximately the body centre of the six nearest b.c.c. cubes surrounding the f.c.c. cube. This multiple splitting is compatible with the small activation volume, and with ΔK, which is larger in lithium than sodium.

Throughout the paper I have wrongly ignored the temperature dependence of the defect enthalpy, ΔH. This is generally done, (e.g. reference 1, equation 15) but it is too crude an approximation. One is compelled to consider momentum transfer from electrons, since it is the source of electromigration, of the residual electrical resistance of the defect, and of electronic entropy. Electronic entropy is temperature dependent, and

$$\left(\frac{\delta \Delta H}{\delta T}\right)_p = T \left(\frac{\delta \Delta S_v}{\delta T}\right)_p. \tag{141}$$

Therefore, one is compelled to include the temperature dependence of ΔH. The following changes are significant. The slope of the Arrhenius plot given by (20) is

$$-kT\frac{\nabla c}{c}\frac{T}{\nabla T} = \mu_0 - \Delta H - T\Delta S_v = kT\ln(c) - 2T\Delta S_v. \quad (142)$$

The flux given by (23) is

$$J = \frac{D}{kT}\cdot T\Delta S_v\cdot\frac{\nabla T}{T}. \quad (143)$$

The exponential factor disappears from (88), leaving the electronic entropy of a vacancy in cobalt still more strongly temperature dependent than indicated by (89). In equations (106) and (115) we must write not $\Delta Q^{*\prime}$ but $-\Delta S_v\cdot\Delta T$.

In the use of equation (106), the new data by Patil and Huntington (*J. Phys. Chem. Solids*, 1970, **31**, 463) supersede that of Ho and Huntington[26]. For the thermomigration of antimony in silver the numerical factor in (111) becomes

$$(17.2 - 6.6 + 1.2\times 2.6) = 13.7. \quad (144)$$

The agreement with the observed value, 13.3, given in (112) is now as good as one could wish. This particular case, in which the data on all the variables are complete, encourages one to believe that the electrical resistivities of adjacent defects may usefully be combined in the proper phase, and that a binary alloy, even when its constituents have very different properties, is amenable to simple treatment.

A final indication of this simplicity is the temperature dependence of the electronic entropy, which is observable in electromigration. By combining the temperature dependent resistivity of a vacancy (3.35 μ ohm cm at 1145°K, from Patil and Huntington) in antiphase with the very nearly temperature independent resistivity of the antimony[40], one expects $T^{-1/4}$ for the impurity–vacancy complex. The experimental temperature dependence, which Doan (*J. Phys. Chem. Solids*, in press) regards as significant, is $T^{-1/4}$.

REFERENCES

1. J.F.D'Amico and H.B.Huntington, *J. Phys. Chem. Solids*, **30**, 2607 (1969).
2. H.B.Huntington, G.A.Shirn and E.S.Wajda, *Phys. Rev.*, **99**, 1085 (1955).
3. H.B.Huntington, *J. Phys. Chem. Solids*, **29**, 1641 (1968).

4. P.Thernquist and A.Lodding, *Proc. Conf. Vacancies and Interstitials in Metals* (KFA: Julich) 55 (1968).

5. A.R.Allnatt and A.V.Chadwick, *Trans. Faraday Soc.*, **63**, 1929 (1967).

6. P.G.Shewmon, *Acta Met.*, **8**, 605 (1960).

7. O.D.Gonzalez and R.A.Oriani, *Trans. Met. Soc. AIME*, **233**, 1878 (1965).

8. P.S.Ho and H.B.Huntington, *J. Phys. Chem. Solids*, **27**, 1319 (1966).

9. L.D.Landau and E.M.Lifshitz, *Statistical Physics* (London: Pergamon Press) (1958).

10. R.E.Howard and A.B.Lidiard, *Rep. Prog. Phys.*, **27**, 161 (1964).

11. S.R. De Groot, *Thermodynamics of Irreversible Processes* (Amsterdam: North Holland) (1952).

12. J.W.S.Rayleigh, *Theory of Sound* (London: MacMillan) (1896).

13. J.M Ziman, *Electrons and Phonons* (Oxford: Clarendon Press) (1960).

14. P.M.Morse, *Vibration and Sound* (New York: McGraw-Hill) (1948).

15. H.J.Lipkin, *Ann. Phys. (N.Y.)* **18**, 182 (1962).

16. S.Takeno, *Prog. Theoretical Physics*, **29**, 191 (1963).

17. R.A.Smith, *Wave Mechanics of Crystalline Solids* (London: Chapman and Hall) (1961).

18. D.N.Vasil'kovskii and I.V.Zakurdaev, *Soviet Physics—Solid State*, **10**, 2886 (1969).

19. D.R.Campbell and H.B.Huntington, *Phys. Rev.*, **179**, 601 (1969).

20. S.Koshino, *Prog. Theoretical Physics*, **24**, 484 (1960).

21. J.N.Mundy, L.W.Barr and F.A.Smith, *Phil. Mag.*, **14**, 785 (1966).

22. A.D.Le Claire, *Phil. Mag.*, **14**, 1271 (1966).

23. A.Lodding, private communication.

24. P.Thernquist and A.Lodding, *Z. Naturforsch.*, **23a**, 627 (1968).

25. A.Ott and A.Lodding, *Proc. Conf. Vacancies and Interstitials in Metals* (KFA: Julich) 43 (1968).

26. P.S.Ho and H.B.Huntington, *J. Phys. Chem. Solids*, **27**, 1319 (1966).

27. A.Seeger and D.Schumacher, *Lattice Defects in Quenched Metals* (New York: Academic Press) 26 (1965).

28. A.Ascoli, E.Germagnoli and G.Guarini, *Acta Met.*, **14**, 1002 (1966).

29. W.C.Mallard, A.B.Gardner, R.F.Bass and L.M.Slifkin, *Phys. Rev.*, **129**, 617 (1963).

30. J.Biersack and W.Diez, *Phys. Stat. Sol.*, **27**, 139 (1968).

31. E.F.Northrup, *J. Franklin Inst.*, **177**, 287; *ibid.* **178**, 85 (1914).

32. P.P.Kuz'menko, *Ukr. Fiz. Zh.*, **7**, 117 (1962).

33. H.Dubler and H.Wever, *Phys. Stat. Sol.*, **25**, 109 (1968).

34. D.Jaffe and P.G.Shewmon, *Acta Met.*, **12**, 515 (1964).

35. R.O.Simmons and R.W.Balluffi, *Phys. Rev.*, **117**, 52 (1960).

36. R.B.McLellan, *Trans. Met. Soc. AIME*, **245**, 379 (1969).

37. L.M.Roberts, *Proc. Phys. Soc.*, **70B**, 744 (1957).

38. P.S.Ho, *J. Phys. Chem. Solids*, **27**, 1331 (1966).

39. N.F.Mott and H.Jones, *The Theory of the Properties of Metals and Alloys* (Oxford: Clarendon Press) (1936).

40. A.N.Gerritsen, *Encyclopedia of Physics*, **19**, 137 (Berlin: Springer) (1956).

41. S.C.Ho, Th.Hehenkamp and H.B.Huntington, *J. Phys. Chem. Solids*, **26**, 251 (1965).

42. W.Biermann, D.Heitkamp and T.S.Lundy, *Acta Met.*, **13**, 71 (1965).

43. F.J. Du Chatenier and J. De Nobel, *Physica*, **28**, 181 (1962).

44. R.W.Dreyfus and A.S.Nowick, *J. Appl. Phys.*, **33**, 473 (1962).

45. F.Bassani and F.G.Fumi, *Nuovo Cimento*, **11**, 274 (1954).
46. R.W.Christy, *J. Chem. Phys.*, **34**, 1148 (1961).
47. K.Wirtz, *J. Am. Chem. Soc.*, **90**, 3098 (1968).
48. K.Wirtz, *Physik Z.*, **44**, 221 (1943).
49. W.Shockley, *Phys. Rev.*, **93**, 345 (1954).
50. C.Bosvieux and J.Friedel, *J. Phys. Chem. Solids*, **23**, 123 (1962).
51. H.B.Huntington, *Trans. Met. Soc. AIME*, **245**, 2571 (1969).
52. N. Van Doan and G.Brebec, Rapport CEA-R-3480 (Paris: Documentation Française) (1968).

3.7

Interdiffusion and Kirkendall shift in binary alloys: test of the Manning theory and influence of the diffusion induced structural defects on the interdiffusion coefficient and Kirkendall shift

A. V. VIGNES and M. BADIA

École Nationale Supériere de la Métallurgie et de l'Industrie des Mines de Nancy, University of Nancy, France

ABSTRACT

Structural changes occurring during the interdiffusion process in pure metal couples have been shown to be a general phenomenon. Local strains due to large initial differences in molar volumes or a high Kirkendall shift may result in progressive recrystallisation and other structural defects. These conditions may lead to an excess of vacancies, or short-circuiting paths when steep concentration gradients exist in pure-metal couples. An effect of the concentration gradient and a time dependence of the interdiffusion coefficient have also been noted by several authors. These anomalous diffusion rates are generally attributed to these short-circuiting paths generated by the diffusion process.

The study of these structural changes on the diffusion process and the Kirkendall shift has been the first aim of this research.

This study was conducted using infinite pure-metal and incremental couples made of very pure materials on seven binary systems Au–Cu, Ni–Pd, Cu–Pd, Nb–V, Nb–Ti, Fe–Ni, Fe–Co. The first four couples have a high size factor, the two last couples have strong Kirkendall shifts. It has been shown that the structural changes do not give rise to anomalously high diffusion rates. In all these systems the interdiffusion coefficients obtained from pure-metal and incremental couples are the same and time independent. The previously contradictory results are analysed.

Only, the porosity when it can appear, can be an important perturbation on the interdiffusion process. In the Fe–Ni system, porosity is important in the iron side when the couple is made from electrolytic and less pure iron, and is reduced with zone-melted iron. The lengths of the diffusion zone and the Kirkendall shifts, even at 1136 °C are bigger in the electrolytic iron–Ni couples than in zone-melted iron–Ni couples. No porosity was found on the other systems where only very pure materials were used.

The application of moderate pressure on electrolytic and zone-melted iron–nickel couples reduces the number of voids, and so the lengths of the diffusion zones and the marker shifts.

With the help of the results on the interdiffusion coefficient and the Kirkendall shift on zone-melted iron–nickel couples, annealed under pressure it has been shown that Manning's equations for the intrinsic coefficients and for the interdiffusion coefficient are very well verified.

INTRODUCTION

In solid binary alloys, the interdiffusion coefficient varies with composition, temperature and pressure. It depends also on the structural state of the materials constituting the diffusion couple, for, in substitutional alloys, one atom can move

by a vacancy mechanism,

along the dislocations and subboundaries,

in the grain boundaries.

A simple analysis of the influence of the short-circuiting paths on the observed value of the diffusion coefficient leads to an expression of the type

$$D_{\text{observed}} = D_v + D_d \cdot f$$

where D_v is the volume diffusion coefficient due to the vacancy mechanism, D_d is the diffusion coefficient in short-circuiting paths and f is the relative surface area. The contribution of these short-circuiting paths decreases when the temperature increases. Thus, with diffusion couples made from work-hardened metals having a high density of dislocations, one notices that the diffusion coefficient is higher than that obtained from couples made from annealed metals. Nonetheless, if the diffusion anneals are of long duration or at high temperatures, the defects anneal and their contribution becomes negligible.

On Fig. 1, we have reproduced Pines'[1] results showing the influence of structural defects introduced by cold work on the length of the diffusion zone in Cu–Ni couples, annealed at 1000 °C. Couple no. 1 was prepared with annealed copper and nickel. The length of the diffusion zone increases parabolically with time. Couple no. 2 was made from workhardened copper and annealed nickel, and couple no. 3 from work-hardened nickel and annealed copper. The two parts of couple no. 4 were work-hardened. We see that the effects of the deformations are important but limited in the first moments of the interdiffusion. Likewise, some authors[2-5] have reported large

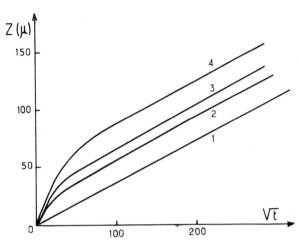

Figure 1 Variation in length of diffusion zone as a function of $\sqrt{t}$ for diffusion in Cu–Ni couples at 1000 °C, with the initial component in different lattice states

enhancements of the diffusion in couples undergoing plastic deformation during their thermal treatment. They have attributed the results to the creation of excess vacancies by concurrent straining, but Sherby and Burke[6] have shown that the strain enhanced diffusion observed is more realistically due to short-circuiting from strain-created subgrain-boundaries and dislocations, because the proportions of excess vacancies are too small to account for such an efficient enhancement. To measure an interdiffusion coefficient representative of diffusion by a vacancy mechanism, our experiments were performed using high purity, well annealed, metals with a low density of dislocations and large grains (to eliminate diffusion along the grain-boundaries) at a sufficiently high temperature ($T > 0.75T_m$: T_m melting temperature of the alloy) for a rapid annihilation of the residual defects.

It has often been noticed that structural changes occur in the diffusion zones. The structure of a diffusion couple after a thermal treatment presents characteristics typical of a metal deformed at high temperature. Rhines and

Mehl[7] have observed grain boundary migration in Cu–α brass couples: appearance of new grains and formation of twins. Sub-structure networks have been observed by X-ray means in Cu–Ni, Ag–Au, Cu–α brass couples by Barnes[8] and Baluffi[9]. In the same way, Doo and Baluffi[10], in a study of diffusion in Cu–α brass couples, have noticed an increase in the density of dislocations, and arrangement of dislocations into subboundaries, including recrystallisation.

These structural modifications show that the whole diffusion zone has been in a stressed state. These stresses arise from either a Kirkendall effect or the variation of the lattice parameter along the concentration gradient.

a) When the rates of diffusion of the two components are unequal, which is proven by the displacement of the markers placed initially at the welding interface (Kirkendall Effect), there is a region of the couple experiencing a net loss of atoms and one other experiencing a gain of atoms. As the diffusion occurs by a vacancy mechanism, vacancies must be annihilated at one side of the couple and generated on the other side. The generation or annihilation of vacancies will cause the lattice to be in a state of compressive or tensile stress if it is prevented from deforming. If, however, the metals can plastically deform, there would be an expansion where vacancies are generated and a contraction where they disappear. The sinks and sources of vacancies are dislocations. Various climb processes increase the dislocation density.

Tensile stresses arise also because of the restraining effect of the bulk of the diffusion couple on the volume contraction in the region losing atoms. As the temperature of the diffusion anneals are relatively high, these dislocations rearrange themselves into subboundaries.

Furthermore, the annihilation of vacancies does not occur only by dislocation climb but in some cases by a precipitation in the form of pores. We shall discuss this point further on.

b) Stresses arise also in the diffusion zone from the change of the lattice parameter along the concentration gradient. High composition gradient variations associated with a high size factor can generate dislocations, the density of which can be found by Friedel's formula[11]:

$$\varrho = \frac{1}{a^2} \frac{da}{dC} \delta \left(\frac{dC}{dz} \right)$$

where a is the mean lattice parameter, da/dC represents the lattice parameter variation with concentration, and $\delta \, (dC/dz)$ represents the composition gradient variation along z. Several authors have revealed by electron microscopy the creation of dislocations due to extremely high concentration gradients (Prussin[12], Queissner[13], Levine[14]). The relative contribution of these two sources of stresses (Kirkendall effect and size factor associated with a strong composition gradient) to the generation of dislocations cannot be evaluated. Since the dislocations and the subboundaries are short-circuiting paths, the diffusion may be enhanced by the diffusion-induced defects.

In fact, several authors have obtained abnormally high values for the interdiffusion coefficient from the pure metal couples.

Reynolds, Averbach and Cohen[15] have noticed that, in the gold–nickel system, the pure-metal couples yielded considerably higher interdiffusion coefficients than did the incremental couples, and that these high values decreased with an increase in the diffusion time.

Goldstein, Hanneman and Ogilvie[16] noticed the same effect in the iron–nickel system: the difference between the D_{AB} values obtained from pure-metal and incremental couples, which is non-existent above 1200°C, increases when the annealing temperature diminishes. Hartley[17] noticed an analogous effect in the Nb–V, Nb–Ti and Mo–Ti systems. In the case of the gold–nickel system, the difference between the D_{AB} values obtained from pure-metal and incremental couples cannot be attributed to short-circuiting paths due to stresses resulting from a Kirkendall effect since this is non-existent in this system. On the other hand, there is a very important concentration gradient in the nickel rich side of the couples, and the size factor is very high, reaching approximately 14%. As we have seen, a high size factor associated with a strong composition gradient in the diffusion zone could be the cause of the production of dislocations which act as short-circuiting paths. In the iron–nickel system, the size factor is weak (about 3%) but there is a marked Kirkendall effect that can be at the origin of stresses that bring about a creation of dislocations acting as short-circuiting paths.

Therfore, we conclude that these two systems constitute typical examples for the study of the influence of diffusion-induced defects on the process of interdiffusion. The temperature, has to be taken into consideration: the efficiency of a diffusional short-circuiting path evidently depends on the temperature. The activation energy of the diffusion all along a short-circuiting path (Q_c) being generally equal to $Q_v/2$ (where Q_v is the activation energy of the volume diffusion). Above a certain temperature the efficiency of the short-circuiting path is low.

c) Another very important structural change is to be found in diffusion couples that are subject to a marked Kirkendall Effect: the appearance of porosity in the part of the couple which is losing atoms. It has been shown[18] that the voids formed during diffusion are heterogeneously nucleated, the nuclei appear to be foreign particles, oxide particles already present in the metals or included in the weld between the metals.

The porosity is therefore connected to the Kirkendall Effect and the presence of particles acting as nucleii. Monty, Zemskoff and Adda[19] have in the same way studied the influence of diffusion-induced defects on the process of interdiffusion in gold–silver couples. According to these authors, the very important porosity found in these couples constitutes the only important perturbation. The voids acting as "short-circuits" by superficial diffusion or evaporation-condensation in the pores.

To study thoroughly the part played by these structural modifications that occur in a diffusion couple on the interdiffusion process, we have studied seven binary systems that present either a high size factor associated to a high gradient (Au–Cu, Ni–Pd, Cu–Pd, Nb–V), or a marked Kirkendall Effect (Fe–Ni, Fe–Co).

System	Au–Cu	Ni–Pd	Nb–V	Cu–Pd	Fe–Ni	Fe–Co	Nb–Ti
Size factor $r_A - r_B/r_A$	10,8%	9,5%	8%	7,1%	3%	1,9%	<1%

(The 4 first systems can also exhibit a Kirkendall Effect, but this has not yet been studied.)

For each system pure-metal and incremental couples were used. The comparison of the interdiffusion coefficients obtained from the two types of couples show the influence of the diffusion-induced defects on the process of diffusion. In incremental couples, the concentration gradient effect and the Kirkendall Effect are smaller, the compositions of the two alloys making up the two parts of the couple being closer to each other. In the Fe–Ni and Fe–Co systems the Kirkendall marker shifts were measured. To study the eventual influence of the porosity on the Kirkendall Effect, some anneals were performed under pressure. These results were also used to test the Darken[31] and Manning[32] relations.

EXPERIMENTAL TECHNIQUES

The materials used were of high purity: the cobalt, nickel, copper and gold are from Johnson-Matthey (99.999). The iron was prepared by zone melting (6 passages) using a bar of purified, electrolytic iron, purified first by electron bombardment. The niobium and vanadium were purified by electron bombardment. The cobalt and nickel were submitted to annealing treatments to obtain large grains (often several millimeters in diameter). The binary alloys used were prepared by induction and annealed several days near the melting point, either in vacuum, or in purified argon: their homogeneity was tested by microprobe analysis. The grain sizes of the alloys (~ 0.4 mm) are generally smaller than those of pure metals. The diffusion couples were prepared from cylindrical pieces 8 mm diameter and 2 to 3 mm in thickness. These pieces were polished through a 25 μ diamond lap, sandwiched together and welded in a little press of stainless steel at 600°C for one hour. The temperature of the welding and its duration were such that no appreciable diffusion occurred during the operation. The measurements of the Kirkendall shift necessitated a special preparation of the couple. The markers used were wires of silimanite with a diameter of 2 to 10 μ. Two types of couples have been prepared.

a) Massive couples of the type $A/B/A$ where the central piece of 1 to 2 mm thickness was made of the metal with the slowest diffusion rate ($D^0_{B(A)} < D^0_{A(B)}$). The markers placed at each interface move outwardly and the eventual porosity that appears in the A rich zones does not interfere with the measurement of the distances separating the two series of markers.

b) One other technique due to Levasseur and Philibert[20] consists of using a reference surface much closer to the marker's plane: for this purpose, one uses the same two pieces A and B, but in between is placed a very thin sheet ($\sim 300 \mu$) of the B metal. The annealing is such that the element A does not diffuse over the interface B/B so that this surface serves as reference. It is simply revealed by a chemical attack. This method yields a higher degree of precision on the measure of the displacement, however the use of a very thin sheet does not allow the best structural conditions required for a correct diffusion experiment. For even after a prolonged anneal, it is not easy to obtain a large grain, which is particularly disturbing at low temperatures. The diffusion anneals were carried out under vacuum at 1250°C or under purified argon above that temperature in furnaces controlled to ± 2°C.

A Cameca electron microanalyser was used to measure the concentrations of two elements across the diffusion couples.

After the anneal, the couples were sectioned and then polished. The preliminary operations, on abrasive papers, are very important for one must utilise a minimum polishing-time. The material to be polished is heterogeneous—especially the couples of pure metals—and the particles of diamond preferentially abrade the softer zones and thereby give rise to grooving. Such a surface is to be avoided for the electron microanalysis. A surface containing a few scratches is to be preferred to a highly polished surface that presents a grooving relief. The specimen is inserted into the specimen chamber of the microanalyser and so oriented that the X-rays received by the crystals come from a zone of homogeneous composition. The conditions of the analyses are given in the following table:

Table 1

Element analysed	Fe	Co	Ni	Cu	Au	Pd	Nb	V	Ti
Voltage kV	25	25	25	30	30	15	20	20	20
X-ray	K_{α_1}	K_{α_1}	K_{α_1}	K_{α_1}	K_{α_1}	K_{α_1}	K_{α_1}	K_{α_1}	K_{α_1}
Peak to background ratio	550	470	400	350	200	100	430	870	770

Each massive couple is simultaneously analysed into its two constituents. The use of a direct method of analysis allows the selection of the zone of analysis: linear scannings parallel and perpendicular to the diffusion direction permit the choice of analysis area and the separation of lattice diffusion from grain-boundary diffusion.

The electron probe analyses were corrected for absorption and fluorescence effects, by the Philibert equation[21] and by using standard alloys.

The Fick equation is valid, only if one uses units of concentration that are either specific or molar. One must, therefore, transform the weight fraction ω_i (obtained from the relative X-ray intensity k_i) into concentrations by using the equation $\varrho_i = \varrho\omega_i$. The density ϱ was computed by the Végard law, from the values of the lattice parameter of pure metals.

The coefficients of interdiffusion were determined by using the Matano method. The limits of these coefficients are determined by using the Hall method (Badia and Vignes[22]).

INTERDIFFUSION IN GOLD–COPPER, COPPER–PALLADIUM, NICKEL–PALLADIUM, NIOBIUM–VANADIUM SYSTEMS

These four systems present high size factors and there is a great variation of the interdiffusion coefficient over the whole of the composition domain.

1 The Gold–Copper System

The diffusion coefficient of gold in copper grain-boundaries between 620 and 760°C was measured by Austin[23]; at 760°C, the value of the ratio $D_J/D_v = 5 \times 10^5$ shows that at this temperature, the efficiency of a short-circuiting path is relatively high. Three couples of pure metals were annealed at 857°C, 770°C and 743°C, and two incremental couples were annealed at 733°C. The variations of the interdiffusion coefficient with the composition are presented on Fig. 2. Variations of D_{AB} with the temperature—for a given composition—are quite linear (Fig. 3). These curves show that there is no difference between the values of D_{AB} obtained from pure-metal and incremental couples. We conclude that there is no effect in the gold–copper system till 733°C.

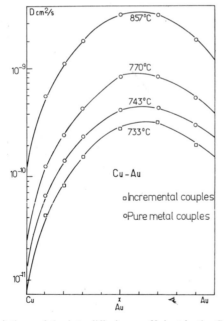

Figure 2 Variations of the interdiffusion coefficient in the Cu–Au system

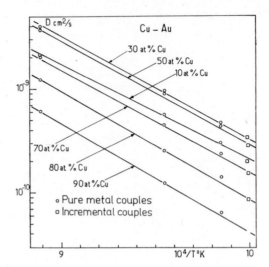

Figure 3   Variations of the interdiffusion coefficient in the Cu–Au system

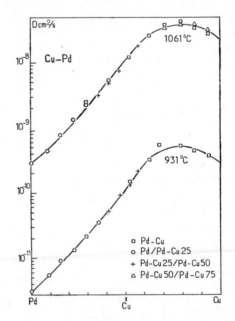

Figure 4   Variations of the interdiffusion coefficient in the Cu–Pd system

2 Copper–Palladium System

Simultaneous anneals at 1061 °C and at 931 °C of pure-metal and incremental couples have been made. At the two temperatures, the three incremental couples (Pd/Pd 75–Cu 25, Pd 75–Cu 25/Pd 50–Cu 50, Pd 50–Cu 50/Cu) yield the same values for D_{AB} as the pure-metal couples (Fig. 4).

3 Nickel–Palladium System

This system has a high size factor: 9.5 % and presents a strong deviation from Végard's law. The calculation D_{AB} is generally made assuming constant partial molar volumes. We have taken into account the variations of the partial molar volumes, in applying the Cohen method[24]. Despite this high size factor, one notices, at 1112 °C, no difference between the interdiffusion coefficient values obtained from pure-metal and incremental couples, as shown on the Fig. 5.

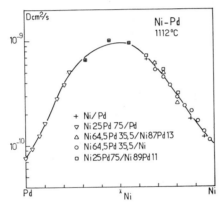

Figure 5 Variation of the interdiffusion coefficient in the Ni–Pd system

4 Niobium–Vanadium System

One simultaneous anneal at 1750 °C of a pure-metal couple and an incremental couple (Nb–Nb 40 V) has been made. Between 20 and 40 % of vanadium, where the variation of the interdiffusion is very great, these two couples yield the same values for D_{AB} (Fig. 6). This result is a part of a complete study of diffusion in the Nb–V system made by Roux[25].

For these four systems, the structural modifications due to a high size factor associated with a strong concentration gradient do not seem sufficient to produce a notable increase of diffusion.

One of the first experiments, that revealed such an enhancement concerns the gold–nickel system (Reynolds[15]) for which the size factor is 14%. In this system, as in the Nb–Ti system, where Hartley[17] noted a similar enhancement of diffusion, there is a great variation of the interdiffusion coefficient.

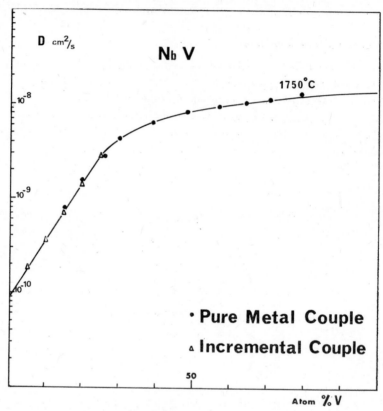

Figure 6   Variation of the interdiffusion coefficient in the Nb–V system
at 1750 °C

At 900 °C, in the Au–Ni system, the limiting values of the interdiffusion coefficient are:

$$D^0_{Ni(Au)} = 10^{-9} \text{ cm}^2/\text{s}. \quad D^0_{Au(Ni)} = 10^{-12} \text{ cm}^2/\text{s}.$$

At 1600 °C, in the Nb–Ti system, the limiting values of the interdiffusion coefficient are:

$$D^0_{Ti(Nb)} = 2 \times 10^{-11} \text{ cm}^2/\text{s}. \quad D^0_{Nb(Ti)} = 1.4 \times 10^{-7} \text{ cm}^2/\text{s}.$$

The great variation of the interdiffusion coefficient involves, in pure-metal couples, the presence of a zone, where the concentration gradient is very high and therefore the calculation of the interdiffusion coefficient by the Matano method is very imprecise.

One can get an idea of the difficulties involved in the calculation of D_{AB} from the concentration-penetration curve of a Nb–Ti couple annealed for 10 hours at 1530°C shown in Fig. 7[25]. The variation of the slope of the con-

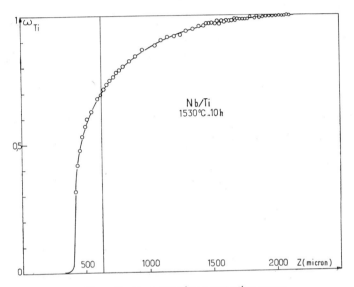

Figure 7 Concentration-penetration curve

centration-penetration curve in the Nb rich side is such that the Matano method cannot be used to determine any value of D_{AB} in this zone. Since the gradient effects were observed in the Ni-rich zone for the Au–Ni system by Reynolds[15] and in the Nb rich zone in the Nb–Ti system by Hartley[27], it is difficult to accept their conclusions.

INTERDIFFUSION IN THE IRON–NICKEL, IRON–COBALT SYSTEMS

The Fe–Ni and Fe–Co systems each have a small size factor; but a marked Kirkendall effect. In addition, for the Fe–Ni system, the preferential diffusion of nickel along the grain boundaries of iron is very high until 1200°C. As has been said, in the Fe–Ni system, below 1200°C, Goldstein, Hanneman and

Ogilvie[16] noted a very definite difference between the values of D_{AB} obtained from pure metal and incremental couples. In these systems, another factor can contribute to the difference in the values obtained for D_{AB} from pure-metal and incremental couples: the iron $\alpha-\gamma$ transformation. It seems, however, that the defects introduced by the transformation are not efficient as diffusional short-circuiting paths unless the diffusional anneal is performed at a temperature very close to the temperature of transformation (Aucouturier[26]).

For the Fe–Ni and Fe–Co couples made with high purity metals, we noticed no difference in the values of D_{AB} obtained from pure-metal couples having undergone long or short anneals, and incremental couples. The Figs. 8 and 9 represent the variations of the interdiffusion coefficient at 1136°C and at

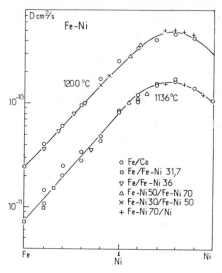

Figure 8   Variations of the interdiffusion coefficient in the Fe–Ni system

1200°C in the iron–nickel, and the iron–cobalt systems. We can conclude that neither the duration of the heat treatment nor the initial gradient influence the values of the interdiffusion coefficient at 1136°C and at 1200°C. Hence, in the Fe/Ni and Fe/Co systems, the structural modifications due to the Kirkendall effect do not give rise to notable acceleration of the interdiffusion process. These results are in contradiction to those of Goldstein et. al.[16].

The only apparent difference between Goldstein's experiments and our own, is in the difference in the purity of the iron used. Goldstein used electrolytic iron (containing 0.007% C, and 0.020% O_2) while we have used zone-

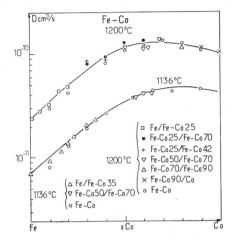

Figure 9 Variations of the interdiffusion coefficient in the Fe–Co system

melted iron. A comparative study of diffusion in the Fe–Ni couples prepared from electrolytic iron was commenced.

In previous experiments with zone-melted iron, three couples were annealed respectively for 48 h, 72 h and 134 h at 1136 °C. Four other couples were

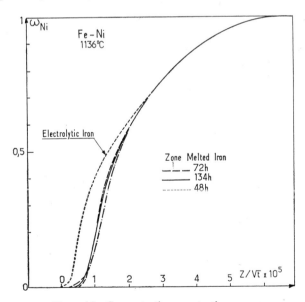

Figure 10 Concentration-penetration curves

made with electrolytic iron and then annealed for 48 h 5, 72 h, 94 h and 116 h, at the same temperature. If one compares the reduced concentration-penetration curves, $C = f(\lambda)$ where $\lambda = z/\sqrt{t}$, for the two series of couples, one sees (Fig. 10) that the length of the diffusion zone in the electrolytic iron couples is greater than those of the zone-melted iron couples. There is a net increase in the diffusion in the iron-rich zone of the elctrolytic iron couples.

A micrographic study was undertaken of the iron-rich zones of both series of couples after a chemical attack that preferentially revealed the iron. Two

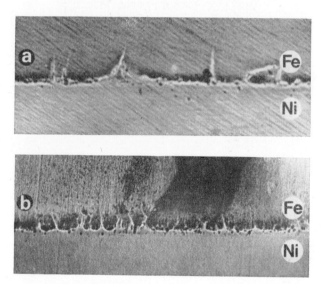

Figure 11   Photomicrographs of iron-rich zones in two Fe/Ni couples:
a—zone-melted iron; b—electrolytic iron ($G = 50$)

pictures (Figs. 11a and 11b) of the diffusion zones of specimens (of both series) showed that in zone-melted iron couples there are few penetrations (clear patches) of nickel into the iron, whilst in the electrolytic-iron couples the penetrations are much more numerous. Some linear scanning by microprobe, perpendicular to the direction of the diffusion, showed the very irregular concentration of the nickel into the iron-rich zone of the couples.

A large number of observations has shown that considerable porosity develops in the iron-rich part of the electrolytic-iron–nickel couples; some pores are also visible in the zone-melted iron–nickel couples.

Another example of irregularity in the diffusion zone was observed in a zone-melted iron couple. We took a Nickel X-ray picture* and an electronic image in a perturbed region of the couple (see Figs. 12 and 13).

One can see pores delimiting a penetration of the nickel in the iron-rich part of the couples.

Figure 12   Nickel X-ray picture of a perturbed region in an Fe/Ni couple
$(100 \, \mu \times 100 \, \mu)$

All these observations suggest that the pores act as diffusional short-circuits. These are particularly numerous in the iron zones of the electrolytic-iron couples.

In conclusion, the differences noted between the lengths of the diffusion zones of the two types of Fe–Ni couples are certainly due to the presence of pores that constitute efficient diffusional short-circuiting paths. In the electrolytic iron, the pores nucleate on impurities and oxide inclusions which are present in much larger quantities than in zone-melted iron.

A study of the Kirkendall effect in the two series confirms these conclusions and allows new observations of diffusional disturbances.

---

* The electron microprobe allows us to take X-ray and electronic images which reflect the various compositions since the second type of photo is sensitive to the difference in he atomic numbers of the diverse constituents.

Figure 13   Electronic image of the perturbed region in the Fe/Ni couple
shown in Fig. 12 (100 $\mu \times$ 100 μ)

STUDY OF THE KIRKENDALL EFFECT IN Fe/Ni COUPLES

The Kirkendall effect in the Fe/Ni system under pressure was studied by Kirwan[27] at temperatures between 1100°C and 1300°C using iron of 99.95% purity and by Levasseur and Philibert[28] at 1200°C. We carried out measurements of Kirkendall shifts in Fe/Ni couples of Series I (zone-melted iron) and series II (Electrolytic iron). In Fig. 14, are gathered together the values obtained for $S/\sqrt{t}$ versus $(1/T °K)$. Those obtained by the authors cited above are also included.

As a consequence of the difficulties of locating markers at high temperatures, the range of temperatures studied is restricted. The results obtained in electrolytic iron–nickel couples are quite dispersed. We can note however that the shifts are greater in these couples than in zone-melted iron couples. The markers move toward the iron-rich zone where, as we have seen, a porosity develops, which is much greater in electrolytic iron couples than in zone-melted couples. We have shown that the difference in the lengths of the diffusion zones were linked to the development of porosity, therefore the difference between the marker shifts obtained in both series of couples is the con-

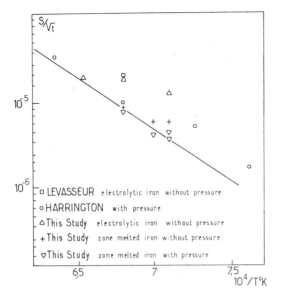

Figure 14 Kirkendall shift in the Fe–Ni system

Figure 15 Nickel penetration in front of a Kirkendall marker in an Fe/Ni couple annealed without pressure (nickel X-ray picture $100\,\mu \times 100\,\mu$)

sequence of a definite increased atom transportation in the electrolytic iron couples.

An additional perturbation is brought about by the markers themselves and shows up locally by an increase in the length of the diffusion zone. Figure 15 shows a characteristic example: it is a nickel X-ray picture in the iron-rich region, beyond a marker (visible in black) in a zone-melted Fe/Ni couple annealed (48 h at 1140 °C) without pressure. The increase in diffusion is due to preferential nucleation of pores on the silimanite/metal interface.

The application of a weak uniaxial pressure is sufficient in numerous cases to reduce the formation of pores. We have made some Kirkendall experiments under uniaxial pressure (~100 kg/cm²) with zone-melted Fe–Ni couples. The results of the displacement measurements are given in Fig. 14 where they can be compared with preceding results obtained without pressure and with those obtained by Levasseur[28] and Harrington[29]. The effect of pressure, in zone-melted iron–nickel couples is very important only at 1140 °C. The shift is reduced by half. In the other couples at 1160 °C and at 1200 °C it is only slightly weaker. In all our experiments under pressure the composition of the marker plane is around 50 %. The composition is much lower in couples without pressure (about 35 % Ni). These results are quite

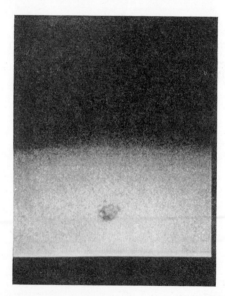

Figure 16   Nickel penetration in front of a Kirkendall marker in an Fe/Ni couple annealed with pressure (nickel X-ray picture 100 μ × 100 μ)

in accordance with Harrington's results also effected under pressure. (The spread at lower temperatures, (1032 and 1132°C) must be connected with abnormal variations of D_{AB} recorded by the author and due to an important intergranular diffusion.) Levasseur's results obtained at 1200°C are equally in accordance with the shifts measured in this study under the same conditions without pressure and using electrolytic iron–nickel couples. In experiments with pressure, we see (Fig. 16) that the diffusion front at the level of the markers is perfectly rectilinear: the pressure thus eliminates the perturbing action of the markers. On the other hand, the Matano interface and the initial welding interface are practically the same, while a weak disparity exists in the preceding couples. The application of the pressure does in effect bring about a reduction of the penetration zone of the order of 5% which has little influence on the interdiffusion coefficient, but this proves that the "normal" porosity is also decreased, a fact which has been observed many times.

The whole of these results relative to the Fe–Ni system confirm the conclusion of Monty et al.[19] in their study on the interdiffusion in the Au–Ag system: "The perturbation which the pores cause is the most important of those parasitic contributions which yield an abnormal increase of the diffusion coefficient and explains Goldstein's[16] results as it does Levasseur's"[28].

CONCLUSION ON THE INFLUENCE OF THE STRUCTURAL MODIFICATIONS CREATED BY THE DIFFUSION PROCESS ON DIFFUSION

We are thus led to draw the following conclusion: the study of the four systems Au–Cu, Ni–Pd and Cu–Pd, Nb–V shows that the structural modifications due to a large size factor associated with a strong gradient are not sufficient to lead to a noteworthy acceleration of the diffusion process at temperatures greater than $0.6T_m$. In Fe–Ni and Fe–Co systems, the structural changes associated with the Kirkendall effect (the size factor is low) are also not sufficient to modify the rate of diffusion. In effect, in all these systems, the values of the interdiffusion coefficient calculated from couples of pure metals are identical to those obtained from incremental couples.

On the other hand, the porosity where it can be developed, brings about a very important perturbation of the diffusion process. In the Fe–Ni system, we have shown that it could play the role of a diffusion short-circuiting path. When we use electrolytic iron, instead of zone melted iron for the preparation of diffusion couples, a great porosity develops on the iron side on account of the presence of numerous nuclei: the lengths of diffusion zones and the marker shifts are greater than those obtained in couples prepared from zone melted iron. We have shown as well that the markers themselves often make up perturbation sources leading to abnormally high shifts. The application of a uniaxial pressure to couples prepared with zone melted iron has little effect on the length of the diffusion zone, but it prevents the formation of pores; the pores are in any case not very numerous in the zone-melted iron– nickel couples. At the lowest temperatures, we observe on the other hand an even more reduced shift of the markers which stems essentially from the elimination of the perturbation (porosity) brought by the marker itself. The fact that perturbations of the interdiffusion process become negligable at high temperatures is explained by the great mobility of the dislocations as well as by a decrease in the efficiency of short-circuits: elimination of the vacancies on the dislocations is facilitated and the pores are of reduced size and fewer in number than at the lowest temperatures.

INTERPRETATION OF THE KIRKENDALL SHIFT RESULTS— TEST OF THE MANNING RELATIONS

The interdiffusion coefficient D_{AB} expresses the distribution of the two constituents of a binary system submitted to a composition gradient. The difference of diffusion rates of the two elements revealed by the Kirkendall effect, cannot be interpreted from the values of the interdiffusion coefficient. Thus the Kirkendall effect has led to the introduction of intrinsic diffusion coefficients, characteristic of the diffusion of each element.

Dehlinger[30] then Darken[31] have set a relation between each intrinsic coefficient and the self-diffusion coefficient corresponding to the same composition:

$$D_A = \alpha_{AB} D_A^*, \quad D_B = \alpha_{AB} D_B^*. \tag{1}$$

The influence of the gradient is expressed by the presence of the thermodynamic factor α_{AB} in these relations. The Darken relations lead to the following relation:

$$D_{AB} = \alpha_{AB} (x_A D_B^* + x_B D_A^*). \tag{2}$$

By statistical analysis of the jump frequencies of the atoms of the two elements situated in the neighbourhood of a vacancy, Manning[32] has introduced a new concept: that of a vacancy flux created by the difference of fluxes of both constituents. This vacancy flux affects the diffusion of different atoms to different extents. It gives rise to an enhancement of the diffusion of the atoms of the fastest element (A for example) and to a slowing down of the diffusion of the atoms of the slowest element (B). According to Manning, the relations between the intrinsic and self-diffusion coefficients are expressed as follows:

$$D_A = D_A^* \alpha_{AB} (1 + V_A)$$
$$D_B = D_B^* \alpha_{AB} (1 - V_A)$$

$$(3)$$

where V_A and V_B are corrective terms which depend on the composition and are function of the ratio D_A^*/D_B^*. Relation (2) is thus written as follows:

$$D_{AB} = (x_A D_B^* + x_B D_A^*) \alpha_{AB} S \qquad (4)$$

where S is a factor due to the vacancy flux, inferior to 1.2.

Even though the verification of these theories is quite difficult given the uncertainty of experimental measurements, it is necessary to try them in the greatest possible number of systems. The results available for the Au–Ag[33], Au–Ni[15], Ag–Cd[34], Cu–Ni[35] systems seem to confirm the validity of the relation (2) quite well. The tests of the validity of the equations (1 or 3) are on the other hand quasi non-existent.

The study of the Kirkendall effect in the Fe–Ni system permits a new test of these theories. We shall compare our conclusions to those of Levasseur and Philibert[36]. Of the results concerning the marker shift, we shall consider only the values obtained in the experiments carried out under pressure and with zone-melted iron–nickel couples. Only these results are affected neither by the porosity due to the Kirkendall effect in electrolytic iron, nor by the porosity induced by the marker. The variations of the self-diffusion coefficients were determined by Wannin and Kohn[37] at 1200°C. We also use values of the thermodynamic factor α_{AB} for the whole composition range at 1100°C[38]. As this factor varies little with temperature in solid solutions, we can use these values at 1200°C. A direct verification of the relation (4) can thus be effected at 1200°C. The parameter S due to the vacancy flux in the whole domain of composition is less than 1.03, so this relation (4) is identical to those of Darken (2). The calculated values of $\alpha_{AB}(x_A D_A^* + x_B D_B^*)$ and the corresponding values of D_{AB} are gathered in Table 2. We note that the agreement is excellent.

The verification of relation (3) between the intrinsic and the self-diffusion coefficient can be done only for the composition of the marker plane (50% Ni)

Table 2   Fe–Ni system—1200 °C—test of the Darken relation (2)

X_{Ni}	$D_{Fe}^* \times 10^{12}$ cm²/s	$D_{Ni}^* \times 10^{12}$ cm²/s	α_{AB}	$(X_{Fe}D_{Ni}^* + X_{Ni}D_{Fe}^*)$ $\alpha_{AB} \times 10^{12}$ cm²/s	$D_{AB\,exp}$ $\times 10^{12}$ cm²/s
0.1	83	49	1	52.4	40
0.2	110	70	1	78	60
0.3	140	90	1.14	111	100
0.4	170	108	1.276	170	160
0.5	200	125	1.46	236	240
0.6	225	140	1.8	344	340
0.7	250	150	2.3	500	460
0.8	260	160	2	480	480
0.9	260	160	1.5	380	420

References	Wannin[37] —1200 °C	Fleisher[38] 1100 °C

and at 1200 °C. The value of the ratio D_{Fe}^*/D_{Ni}^* for this composition (Table 3) leads to: (Manning[32])

$$1 + V_{Fe} \simeq 1.1$$

$$1 - V_{Ni} \simeq 0.9$$

The computed values of $\alpha_{AB}(1 \pm V_i) D_i^*$ and the experimental values of the intrinsic coefficients, are gathered in Table 3. We note a good agreement.

Table 3   Intrinsic coefficients ($\times 10^{12}$ cm²/s in the Fe/Ni system at 1200 °C for 50% Ni). Test of the Manning relations (3)

D_{Fe}^*	$\alpha_{AB}(1 + V_{Fe}) D_{Fe}^*$	D_{Fe}	D_{Ni}^*	$\alpha_{AB}(1 - V_{Ni}) D_{Ni}^*$	D_{Ni}
200	330	353	125	160	146

The experimental results thus confirm the validity of the Manning relations between intrinsic and self-diffusion coefficients.

This important conclusion is in contradiction with that of Levasseur and Philibert[36]. In effect, these authors have carried out their experiments with electrolytic iron and without pressure. They obtained values for the marker shift higher than those which we obtained in zone-melted iron–nickel couples annealed under pressure. We have seen that these differences were due to the porosity which is very important at low temperatures. This led Levasseur and Philibert to evaluate values of D_{Fe} too high and values of D_{Ni} too weak (relative to ours).

REFERENCES

1. B. Ya. Pines and A. F. Sirenko, *Fiz. Metal. Metalloved*, **22**, no. 3, p. 392–399 (1966).
2. K. Hirano, M. Cohen, B. L. Averbach and N. Ujiiye, *Trans. AIME*, **227**, 950 (1963).
3. A. F. Forestieri and L. A. Girifalco, *J. Phys. Chem. Solids*, **10**, 99 (1959).
4. C. H. Lee and R. Maddin, *Trans. AIME*, **215**, 397 (1959), *J. Appl. Phys.*, **32**, 1846 (1961).
5. S. R. Bokhstein, T. E. Gudkova, A. A. Zhukhovitskii and S. T. Kishin, *Dokl. Akad. Nauk. SSSR*, **121**, 1015 (1958).
6. O. D. Sherby and P. M. Burke, *Progress in Materials Science*, **13**, no. 7 (1967).
7. F. N. Rhines and R. F. Mehl, *Trans. AIME*, **128**, 185 (1938).
8. R. S. Barnes, *Proc. Phys. Soc. (London)*, **65B**, 512 (1952).
9. R. W. Baluffi, *J. Appl. Phys.*, **65**, 512 (1952).
10. V. Y. Doo and R. W. Baluffi, *Acta Met.*, **6**, 428 (1958).
11. J. Friedel, *Dislocations*—Pergamon Press (1964).
12. S. Prussin, *J. Appl. Phys.*, **32**, 1876 (1961).
13. H. J. Queissner, *J. Appl. Phys.*, **32**, 1776 (1961).
14. E. Levine, J. Washburn and G. Thomas, *J. Appl. Phys.*, **38**, 1, 81 (1967).
15. J. E. Reynolds, B. L. Averbach and M. Cohen, *Acta Met.*, **5**, 29 (1957).
16. J. I. Goldstein, R. E. Hanneman and R. E. Ogilvie, *Trans. AIME*, **233**, 812 (1965).
17. C. S. Hartley, *Diffusion in b.c.c. metals*—A.S.M., 52 (1965).
18. R. S. Barnes and D. J. Mazey, *Acta Met.*, **6**, 1 (1958).
19. C. Monty, A. Zemskoff and Y. Adda, Rapport C.E.A.—DM/1713, 11 Juin 1968.
20. J. Levasseur and J. Philibert, *Comptes Rend.*, **264**, C, 277 (1967).
21. J. Philibert, *Métaux Corrosion Industries*, no. 465, 466 et 467 (1964).
22. M. Badia and A. Vignes, *Acta Met.*, **17**, 177 (1969).
23. A. E. Austin, N. A. Richard and V. E. Wood, *J. Appl. Phys.*, Vol. 37, 3650 (1966).
24. M. Cohen, C. Wagner and J. E. Reynolds, *Trans. AIME*, **197**, 1534 (1953), **200**, 702 (1954).
25. F. Roux and A. Vignes (to be published).
26. M. Aucouturier, Thèse—Paris 1964.
27. D. J. Kirwan, M.S. Thesis, University of Delaware (1964).
28. J. Levasseur and J. Philibert, *Comptes Rend.*, **264**, C, 380 (1967).
29. J. H. Harrington, M.S. Thesis—University of Delaware.
30. U. Dehlinger, *Z. Physik*, **102**, 633 (1936).
31. L. S. Darken, *Trans. AIME*, **175**, 184 (1948).
32. J. R. Manning, *Acta Met.*, **15**, 817 (1967).
33. T. Heumann, *La diffusion dans les métaux*—Bibliot. Techn. Philips, p. 59 (1957).
34. J. R. Manning, *Phys. Rev.*, **116**, 69 (1959).
35. K. Monma, *J. Jap. Inst. Met.*, **28**, 188 (1964).
36. J. Levasseur and J. Philibert, *Comptes Rend.*, **267**, 1562 (1968).
37. M. Wannin and A. Kohn, *Comptes Rend.*, **267**, 1558 (1968).
38. B. Fleisher and J. F. Elliott, *The physical chemistry of metallic solutions and intermetallic compounds*, **1**, p. 2, F.H.M.S.O. (1959).

Diffusion in metals

Diffusion in metals

J. PHILIBERT

Dept. of Industrial Engineering, Faculté des Sciences,
Orsay & I.R.S.I.D., St. Germain-en-Laye, France

1 INTRODUCTION

Diffusion in metals can be considered from three points of view.

1 The Microscopic Approach

In the physical sense of this term the physicist seeks to express the diffusion coefficient in terms of microscopic quantities such as the lattice parameter, the jump frequency and the geometry of the lattice in the vicinity of the diffusing defects. He should proceed to determine the nature of the defect configuration and hence to calculate the energy associated with its migration through the lattice. Such an approach is however only possible in very simple cases, for example self-diffusion and impurity diffusion at vanishingly small concentrations, and even then only for a small number of elements.

2 The Microstructural Approach

Diffusion studies are not confined to perfect single crystals. All real crystals contain defects; not only thermal agitation and point defects which are essential to all diffusion processes, but other defects such as vacancy clusters, dislocations, sub-boundaries, phase boundaries and grain boundaries. The respective roles of these defects must be understood and evaluated if only to allow the determination of "true" diffusion coefficients, that is to say, those characteristic of the pure lattice diffusion. The physical significance of this approach is very important since these defects are a necessary part of all real

303

crystals. Related to this subject are the unresolved discussions of certain penetration-distance curves which show several successive regions which appear to correspond to different diffusion processes.

Diffusion in grain boundaries and on surfaces is, to a large extent, not understood—even for self-diffusion—mainly as a result of the difficulty in obtaining a pure system. The impurities in the metal have a tendency to segregate to the grain boundaries and to surfaces even at low impurity concentrations in the bulk. Some new analytical techniques, such as L.E.E.D. or Auger electron spectroscopy will probably lead to important progress in this area.

3   The Metallurgical Approach

The metallurgist very often searches for relative results on the diffusion process, in order to analyze, identify or predict the kinetics of metallurgical processes which involve diffusion in the solid or liquid state. In this regard, identification of a process by comparison of the contingent activation energy with tables of experimental results can be misleading, except in the exceptional case of pure metals.

Elsewhere the conditions which hold for metallurgical processes are not necessarily the simple mechanisms which we have referred to above. Let us consider some of the complications of metallurgical processes and their relation to known diffusion phenomena.

1   The presence of numerous elements. A good understanding of ternary diffusion in non-dilute solution is usually a minimum requirement to explain these effects. The existence of strong chemical potential gradients induces drift in addition to other complex effects, due to the different lattice parameters involved and also to the material transport which they cause (vacancy flux or Kirkendall effect). The consequences of these effects are well known in binary diffusion couples but their importance is very much more difficult to appreciate in practical systems.

2   The existence of supersaturations of defects as a result of quenching, plastic deformation or phase transformation.

3   The migration of interfaces—these only result in small atomic displacements, i.e. diffusion over short distances, whereas diffusion measurements result from a much larger number of atomic jumps. A good example of this is the study of diffusion using inelastic relaxation effects, the results of which show some disagreement when compared with those from the conventional methods.

4 The presence of numerous simultaneous diffusion processes—as in the situation where the microstructure is varying e.g. sintering—or the presence of several phases which are not necessarily in thermal equilibrium in the diffusion zone.

Consideration of these phenomena is obviously complicated and in the following we shall limit ourselves to the microscopic and mechanistic approaches in order to show precisely the nature of diffusion in metals in the single crystal state.

2   ELEMENTARY DIFFUSION MECHANISMS

The macroscopic diffusion coefficient can be expressed as a function of the mean square displacement of the diffusing atoms and consequently as a function of the jump frequencies and jump distances for the various types of atomic jumps involved:

$$D_x = \tfrac{1}{2} \cdot \sum_{i=1}^{\zeta} \Gamma_i \cdot \delta x_i^2 \qquad (1)$$

D_x = the diffusion coefficient in the x direction.
ζ = the number of types of elementary jumps
Γ_i = the mean number of jumps of type i per unit time
x_i = the x component of the type i jump distance.

For real cases, this result of the application of random walk theory to diffusion must be corrected by the introduction of a term, $f_x \Gamma$, to replace Γ_1.[1] This term, in which $f_x < 1$, describes the number of effective atom jumps. The correlation factor f_x takes account of the possible non-independence of succesive atomic jumps in some mechanisms. Equation (1) must then be rewritten

$$D_x = \tfrac{1}{2} f_x \sum \Gamma_i \cdot \delta x_i^2 \qquad (2)$$

or for an isotropic crystal

$$D = \tfrac{1}{6} f \sum \Gamma_i \cdot \delta l_i^2 . \qquad (3)$$

f is only a numerical parameter in the case of self-diffusion, otherwise it is a function of the several jump frequencies which can occur in the elementary diffusion process.

The above general expression allows D to be calculated as a function of jump frequency and hence of temperature.

For a vacancy mechanism, Γ will be equal to the product of the vacancy concentration and the jump probability, hence the activation energy for self-diffusion can be written

$$\Delta H = \Delta H_f + \Delta H_m \tag{4}$$

where ΔH_f is the energy for vacancy formation and ΔH_m that for vacancy migration (i.e. the energy corresponding to the jump of an atom into a neighbouring vacancy).

The case of heterodiffusion at infinite dilution, the so-called "impurity diffusion", can be similarly treated, but the expression for the activation energy is more complex since the correlation factor is temperature dependent. It is customary to express the difference in activation energies for heterodiffusion of an impurity B in a metal A and for self-diffusion of A as

$$\Delta Q = \Delta H_B^A - \Delta H_A^A = \Delta H_b + \delta\,(\Delta H_m) - C \tag{5}$$

where ΔH_b is the binding energy between a vacancy and the foreign atom B, $\delta\,\Delta H_m$ the difference between the activation energies for the exchange of a vacancy with atom A in pure A and with a B atom in A matrix, and $C = R\,[d\,(\log f)/d\,(1/T)]$, an expression which accounts for the enthalpies of all types of jumps, because the jump frequencies appearing in f have different activation energies, so that f will vary with temperature.

3 FACE CENTRED CUBIC METALS

In these metals diffusion proceeds via a vacancy mechanism. In addition to theoretical reasons, the experimental proofs are essentially as follows for self-diffusion:

i. Verification of the relationship (4) with good precision for a number of metals (Cu, Ag, Au, Pb, Ni, Al) using values of ΔH from radiotracer studies and values of ΔH_f and ΔH_m from the measurement of physical properties (electrical resistance ...) on quenched or irradiated metals.

ii. Satisfactory agreement of the theoretical value for the correlation factor ($f = 0.78$ in fcc structures) and the experimental value. Actually, experiments do not directly give f but a product $f\Delta K$ (see A.D. LeClaire, this proceedings).

Now let us examine the case of hetero-diffusion. Here one has to consider at least five jump frequencies instead of one frequency W_0 in the pure metal. When a vacancy is in a nearest neighbouring position to a foreign atom B, it is necessary to distinguish between the jumps of frequency W_2, for the exchange vacancy–atom B, W_1 and W_3, for the exchange vacancy–atom A, depending on whether the pair vacancy–atom B is dissociated or not and finally the frequency W_4 which refers to jumps of the vacancy which lead to an association between the vacancy and a foreign atom. This model is obviously simplified since it relies on the hypothesis that interactions between vacancies and foreign atoms are limited to nearest neighbouring positions. Under these conditions it is possible to determine the ratios W_4/W_0, W_3/W_1 and W_2/W_1 experimentally by four different experiments:

1 Isotope effect studies for self-diffusion in A. From this one obtains the quantity $f_0 \Delta K$ and hence knowing the numerical value of f_0, ΔK can be evaluated. It is then assumed that ΔK will be the same for hetero-diffusion as for self-diffusion.

2 Isotope effects for hetero-diffusion. This yields $f_B \Delta K$ and hence f_B. It has been shown that

$$f_B = \frac{1 + \frac{7}{2}(W_3/W_1) F}{1 + (W_2/W_1) + \frac{7}{2}(W_3/W_1) F} \tag{5}$$

where F is a function (calculable) of W_4/W_0.

3 Comparison of the diffusion coefficients D_{B*}^A of tracer B in A (hetero-diffusion) and self-diffusion D_{A*}^A

$$\frac{D_{B*}^A}{D_{A*}^A} = \frac{f_B}{f_0} \frac{(W_2/W_1)}{(W_3/W_1)} \cdot \frac{W_n}{W_0}. \tag{6}$$

4 Variation of self-diffusion coefficient for increasing concentrations of the impurity B. These experiments frequently follow a law

$$D_{A*}^{AB} = D_{A*}^A (1 + b \cdot C_B)$$

where b is determined theoretically as[2]

$$b = -18 + 4 \left(\frac{W_1 + \frac{7}{2} W_3}{W_0} \right) \frac{W_4}{W_3} \tag{7}$$

or more exactly[3]

$$b = -18 + \frac{4W_4}{f_0 W_0}\left(X_1 \frac{W_1}{W_3} + \frac{7}{2}X_2\right) \tag{7b}$$

where X_1 and X_2 are functions of the three ratios W_4/W_1, W_2/W_1 and W_3/W_1.

A complete evaluation using the above relationships has been carried out for the system Ag/Zn by Rothman and Peterson[4]. For other systems for which all the results are unavailable the preceding equations do permit precise limits to be placed on the ratios of the frequencies. These can then be compared with values calculated on the basis of the theory.

4   THEORETICAL CALCULATION OF JUMP FREQUENCY

We will now examine the theoretical calculations which, with the aid of model interatomic potentials, permit the calculation of the energies of formation and migration of vacancies or other point defects in pure metals. We shall limit ourselves to the case of the connections of self-diffusion and hetero-diffusion as expressed in equation 5.

Following Lazarus[5] the simple model has been developed particularly by LeClaire[6] in order to calculate terms ΔH_i from equations of the type:

$$\frac{W_i}{W_0} = \frac{\nu_i}{\nu_0}\exp\left(-\delta\,\Delta Hi/kT\right) \tag{8}$$

This model considers the foreign atom as an ion of charge $+ze$, z being the difference between the valences of the atoms of solute B and solvent A. This charge creates a potential which varies according to $1/r$ but which will be screened by the conduction electrons of the metal. The form of the potential $V(r)$ can be calculated from several approximations (Thomas–Fermi, Hartree, etc.). By considering the vacancy as a charge $-Ze$ (in a metal of valence Z) it is then easy, knowing $V(r)$ to calculate the interaction energy B between a foreign atom and a vacancy. This calculation can be made for the stable configuration—foreign atom/vacancy in nearest neighbouring sites in the lattice and for the metastable, saddle point configuration—foreign atom between two half vacancies. Using the solution of the Thomas–Fermi equation as the potential, LeClaire has evaluated $(\Delta H_{B*}^A - \Delta H_{A*}^A)$ and has found good agreement with experiment for cases in which the solvents are noble metals and the solutes are elements situated to the right of these in the periodic table. In particular, the theory predicts that these elements will diffuse more rapidly than the solvent and that this rate will increase with increasing difference in valence.

LeClaire's theory has been equally successful in its application to close packed bivalent metals such as zinc (hcp) for which he predicts, in addition to the correct differences in activation energy, the sign of the anisotropy[7]. In zinc the slowly diffusing impurities show an anisotropy of the same sign as for the zinc (that is to say diffusion is more rapid in the *c* direction than in the basal plane), the reverse is true for rapidly diffusing impurities.

The theory has been less successful for other close packed metals. In several cases, the theory predicts the wrong sign for the difference in activation energy. This is the case for bivalent (Mg) and trivalent (Al) solvents. The noble metals Cu, Ag, Au behave as rapidly diffusing elements in aluminium[8], whereas the theoretical model predicts the reverse ($Z = -2$, $\Delta Q > 0$ from theory, and approximately -6 kcal mol^{-1} experimentally). This disagreement could result from several causes:

the particular configuration chosen for the saddle point

the form of the perturbation potential

the valence of the foreign atom, in particular with reference to the saddle point.

Several variations have been attempted to allow for these. It is known that the potential initially chosen by LeClaire is a poor approximation for polyvalent metals. More rigorous solutions lead to an oscillating potential[9]. That is to say a potential the sign of which varies periodically with distance. Nevertheless, it appears that in those cases in which there is wild disagreement with the theory of LeClaire, the use of these potentials does not appreciably improve the situation. In those cases where the impurities are transition metals diffusing in normal metals calculations founded on pseudopotentials seem to predict values of ΔQ of the correct sign and magnitude (at least for noble metal solvents). Those cases in which the solvent is a transition metal still resist all theoretical calculations.

Returning to the normal metals, those points on which the validity of theoretical model has been placed in doubt, can often be clarified in part by other experiments such as the effects of hydrostatic pressure and electrical fields on the diffusion process.

5   DIFFUSION UNDER HYDROSTATIC PRESSURE

When diffusion specimens are annealed under hydrostatic pressure one finds that the value of D decreases with increasing pressure. Thermodynamically we can express the variation of free energy with pressure as

$$(\delta G/\delta P)_T = V$$

One can define an activation volume for diffusion ΔV by applying the above relationship to the free energy of diffusion ΔG. It can be shown to a precision of several per cent that for self-diffusion in a pure metal

$$\Delta V = -RT \left[\delta (\log D)/\delta P\right]_T$$

The activation volume thus measured corresponds to the sum of two terms, one related to the formation of a defect ΔV_f, a vacancy for example, the second to its migration ΔV_m.

Attempts have been made to evaluate these terms theoretically. The most important ΔV_f ($\Delta V_m < \Delta V_f$) is not equal to the atomic volume Ω as a consequence of the relaxation of the lattice around the defect. In close packed monovalent metals, the activation volume ΔV is found to be less than the atomic volume in good agreement with theoretical predictions. On the other hand in polyvalent metals such as aluminium or beryllium[11], ΔV is much greater than Ω (1.352 in Al, 1.25 in Be) this leads to the conclusion that ΔV_f is at least equal to Ω.

The question arises, can certain of the theoretical forms of the perturbation potential created in the lattice by a defect explain why the relaxation around a vacancy can be sometimes positive ($\Delta V > \Omega$), sometimes negative ($\Delta V < \Omega$)? This can be answered by use of the oscillating potential of which we have spoken above. One assumes that a vacancy in a crystal of valence Z behaves as if it had a virtual charge Ze at the centre of a screening potential $V(r)$. With this potential is associated an electric field which exerts a force on the ions of the crystal. In the equilibrium state one will find a size effect associated with the vacancy which is large and negative in the close-packed monovalent metals, but which can be positive in the polyvalent metals. This is in good agreement with the experimental results cited above[9].

One can see the considerable interest in continuing this type of experiment with other metals for which the models of the interatomic potential and the mechanism of diffusion are still the subject of discussion and at a later stage to the study of heterodiffusion at infinite dilution.

6 ELECTROTRANSPORT

When a metal is subjected to a high electric field it is found, provided that the temperature is sufficiently high, that a transport of material occurs towards one of the electrodes. This will correspond, from the point of view of diffusion mechanism, to a flux of vacancies in the opposite direction. If the metal contains impurities, these are transported towards one of the electrodes. All this occurs as if the transported atoms were displaced in the field with an effective valence Z^*. One finds that this is generally very different from the true valence Z and does not necessarily have the same sign. In the noble metals the material is transported towards the anode, which corresponds to an effective negative valence of the ions in the crystal. This atomic transport is due to two sorts of forces[10]:

i. an electrostatic force created by the action of the field on the metal ions and directed towards the cathode,

ii. a "frictional force" due to the interaction of the charge carriers (electrons or holes) with the ions; this is directed towards the anode for an electron conductor, the cathode for a hole conductor.

   This second force is usually predominant since it will vary depending upon the position of the ion. It is therefore necessary to consider separately the stable position (ions and vacancies in near neighbouring position) and the saddle point position. For the latter one can assume that the ion behaves like a screened interstitial charge $+ Ze$. Following Friedel and Bosvieux for example[12], the force which will be exerted on this ion is equal to:

$$- Ze\, \vec{\varepsilon}\, \frac{\varrho_d^z}{\varrho_d}\, \frac{n}{n_d^{(z)}}$$

where $\vec{\varepsilon}$ is the applied electric field, n and $n_d^{(z)}$ the concentrations of ions in the stable position and at the saddle point, ϱ_d and $\varrho_d^{(z)}$ the residual resistivities due to defects in the stable position and at the saddle point. From the effective valence of the ions of the solvent or of the solute (electrotransport studies in pure metals, or heterodiffusion under an electric field) one can obtain information on the saddle point configuration and in particular on the electronic structure of the ion in this position. Recent results related to diffusion in silver[13] seem to indicate that when solute and solvent have valences which are equal or almost equal, the electronic characteristics of the ion in the stable position and at the saddle point should be significantly different. If this is correct LeClaire's model for the calulation of heterodiffusion would not be applicable in its original form.

Electrotransport experiments, in addition to their practical importance (metal purification for instance) have important theoretical implications.

7 DIFFUSION IN BODY-CENTRED CUBIC METALS

Body-centred cubic metals have been a centre of interest for several years—an entire conference was devoted to this topic in 1964[14]—the experimental results show that some of these metals behave abnormally. We comment briefly on these problems.

1 Transition Metals

A certain number of transition metals revealed an anomalous behaviour in the sense that the Arrhenius curve of self-diffusion is not linear. We can make a distinction between vanadium and perhaps chromium for which the graph shows a definite break and the elements γTi, βZr, βHf which show a continuous curvature. The values of D_0 and ΔH at low temperature are abnormally small (in comparison with the well-known empirical relations such as $\Delta H = 35T_f$, where T_f is the temperature of fusion in K). It must be noted that V and Cr do not show this latter anomaly. γU and εPu, as a consequence of their low activation energy, can be classified with the anomalous metals of the βZr type. Finally, the diffusion of impurities in these metals also present some anomalies because the heterodiffusion coefficients in transition metals are 10 or 100 times higher than the self-diffusion coefficients—all yield a curvature in the Arrhenius plot.

In general these results have been analysed on the basis of a dual mechanism[15]. For this can be written

$$D = D^{(1)} + D^{(2)}$$
$$= D_0^{(1)} \exp\left(-\Delta H^{(1)}/RT\right) + D_0^{(2)} \exp\left(-\Delta H^{(2)}/RT\right)$$

with $\Delta H^{(2)} < \Delta H^{(1)}$. Thus one determines two activation energies corresponding to the two mechanisms, one of which $D_0^{(2)} \Delta H^{(2)}$ dominates at low temperatures and the other $D_0^{(1)}$, $\Delta H^{(1)}$ at high temperatures.

Amongst the proposed explanations are the following:

i. By analogy with ionic solids the diffusion would be intrinsic at high temperatures, extrinsic at low temperatures, the vacancies being introduced by certain impurities such as oxygen, to which the anomalous metals are particularly sensitive[15].

ii. Volume diffusion at high temperatures, dislocation enhanced diffusion at low temperatures. The abnormally high dislocation content being a consequence of the phase transformation which is present in all these metals at low temperature (*hcp* → *bcc*).

iii. Diffusion by monovacancies at low temperature and by divacancies at high temperature. A model of this type has already been applied to certain *fcc* metals in order to explain a slight curvature in the Arrhenius plots at high temperature[16].

Another completely different type of interpretation is founded on a different approach[17]: it is considered that there is a unique mechanism which is characterised by a temperature-dependent activation energy. This variation is thought to be tied to that of the elastic constants of the crystal; the difference $c' = C_{11} - C_{12}$ would tend towards zero at the temperature of the phase change. If the migration energy is proportional to the elastic constant c' as in the single elastic model (the Zener model) it will tend to zero. Consequently low values will be observed for ΔH at low temperatures.

It would seem at the present time that the mechanisms for diffusion in these metals are not clear. Isotope effect measurements might clarify matters. The effect of hydrostatic pressure is less direct since, in metals which suffer from a phase change, it is not possible to work with single crystalline specimens. Finally it must be noted that for metals such as iron the experimental points for low temperatures (αFe) and at high temperatures (γFe) correspond to a unique mechanism, probably by relaxed vacancies. This relaxation is adduced from the low values of ΔK measured in the two temperature regions[18].

2 Alkali Metals

Diffusion in the alkali metals is of particular interest as a consequence of their loosely packed structure and the possibility of making theoretical calculations with a reasonable accuracy.

The activation energy is in good agreement with empirical relationships of the type $\Delta H = 35 T_f$ but it is almost equal to the formation energy of vacancies in these materials: $\Delta H_m \ll \Delta H_f$. Measurements, under hydrostatic pressure, of self-diffusion and of creep (the kinetics of which is governed by self-diffusion) show that the activation volume is of the order of half the atomic volume for sodium and for potassium[19]. All these indications favour the existence of relaxed vacancies. This deduction is confirmed by measurements of isotope effects for self-diffusion in sodium[20]; the experimental value of $f\Delta K = 0.36$ is compatible with a vacancy mechanism with a strongly relaxed vacancy ($\Delta V_f \approx \Omega/2$). We note however that the last result leaves open the possibility of a mechanism which involves the jumps of two atoms such as that of the indirect interstitial mechanism (interstitialcy).

Theoretical calculations[21] based on an atomic potential between pairs of atoms and deduced from the theory of pseudopotentials allow the degree of relaxation around defect to be calculated as well as the energies of formation and migration. These calculations indicate a strong relaxation around the vacancy in all the alkali metals ($\Delta V_f \sim \Omega/2$) and predict self-diffusion activation energies in good agreement with the experimental values for lithium and sodium (10%) and potassium (15%). However, if the calculations are in favour of a vacancy mechanism in sodium, they do not exclude an interstitial mechanism for lithium. This last mechanism would seem to be reasonable for Li following the high values for the isotope effect which have been determined recently[22]. These results are in accord with a split interstitial mechanism (dumbbell configuration), the importance of the isotope effect ($f\Delta K > 1$) being associated with a lowering of the jump barrier by a tunneling effect, as in the case of the diffusion of hydrogen at low temperatures in transition metals.

Heterodiffusion in the alkali metals also presents anomalies. Whereas lithium is a slowly diffusing element in sodium, rubidium and potassium diffuse rapidly. On the other hand sodium is a rapidly diffusing element in potassium[23]. LeClaire's model founded on the vacancy impurity interactions contradicts this dual behaviour. The possibility of diffusion by interstitials when the atom of the solute is smaller than that of the solvent has not been excluded. This hypothesis becomes more probable when the solute has a very small ionic radius; this is the case for noble metals. At ambient temperatures the heterodiffusion coefficient of gold in sodium is 500 times greater than the self-diffusion coefficient of sodium[24]. Some similar observations have been made in lithium for the heterodiffusion of copper and gold: the copper diffuses twice as quickly as the gold which itself diffuses 10 times faster than the lithium[25]. There again one can consider the diffusion of noble metals by an interstitial mechanism. For, the diffusion of a large number of elements in lithium, the heterodiffusion coefficient decreases as the valence of the solute increases. On the other hand if the solvent is a noble metal one finds, in agreement with the theoretical model of LeClaire, a variation in the opposite direction.

In view of these discrepancies and anomalies it is interesting to remember that a further interesting possibility is the fact that vacancy and interstitial mechanisms can co-exist if the steric and electronic conditions are favourable and that foreign atoms can enter simultaneously in interstitial and substitutional solid solution. If under equilibrium conditions the concentrations of interstitials, N_i, and substitutional atoms, N_s, obey the relationship

$$N_i/N_s \propto \exp\left(-\Delta H_i/kT\right)$$

one will have a composite diffusion coefficient.

$$D = \frac{N_i}{N_i + N_s} D_i + \frac{N_s}{N_i + N_s} D_s$$

$$\approx \frac{N_i}{N_i + N_s} D_i$$

This model was developed 12 years ago by Frank and Turnbull[26] in order to account for the diffusion of copper in germanium. This mechanism, the "dissociative mechanism" has regained prominence, not only on account of the alkali metals but also with regard to metals of groups III B and IV B: Pb, Sn, Tl, In[27]. The interstitial mechanism would seem adequate for heterodiffusion of noble metals in these solvents. The record is held by the system PbCu for which the heterodiffusion coefficient of copper is, depending on the temperature, about 10^4 to 10^5 times higher than the self-diffusion coefficient of lead. In the neighbourhood of the melting point of the solvent one obtains values of the diffusion coefficient as high as 10^{-6} or 10^{-5} cm^2/sec, these correspond to penetrations of the order of millimeters in one hour! Complementary measurements of the variation of self- and heterodiffusion coefficients with impurity content, of the isotope effect, or the effect of hydrostatic pressure or again the electrotransport measurements would be useful to confirm the hypotheses that no single mechanism for diffusion in these solid solutions is adequate.

316 *Diffusion in metals*

CONCLUSIONS

The elementary processes by which atomic jumps take place are still far from being understood in all metals, even if one is limited to pure metals or dilute solid solutions. As we have seen, beyond the measurement of diffusion coefficients, these researches provide confirmation of the theoretical models elaborated by the physicists or alternatively lead to the creation of new models and new theories.

REFERENCES

1. J.R.Manning, *Diffusion Kinetics for Atoms in Crystals* (D. Van Nostrand N.Y.), 1968.
2. A.B.Lidiard, *Phil. Mag.*, **5**, 1171 (1960).
3. R.E.Howard and J.R.Manning, *Phys. Rev.*, **154**, 561 (1967).
4. S.Rothman and N.L.Peterson, *Phys. Rev.*, **154**, 552 (1967).
5. D.Lazarus, *Phys. Rev.*, **93**, 973 (1954).
6. A.D.LeClaire, *Phil. Mag.*, **7**, 141 (1962).
7. P.B.Ghate, *Phys. Rev.*, **133**, A.1167 (1964).
8. F.Maurice and M.Beyeler, Journées d'Automne 1969 de la S.F.M.
9. J.Friedel, *Adv. in Physics*, **3**, 446 (1954).
10. Y.Adda and J.Philibert, *La Diffusion dans les solides*, (P.U.F. Paris) 1966.
11. M.Beyeler and D.Lazarus, Journées d'Automne de la S.F.M. (1969).
12. C.Bosvieux and J.Friedel, *J. Phys. Chem. Solids*, **23**, 123 (1962).
13. N. Van Doan, Journées d'Automne de la S.F.M. Paris (1969).
14. *Diffusion in BCC Metals*, A.S.M. (1965).
15. G.V.Kidson, *Can. J. Physics*, **41**, 1563 (1963).
16. A.Seeger, G.Schottky and D.Schumacher, *Phys. Stat. Sol*, **11**, 363 (1965).
17. H.I.Aaronson and P.G.Shewmon, *Acta Met.*, **15**, 383 (1967).
18. C.M.Walter and W.L.Peterson, *Phys. Rev.*, **178**, 922 (1969)
19. N.H.Nachtrieb, J.A.Weil, E.Catalano and A.W.Lawson, *J. Chem. Phys.*, **20**, 1189 (1952).
20. J.N.Mundy, L.W.Barr and F.A.Smith, *Phil. Mag.*, **14**, 785 (1966).
21. M.Gerl and I.Torrens, *Comptes Rend.*, **269**, B.201 (1969).
22. A.Lodding, J.N.Mundy and A.Ott, to be published (1969).
23. L.W.Barr, J.N.Mundy and F.A.Smith, *Phil. Mag.*, **16**, 1139 (1967).
24. L.W.Barr, J.N.Mundy and F.A.Smith, *Phil Mag.*, **14**, 1299 (1966) and **20**, 389 (1969).
25. A.Ott and A.Lodding, *Vacancies and Interstitials in Metals*, North Holland (1969).
26. F.C.Frank and D.Turnbull, *Phys. Rev.*, **104**, 617 (1956).
27. T.R.Anthony, *Vacancies and Interstitials in Metals*, North Holland (1969).

4.2

The effect of impurities on the diffusion of copper in nickel single crystals

H. HELFMEIER and M. FELLER-KNIEPMEIER

Elektronenmikroskopisches Labor der Fakultät für Bergbau and Hüttenwesen
Technische Universität Berlin

ABSTRACT

The influence of small concentrations (0.1 at.%) of Pd, Au and Ti on the diffusion coefficient of Copper in Nickel single crystals has been measured at temperatures between 843.3 °C and 1050 °C. In this temperature range the diffusion of Cu in Ni crystals of high chemical purity and low dislocation density ($10^3/cm^2$) was found to follow the equation

$$D = D_0 \exp{(Q/RT)}$$

with $D_0 = 0.27 \, cm^2 \, sec^{-1} \pm 22\%$ and $Q = 61.02 \, kcal \, mol^{-1} \pm 1\%$. For the impurities Au and Pd no deviation of the diffusion coefficient of Copper was found between 903.3 °C and 1050 °C, whereas at 843.3 °C the diffusion coefficient for Copper is 17% and 29%, respectively, higher than in pure Nickel. For the low Ti concentration alloy no deviation was found in the temperature range checked.

The effect is discussed using LeClaire's theory and Pauling's classification of metallic valences.

INTRODUCTION

Several calculations have been published of the activation energy for impurity diffusion in metals at small impurity concentrations. The results of these are in reasonable agreement with experiment. Theories of the influence of impurities on self-diffusion data are much more difficult to formulate and thus are less successful in application to experiments. In general the experimental data shows an increase in self-diffusion coefficient, with the addition

317

of small amounts of impurities. If a vacancy mechanism of diffusion is assumed, not only free vacancies share in the transport of material, but also vacancies that are bound to impurity atoms. It seems reasonable to propose that at thermal equilibrium the number of free vacancies is nearly the same in pure and in impure crystals provided that the concentration of impurities is low. The formation energy of a vacancy in a next-neighbour-position to an impurity atom will however be reduced by an amount equal to the binding energy between vacancy and impurity. Thus we envisage that there are two types of vacancies with different formation energies. The relative number of quasi-bound vacancies will increase as the temperature decreases. The influence on the transport mechanism should be to raise the diffusion coefficient at low temperatures to values higher than those predicted by extrapolation of the Arrhenius-line from high temperatures. In selecting materials for self-diffusion experiments to test this speculation, not only suitable isotopes should be available but account should also be taken of the fact that the use of a metal with a high formation energy for vacancies in the pure state will extend the scale of sensitivity of the experiment. That is to say, the combination of an element with a high free vacancy formation energy with an impurity of high binding energy between vacancy and impurity atom will permit a relatively high concentration of bound vacancies at low temperatures. This would thus provide an ideal system for study. On the other hand materials with a high vacancy-formation energy are characterised by low self-diffusion coefficients which is a severe handicap to measurement at low diffusion temperatures. This difficulty can be overcome by studying hetero-diffusion. In this case, a matrix metal with a high vacancy-formation energy may be chosen, but a diffusing element is selected with a high diffusion coefficient in the matrix metal. The diffusing element thus works as indicator of vacancy concentration with a higher sensitivity than the matrix metal itself. As a consequence of these considerations, nickel was chosen as matrix material and copper as diffusing partner. This system is also ideal for study by electron microprobe analysis which was the chosen technique.

EXPERIMENTAL

The starting materials were nickel and copper rods of 5 mm diameter × 150 mm long with an impurity content of less than 10 ppm. After three times zone melting in an electron beam furnace the nickel was degassed; a single crystal was obtained, about 70 mm in length. These crystals were spark cut into 8 mm pieces, ground and polished with diamond paste. The deformed layer was electrolytically removed. In this way the enhancement of volume diffusion measurements by grain boundaries could be eliminated. The possible influence of dislocations was reduced by annealing the crystals under a vacuum of $5 \cdot 10^{-8}$ Torr in three stages of 36 h each, a linear warm up to 900 °C, a stay and a linear decrease. This procedure reduced the dislocation concentration to about 10^3–10^4 dislocations per square centimeter.

For the diffusion experiments, the thin film method was used. To avoid the high impurity content in the copper layer which can result from electrolytic deposition we chose to evaporate the copper on to a cooled nickel crystal from a high purity tungsten crucible under a vacuum of less than $5 \cdot 10^{-8}$ Torr. The vacuum apparatus (Fig. 1) was degassed at 150 °C. During that time the specimen was kept at a temperature of 250 °C to produce a clean surface. After evaporation, the specimen was sealed under the same vacuum in a small quartz tube.

The diffusion annealing was done in a furnace controlled to ± 0.2 °C. Taking into account the aging of the PtRh/Pt thermocouple and the drift in the

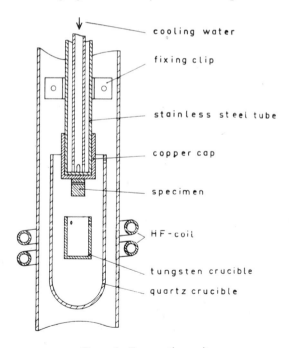

Figure 1 Evaporation unit

electronic device, the temperature constancy was better than ±0.5°C per month or ±0.3°C per day. The absolute temperature specification better than ±1.5°C.

In order to measure the gradient of copper in nickel with a microprobe analyser, the specimens were prepared in the following way. First a 10 µm

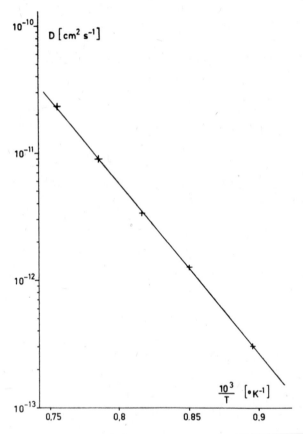

Figure 2 Diffusion coefficient of copper in pure nickel single crystals

thick nickel layer was electroplated on the crystal followed by a 200 µm thick copper layer. After grinding a right angle between the front side and a plane surface parallel to the cylinder axis, the specimen was embedded in an electrically conducting resin and polished. The boundaries specimen/nickel and nickel/copper were used for calibration and the electroplated copper which was always in the field of view during the measurement served as standard.

RESULTS

Initially the diffusion coefficient of copper in pure nickel was investigated in a temperature range from 843 °C to 1050 °C. In this region the results show the usual Arrhenius type dependence. The measured values lay within a region of $\pm 3\%$ about the least-square line, as can be seen in Fig. 2. The acti-

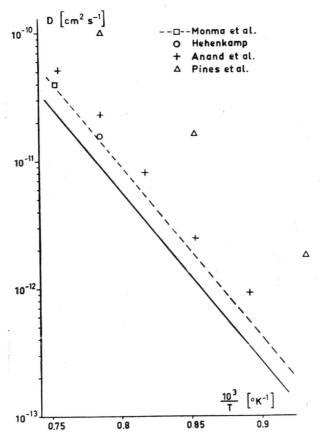

Figure 3 Diffusion coefficient of copper in nickel (the unbroken line shows our own measurements)

vation energy was found to be, $Q = 61.02 \text{ kcal mol}^{-1} \pm 1\%$, the frequency factor $D_0 = 0.27 \text{ cm}^2 \text{ s}^{-1} \pm 22\%$.

The necessity of these measurements is made evident in Fig. 3; the wide spread of previous literature values[1-4] precluding their use as reference values for our subsequent calculations. Former experiments were carried out with polycrystalline material of low purity. Because of this and the method used (integral measurement with radioactive isotopes over the cross-section) they probably yielded enhanced diffusion coefficients. An experiment by ourselves using polycrystalline material showed the influence of grain boundary diffusion. The diffusion coefficient was in some cases as much as three times as high as the value for a single crystal, depending on

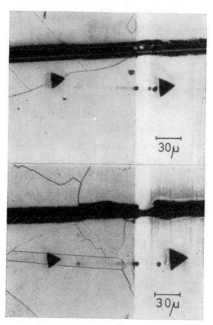

Figure 4 Diffusion coefficient of copper in polycrystalline nickel: top—$D = 3.45 \times 10^{-12} \text{ cm}^2 \text{ s}^{-1}$; bottom—$D = 9.02 \times 10^{-12} \text{ cm}^2 \text{ s}^{-1}$

the distance between the measured line and the grain boundary. Figure 4 shows an example of a polycrystalline specimen annealed at 952 °C. The black lines and triangles are marks on the specimen.

For alloying, a thread of the material concerned was fixed into a slot in a prepared nickel rod. By zone melting in an electron beam furnace in the former manner, alloyed single crystals were produced directly. The elements which were examined for their influence on the diffusion of copper in nickel are gold (0.17 at %), palladium (0.27 at %) and titanium (0.18 at %). The titanium-alloyed specimens showed no deviation from the pure specimens over the whole temperature range. The specimens with gold and palladium agreed with the pure crystals at 1000 °C and 903 °C but not at 843 °C. The diffusion coefficients were raised by 29% and 17% respectively. In a preliminary test one specimen with 2.7 at % arsenic showed the biggest increase of the diffusion coefficient related to the impurity concentration. Figure 5 gives a general picture of the results.

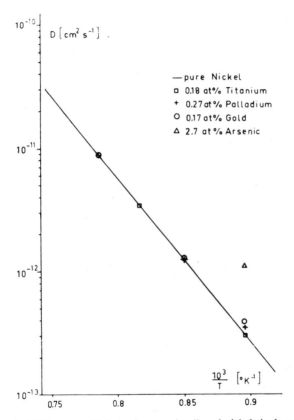

Figure 5 Diffusion coefficient of copper in alloyed nickel single crystals

DISCUSSION

The present work represents the first known on this subject. No theories are available to test the results. A quantitative calculation of the effect is much more difficult than in the case of self-diffusion. However, by using the concept of metal valence from Pauling[5] and the statements on the potentials from LeClaire[6,7], a semi-quantitative interpretation of the observed variations can be given. The result of this is, that titanium, an element with a lack of electrons compared with the transition metals, would not be expected to alter the diffusion coefficient of copper in nickel. This parameter should then increase from palladium through gold to arsenic corresponding to the increasing differences of valencies between nickel and these elements with a surplus of electrons. It is assumed here that the total number of vacancies (free and bound) is dominant in the enhancement of diffusion and no consideration is taken of the different consequences of tight and loosely bound vacancies and moving or fixed impurity atoms.

The experiments are to be continued to lower temperatures to find if there is a constant activation energy in the low temperature region and to determine if the dependence upon the concentration of the impurity is linear or exponential. Following this it is proposed to carry out quenching experiments with the same alloys to determine the real vacancy concentrations at different temperatures.

REFERENCES

1. K. Monma, H. Suto and H. Oikawa, *J. Inst. Metals*, **28**, 192 (1964).
2. T. Hehenkamp, *Z. Naturforschung*, **23a**, 229 (1968).
3. M. S. Anand, S. P. Murarka and R. P. Agarwala, *J. Appl. Phys.*, **36**, 3860 (1965).
4. B. Ya. Pines, I. G. Ivanov and I. V. Smushkov, *Soviet Phys.—Solid State*, **4**, 7, 1380 (1963).
5. L. Pauling, *Die Natur der chemischen Bindung;* Weinheim (1968).
6. A. D. LeClaire, *Phil. Mag.*, **7**, 73, 141 (1962).
7. A. D. LeClaire, *Phil. Mag.*, **10**, 106, 641 (1964).

Interpretation of self-diffusion in copper

H. MEHRER

Institut für theoretische und angewandte Physik der Universität Stuttgart

and

A. SEEGER

Max-Planck-Institut für Metallforschung,
Institut für Physik, Stuttgart

EXTENDED ABSTRACT*

For an empirical description of tracer self-diffusion coefficients D^T in a limited temperature region in face-centred cubic metals it may in many cases suffice to use a simple Arrhenius law

$$D^T = D_o^T \exp\left(-\frac{Q^{SD}}{kT}\right). \qquad (1)$$

Here Q^{SD} is the activation energy for self-diffusion, D_o^T a pre-exponential factor, k is Boltzmann's constant, and T the absolute temperature.

The quantities Q^{SD} and D_o^T should not be interpreted simply in terms of monovacancies. When one wants to proceed to an interpretation of the data in terms of vacancy properties one has to consider the contribution of the next important mechanism and a possible temperature dependence of the activation parameters. Our knowledge of vacancies in copper indicates that the mechanism likely to give the second largest contribution to the measured tracer self-diffusion coefficient at high temperatures is the divacancy mechanism.

---

* A full account of this paper has appeared elsewhere.[4]

The diffusion coefficient due to mono- and divacancies may be written as

$$D^T = D_{10}\, e^{-Q_1/kT}\, e^{-\frac{2\alpha(T-T_0)}{T}}\, e^{\frac{2\alpha\ln(T/T_0)}{T}} \left[1 + D_{21}\, e^{-Q_{21}/kT}\right]. \qquad (2)$$

This equation contains five parameters: The monovacancy self-diffusion energy

$$Q_1 = E_{1V}^F + E_{1V}^M, \qquad (3)$$

as usual, is the sum of the formation energy E_{1V}^F and the migration energy E_{1V}^M of the monovacancy. The pre-exponential factor for monovacancies

$$D_{10} = a^2 f_{1V} \nu_{1V}^0 \exp\left(\frac{S_{1V}^M + S_{1V}^F}{k}\right) \qquad (4)$$

contains the cubic lattice constant a, the geometrical correlation factor for monovacancies f_{1V}, the monovacancy frequency factor $\nu_{1V}^0 \exp\left(S_{1V}^M/k\right)$, and the monovacancy formation entropy S_{1V}^F.

$$Q_{21} = E_{1V}^F - E_{1V}^M + E_{2V}^M - E_{2V}^B \qquad (5)$$

is an abbreviation for the difference of the di- and monovacancy self-diffusion energies, where E_{2V}^M and E_{2V}^B are the divacancy migration and binding energy. Finally the ratio of the pre-exponential factors for di- and monovacancies in terms of vacancy properties is given by

$$D_{21} = 4\, \frac{f_{2V}}{f_{1V}}\, \frac{\nu_{2V}^0}{\nu_{1V}^0}\, \exp\left(\frac{S_{2V}^M + S_{1V}^F + \Delta S_{2V} - S_{1V}^M}{k}\right). \qquad (6)$$

Here f_{2V} is the geometrical correlation factor for divacancies, $\nu_{2V}^0 \exp\left(S_{2V}^M/k\right)$ their frequency factor, and ΔS_{2V} the divacancy association entropy.

The quantities Q_1, D_{10}, Q_{21}, and D_{21} refer to the reference temperature T_0. The temperature dependence of the energies and entropies of formation and migration has been recently discussed in detail[1]. The result is that in the case of the noble metals the simplest possible dependence

$$E^i = E^i(T_0) + \alpha^i k\,(T - T_0) \qquad (7a)$$

$$S^i = S^i(T_0) + \alpha^i k\,\ln T/T_0 \qquad (7b)$$

should be adequate. In equation (7) the superscript i stands for F (formation) or M (migration). The reference temperature should be chosen near or above the Debye temperature. In the present case we take $T_0 = 400\,°K$. For copper we expect the temperature coefficients α^i to be small, i.e.,

$$0 < \alpha = (\alpha^F + \alpha^M)/2 \leqslant 0.5.$$

Since the divacancy corrections in (2) are also small, we presumably introduce a negligible error if we use the same α-values for di- and monovacancies.

Recent tracer measurements of the self-diffusion coefficient by Rothman and Peterson[2] are analysed in terms of equation (2). The analysis is performed in conjunction with the determination of equilibrium concentration of vacant lattice sites by Simmons and Balluffi[3] and is corroborated by measurements of the isotope effect on self-diffusion[2]. The values listed in Table 1 for self-diffusion parameters and vacancy properties are obtained. For a discussion of these values and for a detailed description of the analysis we refer the reader to a recent paper[4] of the authors.

Table 1 Consistent set of self-diffusion parameters and vacancy properties for copper. (The underlined value has been determined by quenching experiments[5])

Self-diffusion	Monovacancies		Divacancies
$Q_1 = 2.09$ eV	$E_{1V}^F = 1.04$ eV $\left.\right\}$	$T = T_m$	$E_{2V}^M - E_{2V}^B \approx 0.54$ eV
$D_{10} = 0.19$ cm²/s	$S_{1V}^F \approx 0.4k$	$(\alpha T = 0.1)$	$E_{2V}^B = 0.12$ eV
$Q_{21} = 0.51$ eV	$E_{1V}^F = 1.03$ eV $\left.\right\rfloor$		$\Delta S_{2V} \approx 2k$
$D_{21} = 28$	$S_{1V}^F \approx 0.3k$		$\nu_{2V}^0 \exp \dfrac{S_{2V}^M}{k} = 1.6 \cdot 10^{14}$
$2\alpha = 0.5$	$E_{1V}^M = 1.06$ eV $\left.\right\}$	$T = T_0$	s^{-1}
$(T_0 = 400\,^\circ\text{K})$	$\nu_{1V}^0 \exp \dfrac{S_{1V}^M}{k} = 1.4 \cdot 10^{14}\ \text{s}^{-1}$		

REFERENCES

1. A. Seeger and H. Mehrer, *Vacancies and Interstitials in Metals*, Conf. Jülich 1968, Eds.: A. Seeger, D. Schumacher, J. Diehl and W. Schilling, North-Holland Publ. Company; Amsterdam 1970.
2. S. J. Rothman and N. L. Peterson, *Phys. Stat. Sol.*, **35**, 305 (1969).
3. R. O. Simmons and R. W. Balluffi, *Phys. Rev.*, **129**, 1533 (1963).
4. H. Mehrer and A. Seeger, *Phys. Stat. Sol.*, **35**, 313 (1969).
5. F. Ramsteiner, G. Lampert, A. Seeger and W. Schüle, *Phys., Stat. Sol.*, **8**, 863 (1965).

4.4

Self-diffusion and the diffusion
mechanism in b.c.c. transition metals

G. M. NEUMANN

OSRAM-Studiengesellschaft, München, Germany

ABSTRACT

Self-diffusion in b.c.c. transition metals is reviewed and a simple theory for diffusion in these metals proposed. Literature data are critically re-analyzed and appropriate Arrhenius parameters evaluated. In order to check the data, vacancy and divacancy properties for b.c.c. metals are estimated on the basis of an appraisal of the corresponding point defect properties of f.c.c. metals. It is shown that non-linear Arrhenius relationships in the metals of the Vth and VIth sub-groups—V, Nb, Ta, Cr, Mo, W—are consistent with a mechanism of vacancy-divacancy diffusion. Non-linear Arrhenius relationships in the metals of the IVth sub-group—Ti, Zr, Hf—can then be attributed to a more complex mechanism of diffusion including short-circuit and dislocation contributions.

1 INTRODUCTION

In the past ten years knowledge regarding diffusion, and transport phenomena as well as that regarding point-defect properties in metals has increased considerably.

More recently the main interest in the field of diffusion-processes has shifted toward the behaviour of body-centred cubic metals, partly because of their increasing importance in high-temperature applications, partly because of a natural interest in the fundamentals of these systems. In 1964 a conference devoted entirely to diffusion in body-centred cubic metals included considerable discussions of the results of measurements. The amount of new work reported since then has been relatively small and the situation

329

concerning the interpretation of the results has not changed very much. The theoretical information is only scant. The present paper, therefore, is not primarily an attempt to give an exhaustive survey of the literature, but rather an attempt to consider the results in terms of current ideas concerning diffusion in metals. It is written to stimulate further experimental investigations and theoretical treatments in this field. In the first part, the paper gives a brief review on the subject of self-diffusion in body-centred cubic transition metals—mainly with the metals of the Vth, VIth, and IVth subgroup of the periodic table. The compiled data are critically reconsidered and best sets of diffusion parameters are evaluated for each system. In the second part of the paper the results of this analysis are compared with the known data of point-defect properties in body-centred cubic metals. The concept of a combined vacancy-divacancy diffusion mechanism as was developed from the study of point-defects and diffusion processes in face-centred cubic metals— is applied to the interpretation of the diffusion parameters.

2   SELF-DIFFUSION IN BODY-CENTRED CUBIC TRANSITION METALS

The relatively large amount of experimental data on diffusion in body-centred cubic metals shows a wide scatter for individual systems. There are also some unusual features of diffusion in these metals relative to diffusion in face-centred cubic metals. From the experimental evidence it appears that the body-centred cubic metals can be classified into two distinct groups with regard to the temperature dependence of their diffusion behaviour. There are those that behave quite normally in the sense that the temperature dependence of the diffusion process is of the form

$$D(T) = D_0 \exp(-\Delta H_V / RT), \qquad (1)$$

where D_0 is a temperature-independent frequency factor and ΔH_V is the activation energy for diffusion. Others are "anomalous" in the sense that their Arrhenius-type temperature dependence shows a more or less pronounced nonlinear behaviour that can be described by a sum of two exponential terms of the form

$$D(T) = D_{01} \exp(-\Delta H_{V1}/RT) + D_{02} \exp(-\Delta H_{V2}/RT). \qquad (2)$$

This resolution of the measured Arrhenius-curve into two separate contributions was first applied by Kidson[1] and is based on the following assumptions.

First, that there are two separate mechanisms of diffusion each of which dominates in a certain temperature range. Second, that at least one of these mechanisms is a vacancy mechanism. If it is further assumed that the contribution of the high-temperature diffusion-mechanism to the measured diffusion coefficient becomes negligibly small at the lower temperatures, the Arrhenius-curves may then be resolved. Values of the true diffusion coefficient at higher temperatures can be obtained by extrapolating the low temperature diffusion coefficients to higher temperatures and subtracting them from the observed diffusion data. This procedure—the so-called Kidson-analysis—is applied in the reanalysis of the diffusion data of nearly all of the body-centred cubic metals.

2.1 Vanadium

There are three investigations of the self-diffusion in vanadium. The measurements of Lundy and McHargue[2] and of Peart[3] that cover a fairly wide temperature range (840° to 1890°C) and the more recent work of Agarwala, Murarka and Anand[4] that covers only the lower temperature range (700° to 1400°C). Furthermore there is a reanalysis of the data of Peart by Lazarus[5].

The results of these investigations which all show a curvature in the Arrhenius-diagram, are tabulated in Table 1 and shown in graphical form in Fig. 1.

The diffusion-coefficients from the separate investigations in the high-temperature region are in good agreement. They can be fitted by two nearly

Table 1   Self-diffusion data for vanadium

Arrhenius-parameters		Temperature range (°C)	Method of analysis	Reference
D_0 (cm²/s)	ΔH_v (kcal/mole)			
1.1×10^{-2}	61.0	1000–1400	Lathe-sectioning and grinding	2
58	91.5	1400–1890		
0.36	73.6	840–1356	Lathe-sectioning	3
14	94.1	1356–1830		
0.35	72.5		Reanalysis of data of Peart	5
130	92.5			
1.07×10^{-1}	64.6	700–1300	Residual activity	4
10.45	76.8	1300–1400		

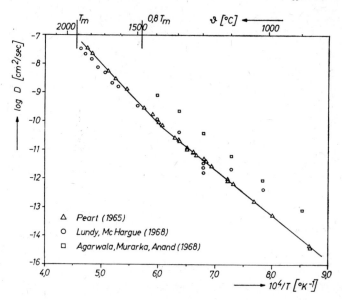

Figure 1 Arrhenius-diagram for self-diffusion in vanadium

parallel curves. In the low-temperature region, however, the data show some deviations and the curves are not parallel. This difference in diffusivity is attributed by Agarwala *et al.*[4] to the varying impurity content of the mate-rial (Table 2), which is reasonable if we accept an increasing enhancement of diffusion with increasing impurity content.

Table 2 Impurity content of vanadium

Author	Lundy[2]	Peart[3]	Agarwala[4]
Element			
O_2	880 ppm	5 ppm	800 ppm
C	600 ppm	–	–
N_2	800 ppm	5 ppm	< 10 ppm
Metals	360 ppm	95 ppm	60 ppm

On the whole it seems that the diffusion behaviour of vanadium may best be represented by the data of Peart[3] which give for the temperature depen-dence of the diffusion coefficient in the low-temperature and high-temperature region the equations

$$880° < T < 1230°C$$

$$D = 0.39 \exp(-73, 820/RT) \tag{3}$$

and

$$1460° < T < 1830°C$$

$$D = 204 \exp(-94, 100/RT) \tag{4}$$

while the Kidson-analysis results in the relation

$$D = 0.39 \exp(-73, 800/RT) + 194.4 \exp(-94, 900/RT). \tag{5}$$

2.2 Niobium

The self-diffusion of niobium has been the subject of four investigations, which agree quite well with each other.

While the studies by Resnick and Castleman[6], Peart, Graham and Tomlin[7] and Lyubimov, Geld and Shveikin[8] cover only a small temperature region (1600° to 2100°C), the measurements of Lundy, Winston, Pawel and McHargue[9,10] cover a wide temperature interval (880° to 2400°C).

The results of the investigations are tabulated in Table 3 and shown graphically in Fig. 2.

Table 3 Self-diffusion data for niobium

Arrhenius-parameters		Temperature range (°C)	Method of analysis	Reference
D_0 (cm²/s)	ΔH_V (kcal/mole)			
12.4	105.0	1535–2120	Sectioning	6
1.3	95.0	1700–2100	Autoradiography	7
49.0	115.0	1700–2100	Residual activity	8
1.1	96.0	880–2400	Lathe-sectioning	9
1.94	98.0	880–2400	grinding and anodizing technique	10

While the data of Lyubimov *et al.*[8] are in excellent agreement with those of Lundy *et al.*[9] the previous results of Resnick *et al.*[6] and Peart *et al*[7] are somewhat higher by a factor of 2 to 3. The data of Lundy *et al.*[9] do show however upward deviations in some data points which are of this order of magnitude. These deviations are explained by short-circuit diffusion, which can be recognized by tailing of the penetration plots. The depth at which this tailing occurred and its magnitude varied from sample to sample. Generally, it was more prominent at lower temperatures. Thus the diffusion-coefficients for each temperature interval that lay deepest should be the most accurate data.

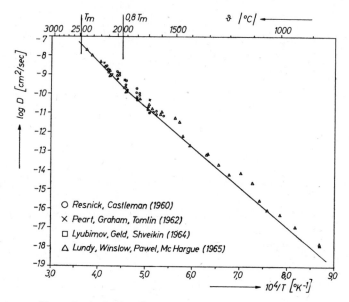

Figure 2   Arrhenius-diagram for self-diffusion in niobium

An analysis with respect to these considerations yields a slightly curved Arrhenius-plot, the curvature beginning in the temperature region of $0.8T_m$. Hence, the temperature dependence of self-diffusion in niobium can be described in the low-temperature and high-temperature region by the equations

$$900° < T < 1900°C$$

$$D = 0.47 \exp(-95,100/RT) \tag{6}$$

and

$$1900° < T < 2400°C$$

$$D = 29.5 \exp(-112,000/RT); \tag{7}$$

the Kidson-analysis giving the relation

$$D = 0.47 \exp(-95,100/RT) + 200 \exp(-125,100/RT). \tag{8}$$

2.3 Tantalum

The self-diffusion of tantalum has been studied by Langmuir and Eager[11] and Gruzin and Meshkov[12] and more recently by Pawel and Lundy[13].

The measurements of Langmuir and Eager[11] were made in the temperature region 1830° to 2530°C, but unfortunately no data points are reported. The Arrhenius-plot, however, lies well within the order of magnitude of the data from the work of Pawel and Lundy[13]. The data of Gruzin and Meshkov[12] are of little value, since they were measured over an extremely small temperature-interval. The data are influenced considerably by grain-boundary diffusion.

Thus our information is restricted to the investigation of Pawel and Lundy[13] whose study covers the temperature range 1200° to 2300°C.

The results of the measurements are given in Table 4 and shown graphically in Fig. 3.

Table 4 Self-diffusion data for tantalum

Arrhenius-parameters		Temperature range (°C)	Method of analysis	Reference
D_0 (cm^2/s)	ΔH_V (kcal/mole)			
3×10^{-2}	89.5	1830–2530	–	11
1.3×10^2	110.0	1200–1300	Residual activity	12
0.124	98.7	1200–2300	Sectioning	13

The data for self-diffusion in tantalum can be represented by a linear Arrhenius-relationship, but unfortunately they extend only up to temperatures at which deviations from linearity should be just noticeable. If one is searching for deviations in this sense, however, one can evaluate this tendency by comparing the highest data points with the calculated data from Langmuir's analytical expression.

An analysis then gives for the temperature dependence of the diffusion in tantalum for the low temperature and the high-temperature region the relations

$$1200° < T < 2200°C$$

$$D = 0.10 \exp(-99, 100/RT) \tag{9}$$

and

$$T > 2200\,°C$$

$$D = 1.218 \exp\left(-111, 100/RT\right) \tag{10}$$

while the Kidson-analysis results in

$$D = 0.10 \exp\left(-99, 100/RT\right) + 16.3 \exp\left(-132, 600/RT\right). \tag{11}$$

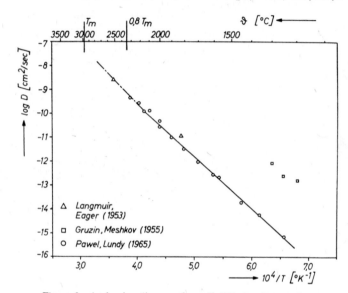

Figure 3   Arrhenius-diagram for self-diffusion in tantalum

2.4   Chromium

For self-diffusion in chromium there are a great number of investigations, the data points exhibiting a wide scatter over a range of almost a factor 10^2 to 10^3.

Owing to high data scatter and to the measurement at only three intermediate temperatures the results of the investigations by Paxton and Gondolf[15] and those by Bogdanov[17] are essentially inconclusive and can be neglected; especially since the low values of the pre-exponential factor as reported by Paxton and Gondolf[15] indicate preferential diffusion along grain-boundaries or other short-circuiting diffusion paths.

On the other hand the results of the investigations by Bokshtein, Kishkin, and Moroz[14b], Hagel[18] and Askill[20,21] are in excellent agreement. The data points of these three studies do not differ by as much as a factor of 0.5 thus confirming the evaluated Arrhenius-parameter. It is unfortunate that these studies were not extended to higher temperature regions.

The diffusion coefficients as measured in the experiments of Gruzin[16] and Ivanov, Matveeva, and Morozov[19] are higher by a factor of 10 to 30, com-

Table 5   Self-diffusion data for chromium

Arrhenius-parameters		Temperature		
D_0 (cm²/s)	ΔH_V (kcal/mole)	range (°C)	Method of analysis	Reference
–	76.0	1100–1350	Autoradiography	14a
1.5×10^{-4}	52.7	950–1250	Sectioning	15
2.3×10^{-5}	51.3		Absorption	
45	85.0	1000–1350	Sectioning	16
1.65×10^{-3}	62.4	1080–1340	Autoradiography	17
0.4	76.0	700–1350	Autoradiography	14b
0.28	73.2	1200–1600	Absorption and residual activity	18
6.47×10^{-2}	59.2	1050–1400	Residual activity	19
0.2	73.7	1030–1550	Lathe-sectioning and serial grinding	20, 21

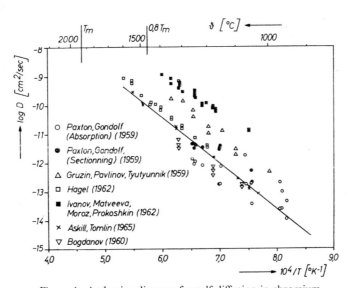

Figure 4   Arrhenius-diagram for self-diffusion in chromium

pared to the former data points. Allowing for upward deviations due to short-circuiting the results may be adjusted, nevertheless, to analytical expressions including the same activation energy for diffusion and differing only in the pre-exponential factor.

The results of the individual investigations are tabulated in Table 5 and shown in graphical form in Fig. 4.

The temperature dependence of the self-diffusion in chromium follows a linear Arrhenius-relationship of the form

$$1000° < T < 1500°C$$

$$D = 0.2 \exp(-73, 700/RT). \tag{12}$$

Because of the scatter in experimental data the Kidson-analysis was not applied to chromium.

2.5 Molybdenum

The investigations on self-diffusion in molybdenum can be separated into two groups, the data points of which differ by a factor of 4 to 5. The results of these are tabulated in Table 6 and shown in graphical form in Fig. 5.

The measurements of three goups of Russian workers Borisov, Gruzin, Pavlinov, and Fedorov[22], Gruzin, Pavlinov, and Tyutyunnik[16] and Pavlinov and Bykov[26] cover the temperature range 1880° to 2270°C and can best be represented by a linear Arrhenius-relationship

$$D = 0.243 \exp(-90, 150/RT). \tag{13}$$

Table 6   Self-diffusion data for molybdenum

Arrhenius-parameters		Temperature		
D_0 (cm^2/s)	ΔH_V (kcal/mole)	range (°C)	Method of analysis	Reference
4.0	115.0	1860–2180	Residual activity	22
4.0	114.0	1800–2200	Residual activity	16
2.77	111.0	1700–1920		23
0.38	100.8	1790–2155	Sectioning	24
0.1	92.2	1850–2345	Sectioning	25
1.8	110.0	1880–2270	Residual activity	26

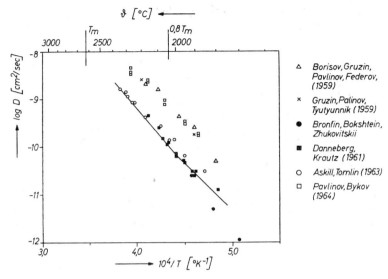

Figure 5 Arrhenius-diagram for self-diffusion in molybdenum

The second group of diffusion-coefficients consists of the measurements of Danneberg and Krautz[24] and Askill and Tomlin[25]. If these data are combined, they show a curved Arrhenius behaviour, the data of Danneberg and Krautz[24] resulting in the low-temperature term and part of the data of Askill and Tomlin in the high-temperature term. The corresponding analytical expressions are

$$1800° < T < 2000°C$$

$$D = 3.06 \times 10^{-2} \exp(-89,750/RT) \tag{14}$$

and

$$2000° < T < 2500°C$$

$$D = 2.06 \exp(-108,650/RT) \tag{15}$$

the Kidson-analysis giving

$$D = 3.06 \times 10^{-2} \exp(-89,750/RT) + 63 \exp(-131,100/RT). \tag{16}$$

2.6 Tungsten

The diffusion behaviour of tungsten has been the subject of several experimental and theoretical investigations, a critical discussion of which has been given by Neumann[27].

The results of the measurements are tabulated in Table 7 and shown in graphical form in Fig. 6.

Table 7　Self-diffusion data for tungsten

Arrhenius-parameters		Temperature range (°C)	Method of analysis	Reference
D_0 (cm²/s)	ΔH_V (kcal/mole)			
6.3×10^7	135.8	1290–1450	Absorption	28
0.257	120.5	2000–2700	Sectioning	29
42.8	153.1	2700–3230	Residual activity	30
4.64×10^{-2}	113.8	2000–2700	Estimation from data of Danneberg	27
1.88	140.3	1800–2400	Anodizing	31

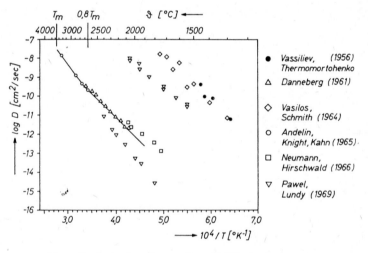

Figure 6　Arrhenius-diagram for self-diffusion in tungsten

The temperature dependence of the diffusion in tungsten is described for the low-temperature and high-temperature region by the expressions

$$2000° < T < 2750°C$$

$$D = 4.64 \times 10^{-2} \exp\left(-113, 800/RT\right) \tag{17}$$

and

$$2750 < T < 3300\,°C$$

$$D = 42.8 \exp\left(-153,\,100/RT\right) \tag{18}$$

the Kidson-analysis resulting in the equation

$$D = 4.64 \times 10^{-2} \exp\left(-113,\,800/RT\right) + 223 \exp\left(-167,\,600/RT\right) \tag{19}$$

More recently an investigation of self-diffusion in tungsten has been reported by Pawel and Lundy[31] but unfortunately the temperature range is too low to indicate the presence of any curvature.

2.7 β-Titanium

The self-diffusion behaviour of β-titanium has been studied in three investigations, but only two sets of data points from the measurements of Murdock, Lundy, and Stansbury[23] and DeReca and Libanati[35] are known.

The results of the investigations are tabulated in Table 8 and shown in graphical form in Fig. 7.

Table 8   Self-diffusion data for β-titanium

Arrhenius-parameters		Temperature range (°C)	Method of analysis	Reference
D_0 (cm²/s)	ΔH_V (kcal/mole)			
2×10^{-4}	28.6	900–1300		32
3.58×10^{-4}	31.2	900–1500	Lathe-sectioning	33
1.09	60.0			
1.63×10^{-3}	35.0	890–1580	Residual activity	34
1.9×10^{-3}	36.5	900–1580	Sectioning Gruzin-method	35

Although the diffusion coefficients are similar there is a severe distinction in the observed temperature dependence of the data. While the results of Murdock *et al.*[33] exhibit a curvature in the Arrhenius-diagram the results of DeReca *et al.*[35] imply a linear Arrhenius-behaviour. The linearity, however, is achieved by correcting the measured diffusion coefficients at high temperatures with respect to their annealing time. As this procedure is not quite clear it is assumed that the data of Murdock *et al.*[33] are more representative of the diffusion behaviour of β-titanium.

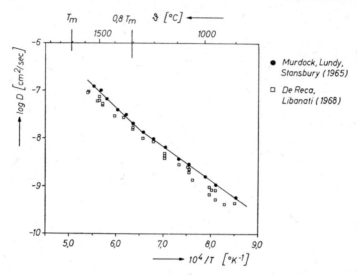

Figure 7   Arrhenius-diagram for self-diffusion in β-titanium

Furthermore Askill and Gibbs[36] have demonstrated a curvature in the Arrhenius-diagram for the diffusion of many metallic impurities in β-titanium. Thus the self-diffusion in β-titanium is represented by the Arrhenius-relationships

$$900° < T < 1230°C$$

$$D = 3.58 \times 10^{-4} \exp(-31, 200/RT) \qquad (20)$$

and

$$1230° < T < 1550°C$$

$$D = 4.6 \times 10^{-2} \exp(-45, 910/RT) \qquad (21)$$

the Kidson-analysis, as reported by Murdock *et al.*[33], resulting in the relation

$$D = 3.58 \times 10^{-4} \exp(-31, 200/RT) + 1.09 \exp(-60, 000/RT). \qquad (22)$$

2.8   β-Zirconium

The diffusion behaviour of β-zirconium has been studied by several authors. Except for the work of Volokoff, May, and Adda[40], whose data are obviously influenced by grain-boundary or short-circuit diffusion processes, the diverse investigations match fairly well.

Unfortunately only one study extends to sufficiently high temperatures that a curvature in the Arrhenius-plot becomes obvious.

The results of the measurements are tabulated in Table 9 and shown graphically in Fig. 8.

Table 9 Self-diffusion data for β-zirconium

Arrhenius-parameters		Temperature range (°C)	Method of analysis	Reference
D_0 (cm²/s)	ΔH_V (kcal/mole)			
10^{-4}	27.0	900–1500	Residual activity	37
2.4×10^{-3}	38.0	900–1500	Residual activity	38
4.0×10^{-5}	26.0	1000–1250	Residual activity	39
4.2×10^{-5}	24.0	900–1240	Sectioning	40
2.4×10^{-4}	30.1	1100–1500	Sectioning	41
1.34	65.2		Analysis of data of Federer	1, 43
8.5×10^{-5}	27.7			
4.8×10^{-6}	20.7	900–1750	Sectioning	42
2.5×10^{-2}	46.9			

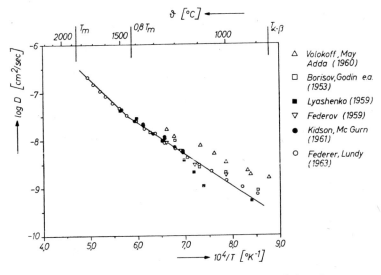

ϑ [°C] ←

△ Volokoff, May Adda (1960)
□ Borisov, Godin e.o. (1953)
■ Lyashenko (1959)
▽ Federov (1959)
● Kidson, Mc Gurn (1961)
○ Federer, Lundy (1963)

Figure 8 Arrhenius-diagram for self-diffusion in β-zirconium

The temperature dependence of diffusion in β-zirconium in the low-temperature and the high-temperature regions is described by the expressions

$$900° < T < 1450°C$$

$$D = 2.52 \times 10^{-4} \exp\left(-30,700/RT\right) \tag{23}$$

and

$$1500° < T < 1750°C$$

$$D = 1.55 \times 10^{-2} \exp\left(-45,200/RT\right) \tag{24}$$

the Kidson-analysis resulting in the relation

$$D = 2.52 \times 10^{-4} \exp\left(-30,700/RT\right) + 0.37 \exp\left(-61,600/RT\right). \tag{25}$$

2.9 β-Hafnium

The diffusion behaviour of β-hafnium has been investigated only to a small degree in a limited temperature region (1800° to 2200°C). The results of the studies are tabulated in Table 10 and shown graphically in Fig. 9.

Table 10 Self-diffusion data for β-hafnium

Arrhenius-parameters D_0 (cm²/s)	ΔH_V (kcal/mole)	Temperature (°C)	Method of analysis	Reference
1.2×10^{-3}	38.7	1800–2000	Sectioning	44
4.8×10^{-3}	43.8	1800–2160	Sectioning— Gruzin-method	35

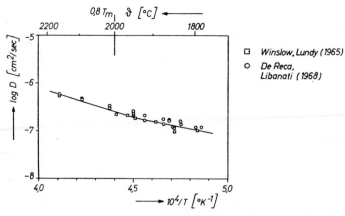

Figure 9 Arrhenius-diagram for self-diffusion in β-hafnium

Assuming the data of Winslow and Lundy[44] to be representative for the low-temperature region, the data of DeReca and Libanati[35] yielding a slow curvature in the Arrhenius-plot, the Kidson-analysis results in the relation

$$D = 1.2 \times 10^{-3} \exp\left(-38, 700/RT\right) + 0.33 \exp\left(-68, 400/RT\right). \quad (26)$$

2.10 Summary of Diffusion Data in BCC-transition Metals

Normal self-diffusion behaviour is commonly characterized by three essential features:

1 The temperature dependence of the diffusion coefficient is nearly always accurately described by the Arrhenius equation

$$D(T) = D_0 \exp\left(-\Delta H_V/RT\right) \quad (27)$$

2 The activation energy is roughly proportional to the melting-temperatures of the metal by the empirical relation

$$\Delta H_V = 34T_m \quad (28)$$

which seems to be accurate to a degree of 20%.

3 The pre-exponential factor seems always to be within an order of magnitude of 0.5, that is, it is in the range 0.05 and 5.0 cm^2 s^{-1}. These three properties have come to be regarded as characterizing "normal" diffusion behaviour because they are common to self-diffusion in such a wide range of face-centred cubic, body-centred cubic, hexagonal close-packed, and other structured metals.

The results of the present reanalysis of the diffusion data in b.c.c.-transition metals are summarized in Table 11.

The reliability of the empirical melting-point rule can be inferred from the Table 12.

From the evidence thus available, it would appear that the bcc-transition metals can be classified into two fairly distinct groups with regard to their diffusion behaviour. Both groups having in common the remarkable feature that their Arrhenius-type temperature dependence is more or less nonlinear.

The first group, consisting of the elements of the Vth and VIth sub-group of the periodic table, show normal diffusion behaviour in the low-tempera-

Table 11 Arrhenius-parameters for self-diffusion in b.c.c.-metals as reanalyzed
applying the Kidson-model to literature data

Metal	Diffusion-parameters			
	D_{01} (cm²/s)	H_{V1} (kcal/mole)	D_{02} (cm²/s)	ΔH_{V2} (kcal/mole)
Vanadium	0.39	73.8	194	94.9
Niobium	0.47	95.1	200	125.1
Tantalum	0.10	99.1	16.3	132.6
Chromium	0.20	73.7	–	–
Molybdenum	3.06×10^{-2}	89.8	63	131.1
Tungsten	4.64×10^{-2}	113.8	223	167.6
β-Titanium	3.58×10^{-4}	31.2	1.09	60.0
β-Zirconium	2.52×10^{-4}	30.7	0.37	61.6
β-Hafnium	1.2×10^{-3}	38.7	0.33	68.4

Table 12 Relation between activation energy of diffusion and melting
temperature for b.c.c.-metals[a]

Metal	Melting-temperature T_m (°K)	$\Delta H_V/T_m$	
		Low-temperature region	High-temperature region
Vanadium	2190	0.337	0.433
Niobium	2770	0.343	0.451
Tantalum	3270	0.303	0.406
Chromium	2176	0.338	–
Molybdenum	2890	0.311	0.454
Tungsten	3655	0.312	0.458
β-Titanium	1980	0.157	0.302
β-Zirconium	2125	0.145	0.290
β-Hafnium	2250	0.172	0.304

[a] Melting-temperatures according to Nesmeyanov[45]

ture region and deviations from the melting point rule at temperatures near
the melting point.

The second group, consisting of the elements of the IVth sub-group of the
periodic table, show normal diffusion behaviour in the high-temperature
region and deviations from the melting point rule at lower temperatures.

3 MECHANISM OF DIFFUSION IN BODY-CENTRED CUBIC TRANSITION METALS

There are a certain number of theoretical investigations concerning the mechanism of diffusion in metals, but there is no theoretical treatment that recognizes any fundamental difference between diffusion in body-centered cubic metals and diffusion in metals of other structures.

Inasmuch as metals that are strongly anisotropic often show a nearly iso-tropic diffusion behaviour there is no argument against not applying current ideas on the mechanism of diffusion in face-centered cubic metals to body-centered cubic metals.

Consequently, the Arrhenius-parameters as evaluated in the first part of this paper are discussed in terms of the recently well established concept of a combined vacancy-divacancy mechanism of diffusion.

3.1 Mechanism of Diffusion and Point-defect Properties in Face-centred Cubic Metals

It is generally accepted and supported by considerable experimental evidence that in nearly all metals self-diffusion occurs by a vacancy mechanism. Accordingly, the parameters of the Arrhenius-relation

$$D = D_{01} \exp\left(-\Delta H_{V1}/RT\right)$$

are interpreted as

$$D_{01} = a^2 \nu \gamma f \exp\left(\frac{\Delta S_f^\square + \Delta S_m^\square}{R}\right) \tag{29}$$

and

$$\Delta H_{V1} = \Delta H_f^\square + \Delta H_m^\square \tag{30}$$

where

a = lattice parameter
ν = atomic vibration frequency
γ = geometric constant[46]
f = correlation factor[47]
$\Delta S_f^\square$ = formation entropy of single vacancy
$\Delta S_m^\square$ = migration entropy of single vacancy
R = gas-constant
$\Delta H_f^\square$ = formation enthalpy of single vacancy
$\Delta H_m^\square$ = migration enthalpy of single vacancy

The results of recent investigations on point-defect properties of solids (see Seeger[48]) have confirmed the existence of divacancies in addition to monovacancies, the concentration of the former increasing with increasing temperature. In processes which are dependent on vacancies, an influence of divacancies should be observed at high temperatures. The effects of divacancies on the diffusion behaviour of metals has been discussed by Seeger, Schottky and Schumacher[49] for the case of the self-diffusion in nickel. By a detailed analysis of literature data they demonstrated that the observed deviations from a straight Arrhenius-behaviour at high temperatures may be accounted for by contributions of di- and trivacancies.

This concept of a contribution of multivacancies to diffusion processes in face-centred cubic metals and of a resulting diffusion mechanism involving combined mono- and divacancies has been confirmed further by several consecutive treatments. The effect has been established by Schumacher, Seeger, and Harlin[50] for the case of platinum, by Seeger and Mehrer[51] for the case of gold, by Stoebe and Dawson[52] for the case of aluminium, and now by Mehrer and Seeger[53] for the case of copper.

While these treatments are based on an analysis of different literature data, Reimers[54] recently reported very accurate measurements of self-diffusion in silver up to temperature just 2° below the melting point. The observed deviation from linear Arrhenius-behaviour was the first direct experimental verification of the contribution of divacancies to self-diffusion in metals, thus giving strong evidence for the reality of this effect.

The temperature dependence of the diffusion-coefficient can then be expressed by a sum of two Arrhenius-type exponential terms of the form

$$D = D_{01} \exp\left(-\Delta H_{V1}/RT\right) + D_{02} \exp\left(-\Delta H_{V2}/RT\right).$$

The meaning of the Arrhenius-parameters being as for the monovacancy mechanism in equations (29) and (30) and

$$D_{02} = a_2 v' \gamma' f' \exp\left(\frac{\Delta S_f + \Delta S_m}{R}\right) \qquad (31)$$

and

$$\Delta H_{V2} = \Delta H_f^{\square} + \Delta H_m^{\square} = 2\Delta H_f^{\square} + \Delta H_m^{\square} - 2B^{\square} \qquad (32)$$

where

v' = relevant atomic vibration frequency
γ' = relevant geometric factor[55]
f' = relevant correlation factor[55]
$\Delta S_f^{\square}$ = formation entropy of divacancies

$\Delta S_m^{\square\square}$ = migration entropy of divacancies
$\Delta H_f^{\square\square}$ = formation enthalpy of divacancies
$\Delta H_m^{\square\square}$ = migration enthalpy of divacancies
$B^{\square\square}$ = binding energy of divacancies

The consistency of the experimental data can be checked by comparing the data from the measurement of diffusion properties and point-defect properties of the corresponding metals.

The data for the vacancy and divacancy formation- and migration-energies as evaluated for face-centred cubic metals are tabulated in Table 13 and Table 14.

Table 13   Properties of monovacancies in f.c.c.-metals

Metal	Formation-energy $\Delta H_f^{\square}$		Migration-energy $\Delta H_m^{\square}$		Reference
	(eV)	(kcal)	(eV)	(kcal)	
Copper	1.03	23.8	1.06	24.5	53
Silver	1.10	25.4	0.87	20.1	54
Gold	0.97	22.4	0.89	20.6	51
Aluminum	0.76	17.6	0.72	16.6	52
Nickel	1.35	31.2	1.46	33.7	49
Platinum	1.49	39.4	1.38	31.9	50

Table 14   Properties of divacancies in f.c.c.-metals

Metal	Formation-energy $\Delta H_f^{\square\square}$		Migration-energy $\Delta H_m^{\square\square}$		Binding-energy $B^{\square\square}$		References
	(eV)	(kcal)	(eV)	(kcal)	(eV)	(kcal)	
Copper	1.94	44.9	0.66	15.3	0.12	2.8	53
Silver	2.07	47.8	0.57	13.2	0.13	3.0	54
Gold	1.78	41.1	0.79	18.3	0.16	3.7	51
Aluminum	1.42	32.8	0.38	8.8	0.10	2.3	52
Nickel	2.40	55.4	0.82	18.9	0.3	6.9	49
Platinum	2.87	66.3	1.11	25.6	0.11	2.5	50

3.2   Point-defect Properties in Body-centred Cubic Metals

The knowledge of point-defect properties in body-centred cubic metals is limited. A review of the present situation in this field of research has been recently given by Schultz[56].

In order to discuss the diffusion-parameters of these metals in terms of a combined vacancy-divacancy mechanism the knowledge of the formation- and migration-energy of vacancies and divacancies as well as the binding-energy of divacancies would be quite useful. Unfortunately the data are restricted to the knowledge of some monovacancy-properties.

The experimental data of the corresponding formation- and migration-energy are tabulated in Table 15.

Table 15 Experimental data for the formation- and migration-energies of vacancies in b.c.c.-metals[a]

Metal	Sublimation-enthalpy ΔH_{sub}		Formation-energy $\Delta H_f^\square$		$\dfrac{\Delta H_f^\square}{\Delta H_{sub}}$	Ref.	Migration-energy $\Delta H_m^\square$		$\dfrac{\Delta H_m^\square}{\Delta H_{sub}}$	Ref.
	(eV)	(kcal)	(eV)	(kcal)			(eV)	(kcal)		
Vanadium	5.32	122.9	–	–	–		–	–	–	
Niobium	7.46	172.5	2.03	47.0	0.273	57	–	–	–	
Tantalum	8.08	186.5	2.91	67.2	0.360	58	1.35	31.2	0.167	59
Chromium	4.04	93.3	1.48	34.2	0.366	60	–	–	–	
Molybdenum	6.85	158.3	2.40	55.2	0.350	61	–	–	–	
Tungsten	8.73	201.8	3.14	72.5	0.360	62	1.3	30.0	0.190	61
β-Titanium	4.90	113.2	1.55	35.8	0.316	64	1.7	39.3	0.198	63
β-Zirconium	6.31	145.8	1.75	40.4	0.278	65	–	–	–	
β-Hafnium	6.71	155.0	–	–	–		–	–	–	

[a] The sublimation enthalpy is taken from Nesmeyanov[45]

If one considers the sublimation-enthalpy of a metal to be a true measure of the lattice energy one may estimate the formation- and migration-energy of a vacancy in the corresponding metal from its sublimation-enthalpy, for these energies are closely related to the lattice energy as well.

The most reliable data of the point-defect energies are plotted in Fig. 10 and Fig. 11 as a function of the sublimation-enthalpy of the metals.

As can be inferred from the diagrams the expected relationship between these parameters may be approximated by straight lines, resulting in the equations

$$\Delta H_f^\square = 0.360 \, \Delta H_{sub}^{289} \tag{33}$$

and

$$\Delta H_m^\square = 0.200 \, \Delta H_{sub}^{298}. \tag{34}$$

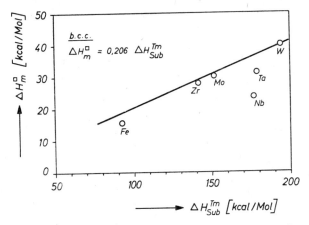

Figure 10   Formation energy for vacancies in b.c.c.-metals as a function of their sublimation-enthalpy

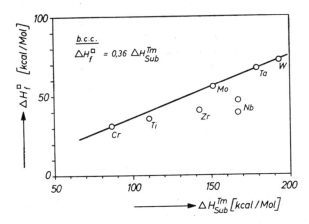

Figure 11   Migration energy for vacancies in b.c.c.-metals as a function of their sublimation-enthalpy

The point-defect data, calculated according to these empirical relations, are tabulated in Table 16.

As for the divacancy-properties in body-centred cubic metals there are no data reported.

A reconsideration of the situation in face-centred cubic metals with regard to the ratio of the monovacancy-energies to the divacancy-energies, and assuming that the same ratios will hold for the body-centred cubic metals,

Table 16 Estimated values for the formation- and migration-energy of vacancies in b.c.c.-metals

Metals	Formation-energy $\Delta H_f^{\square}$		Migration-energy $\Delta H_m^{\square}$	
	(eV)	(kcal)	(eV)	(kcal)
Vanadium	1.91	44.2	1.05	24.3
Niobium	2.68	62.0	1.47	34.1
Tantalum	2.91	67.0	1.60	36.9
Chromium	1.45	33.6	0.80	18.5
Molybdenum	2.47	56.9	1.36	31.3
Tungsten	3.14	72.5	1.73	39.3
β-Titanium	1.76	40.8	0.97	22.4
β-Zirconium	2.27	52.4	1.25	28.9
β-Hafnium	2.41	55.8	1.33	30.7

enables an estimate to be made of the divacancy-properties in these metals. The ratio of the corresponding formation- and migration-energies for mono- and divacancies are shown in graphical form in Fig. 12 and Fig. 13.

From the diagrams the following correlations can be inferred:

$$\Delta H_f^{\square\square} = 1.88 \, \Delta H_f^{\square} \tag{35}$$

$$B^{\square\square} = 0.12 \, \Delta H_f^{\square} \tag{36}$$

$$\Delta H_m^{\square\square} = 0.68 \, \Delta H_m^{\square}. \tag{37}$$

While for the formation-energy of point-defects a very accurate relationship is found, the data for the migration-energy are scattered. Nevertheless, both these empirical relationships found for face-centred cubic metals will be used in estimating the corresponding data for body-centred cubic metals.

By combining the equations (33) and (34) with the equations (35), (36), and (37), the equations

$$\Delta H_f^{\square\square} = 0.677 \, \Delta H_{sub}^{298} \tag{38}$$

$$B^{\square\square} = 0.051 \, \Delta H_{sub}^{298} \tag{39}$$

$$\Delta H_m^{\square\square} = 0.136 \, \Delta H_{sub}^{298} \tag{40}$$

are obtained, which are valid for body-centred cubic metals. The divacancy-properties as estimated by this hypothesis are tabulated in Table 17.

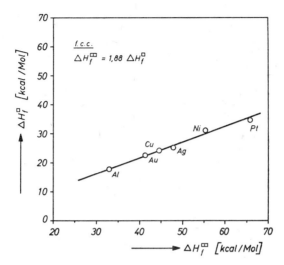

Figure 12   Ratio of mono- and divacancy formation-energy in f.c.c.-metals

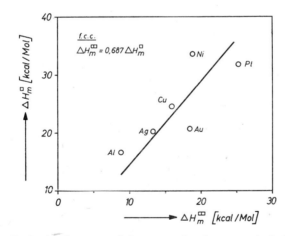

Figure 13   Ratio of mono- and divacancy migration-energy in f.c.c.-metals

Table 17 Estimated values for divacancy-properties in b.c.c.-metals

Metal	Formation-energy $\Delta H_f^{\square\square}$ (eV)	(kcal)	Binding-energy $B^{\square\square}$ (eV)	(kcal)	Migration-energy $\Delta H_m^{\square\square}$ (eV)	(kcal)
Vanadium	3.60	83.2	0.27	6.3	0.72	16.7
Niobium	5.05	116.6	0.38	8.8	1.01	23.4
Tantalum	5.47	126.3	0.41	9.5	1.10	25.4
Chromium	2.74	63.2	0.20	4.8	0.55	12.7
Molybdenum	4.64	107.1	0.35	8.1	0.93	21.5
Tungsten	5.91	136.5	0.45	10.3	1.19	27.4
β-Titanium	3.32	76.6	0.25	5.8	0.67	15.4
β-Zirconium	4.27	98.7	0.32	7.5	0.86	19.6
β-Hafnium	4.52	104.9	0.34	7.9	0.91	21.1

2.3 Vacancy-divacancy Mechanism of Diffusion in BCC-metals

The concept of a combined vacancy-divacancy mechanism of self-diffusion in body-centred cubic metals was first adopted for the interpretation of the diffusion behaviour of tungsten by Neumann and Hirschwald[66]. Later, it was applied by Neumann[67] and more extensively by Peart and Askill[68] to the discussion of the diffusion behaviour of body-centered cubic metals in general. The treatments, however, differ in some aspects. In the approach of Peart and Askill[68] the main points in the evaluation of the activation energy for the divacancy-mechanism according to the relation

$$\Delta H_{V2} = \Delta H_f^{\square\square} + \Delta H_m^{\square\square} \tag{32}$$

where

$$\Delta H_f^{\square\square} = 2\,\Delta H_f^{\square} - B^{\square\square} \tag{41}$$

are the following assumptions:

1 In the evaluation of the formation-energy of divacancies according to the above equation (41) the formation-energy of monovacancies can be inferred from the measured activation-energy of self-diffusion using the empirical relation

$$\Delta H_f^{\square} = k_1 \cdot \Delta H_{V1} \tag{42}$$

the constant k_1 being 0.6, and

2 The migration energies of monovacancies and divacancies are related by

$$\Delta H_m^{\square\square} = k_2 \cdot \Delta H_m \tag{43}$$

the constant k_2 being unity.

While the first assumption is sustained by the experimental data, the second assumption should be regarded with some doubts.

In several theoretical treatments by Damask, Dienes and Weizer[69], Schottky, Seeger and Schmid[70] and Johnson[71] it has been established that divacancies are more mobile than are monovacancies

$$\Delta H_{m\,\Box}^{\Box} < \Delta H_m^{\Box}. \tag{44}$$

This relation is valid at least for face-centred cubic metals, and there are no reasons to assume that the situation in body-centered cubic metals is not the same.

Thus the data of Peart *et al.*[68] for the activation energy of self-diffusion for the case in which a divacancy-mechanism is operating, have to be corrected with respect to this point. On the other hand the data of Neumann[67] have been obtained from an estimate of the properties of point-defects as evaluated in the preceeding part of this paper. Since this information is derived from experimental studies which are independent of the investigation of diffusion behaviour, these data seem to be more realistic.

Accordingly, the activation-energies of self-diffusion were estimated for the case of a monovacancy-mechanism and a divacancy-mechanism making use of the equations

$$\Delta H_{V1} = \Delta H_f^{\Box} + \Delta H_m^{\Box}$$

and

$$\Delta H_{V2} = \Delta H_f^{\Box} + \Delta H_m^{\Box} = 2\,\Delta H_f^{\Box} - B^{\Box} + \Delta H_m^{\Box}. \tag{32}$$

The resulting data are tabulated in Table 18.

These estimates may be refined to a certain degree if consideration is made of the ratio of monovacancies to divacancies.

Undoubtedly the concentration of monovacancies exceeds the concentration of divacancies in the diffusion specimens by a factor of almost 10. But, the diffusion process is occurring, due to their mobility, mainly by the migration of divacancies. As has been discussed by Schottky[72] for the case of the annealing of quenched-in point defects, the migration of divacancies is represented by an energy which is lower by the binding-energy of divacancies

$$\overline{\Delta H_m^{\Box}} = \Delta H_m^{\Box} - B^{\Box}. \tag{45}$$

The activation-energy of self-diffusion, when inserting (45) and (41) in (32), is then given by

$$\Delta H_{V2} = 2\,\Delta H_f^{\Box} + \Delta H_m^{\Box} - 2B^{\Box}. \tag{46}$$

Table 18 Activation energy for self-diffusion in b.c.c.-metals

Metal	Monovacancy-mechanism				Divacancy-mechanism					
	$\Delta H_{V1}=\Delta H_f^\square + \Delta H_m^\square$ calculated		$\Delta H_V^{exp.}$ experimental		$\Delta H_{V2}=2\Delta H_f^\square + \Delta H_m^{\square\square}-B^{\square\square}$	$\Delta H_{V2}=2\Delta H_f^\square + \Delta H_m^{\square\square}-2B^{\square\square}$ calculated			$\Delta H_V^{exp.}$ experimental	
	(eV)	(kcal)	(eV)	(kcal)	(eV)	(kcal)	(eV)	(kcal)	(eV)	(kcal)
Vanadium	2.97	68.6	3.19	73.8	4.32	99.9	4.05	93.6	4.10	94.9
Niobium	4.16	96.2	4.11	95.0	6.08	140.4	5.70	131.6	5.41	125.0
Tantalum	4.50	104.1	4.29	99.1	6.57	151.7	6.16	142.2	5.74	132.6
Chromium	2.26	52.1	3.18	73.6	3.28	75.9	3.08	71.1	–	–
Molybdenum	3.82	88.3	3.88	89.8	5.57	128.6	5.22	120.6	5.68	131.1
Tungsten	4.87	112.5	4.93	113.8	7.32	169.2	6.86	158.5	7.25	167.5
β-Titanium	2.74	63.2	2.60	60.0	3.99	92.2	3.73	86.2	–	–
β-Zirconium	3.52	81.3	2.67	61.6	5.13	118.5	4.81	111.1	–	–
β-Hafnium	3.74	86.5	2.96	68.4	5.45	126.0	5.11	118.1	–	–

This influence of the binding-energy is the result of local equilibrium between mono- and divacancies. An increase in temperature, which results in an increase in mobility of the divacancies gives rise at the same time to a certain dissociation of the divacancies, i.e. the increase in mobility of the divacancies is counteracted by the decrease of the divacancies due to dissociation.

The data of the diffusion-energies, corrected with respect to these considerations, are also tabulated in Table 18.

The agreement between calculated and experimentally observed values of the activation-energy of self-diffusion is reasonable if allowance is made for the assumptions applied. The only exception is chromium. The reason for disagreement in this case more probably arises from the experimental uncertainty in the diffusion data than from any real difference in physical behaviour. The coincidence of experimental activation energy for diffusion and that evaluated for divacancy-diffusion is possibly fortuitous.

Thus the present considerations and calculations show that it is reasonable to interpret the diffusion behaviour of body-centred cubic metals in the way as represented in Table 19.

The body-centred cubic elements of the Vth and VIth subgroup of the periodic table exhibit the same diffusion behaviour as do the face-centred cubic metals: A monovacancy-mechanism of diffusion operating in the low-temperature region and a divacancy-mechanism of diffusion operating in the high-temperature region.

On the other hand the elements of the IVth subgroup of the periodic table exhibit a so-called "anomalous" diffusion-behaviour, inasmuch as a mono-vacancy-mechanism of diffusion is operating in the high-temperature region leaving the diffusion mechanism in the low-temperature region open to discussion. There have been a considerable number of attempts to provide an interpretation for the diffusion behaviour of the latter metals. A detailed

Table 19   The probable diffusion mechanisms in b.c.c.-metals in the different temperature ranges

Metal	Low-temperature region	High-temperature region
Vanadium Niobium Tantalum Chromium Molybdenum Tungsten	Monovacancy-mechanism	Divacancy-mechanism
β-Titanium β-Zirconium β-Hafnium	Dislocation-mechanism	Monovacancy-mechanism

discussion of these was given at the Gatlinberg conference on diffusion in b.c.c.-metals[5,10,43]. Since the present paper is devoted mainly to the concept of a combined vacancy-divacancy mechanism of diffusion, there will be no further discussion of these other proposed models of diffusion in these systems. The three most striking features are however:

1   The values of the pre-exponential factor D_0 are very low, indicating a remarkable contribution of diffusion via dislocations and other short-circuit paths of diffusion in these metals.

2   The absolute values of the diffusion-coefficients in these systems are one or two orders of magnitude higher than those in normal metals and approach those for liquid metals.

3   There is little difference in the magnitude of the activation-energy of self-diffusion, in the low-temperature and in the high-temperature region.

According to Peart *et al.*[68] the diffusion behaviour of the metals of the IVth subgroup may be explained by a vacancy-divacancy mechanism too, taking into account an influence of short-circuit diffusion as a third contribution. Although the agreement of this approach with experimental values is fairly good, it seem that the diffusion behaviour in these systems is quite inexplicable on any simple model.

4   CONCLUSIONS

Although the results of the present concept of a combined vacancy-divacancy mechanism for the interpretation of the diffusion behaviour in body-centred cubic metals seem to be quite reasonable, one has to remember the uncertainties in the values of several of the parameters used in the estimations. In spite of the hypothetical character of the treatment it does provide some evidence for the reality of this diffusion model.

There is a considerable deal of theoretical and experimental work to be done on diffusion and on point-defect properties in body-centred cubic metals before a unique understanding of diffusion behaviour can be achieved. It should be emphasized that it is no longer sufficient to make measurements in limited temperature ranges but that they should be extended to regions just below the melting point of the metals in order to obtain meaningful diffusion parameters and internally consistent sets of point-defect properties.

ACKNOWLEDGMENT

I wish to thank Mr. W. Knatz for assistance in programming and computing the experimental data.

REFERENCES

1. G. V. Kidson, *Can. J. Phys.*, **41**, 1563 (1963).
2. T. S. Lundy and C. J McHargue, *Trans. AIME*, **233**, 243 (1965).
3. R. F. Peart, *J. Phys. Chem. Solids*, **26**, 1853 (1965) and in *Diffusion in Body-Centred Cubic Metals*, American Society for Metals, Metals Park, Ohio, 1965, p. 235.
4. R. P. Agarwala, S. P. Murarka and M. S. Anand, *Acta Met.*, **16**, 61 (1965).
5. D. Lazarus, in *Diffusion in Body-Centred Cubic Metals*, American Society for Metals, Metals Park, Ohio, 1965, p. 156.
6. R. Resnick and L. S. Castleman, *Trans. AIME*, **218**, 307 (1960).
7. R. F. Peart, D. Graham and D. H. Tomlin, *Acta Met.*, **10**, 519 (1962).
8. V. D. Lyubimov, P. V. Geld and G. P. Shveikin, *Izvest. Akad. Nauk SSSR, Met., Gorn Delo*, **5**, 137 (1964).
9. T. S. Lundy, F. R. Winslow, R. E. Pawel and C. J. McHargue, ORNL-3617 (1964) and *Trans. AIME*, **233**, 1533 (1965).
10. T. S. Lundy, J. I. Federer, R. E. Pawel and F. R. Winslow, in *Diffusion in Body-Centred Cubic Metals*, American Society for Metals, Metals Park, Ohio, 1965, p. 35.

11. D. B. Langmuir and R. L. Eager, *Phys. Rev.*, **89**, 911 (1953).

12. P. L. Gruzin and V. I. Meshkov, *Vorp. Fiz. Met.; Met. Akad. Nauk Ukr. SSR, Sbor Nauchn. Rabot*, **4**, 570 (1955), and AEC-tr-2926.

13. R. E. Pawel and T. S. Lundy, *J. Phys. Chem. Solids*, **26**, 937 (1965).

14. S. Z. Bokshtein, S. T. Kishkin and L. M. Moroz, *Zavod Lab.*, **23**, 316 (1957) and AEC-tr-4505.

15. H. W. Paxton and E. G. Gondolf, *Archiv Eisenhüttenwesen*, **30**, 55 (1959).

16. P. L. Gruzin, L. V. Pavlinov and A. D. Tyutyunnik, *Izvest. Akad. Nauk SSSR, Ser. Fiz.*, **5**, 155 (1959).

17 N. A. Bogdanov, *Izvest. Akad. Nauk SSSR, Met., Toplivo*, **3**, 95 (1960).

18. W. C. Hagel, *Trans. AIME*, **224**, 430 (1962).

19. L. I. Ivanov, M. P. Matveeva, V. A. Morozov and D. A. Prokoshkin, *Izvest. Akad. Nauk SSSR, Met., Toplivo*, **2**, 104 (1962).

20. J. Askill and D. H. Tomlin, *Phil. Mag.*, **11**, 467 (1965).

21. J. Askill, in *Diffusion in Body-Centred Metals*, American Society for Metals, Metals Park, Ohio, 1965, p. 253.

22. E. V. Borisov, P. L. Gruzin, L. V. Pavlinov and G. B. Fedorov, *Met. i Metalloved. Chist. Met.*, **1**, 213 (1959).

23. M. B. Bronfin, S. Z. Bokshtein and A. A. Zhukhovitskii, *Zavod. Lab.*, **26**, 828 (1960).

24. W. Danneberg and E. Krautz, *Z. Naturforsch.*, **16a**, 854 (1961).

25. J. Askill and D. H. Tomlin, *Phil. Mag.*, **8**, 997 (1963).

26. P. L. Pavlinov and V. N. Bykov, *Fiz. Met. i Metalloved.*, **18**, 459 (1964).

27. G. M. Neumann, 6th International Plansee Seminar "De Re Metallica", Reutte/Tirol, 1968.

28. V. P. Vasiliev and S. G. Thermomortohenko, *Zavod. Lab.*, **22**, 688 (1956).

29. W. Danneberg, *Metall*, **15**, 977 (1961).

30. R. L. Andelin, J. D. Knight and M. Kahn, USAEC-LA-2880 (1963) and *Trans. AIME*, **233**, 19 (1965).

31. R. E. Pawel and T. S. Lundy, *Acta Met.*, **17**, 979 (1969).

32. R. P. Elliot, AD 290 336 (1962).

33. J. F. Murdock, T. S. Lundy and E. E. Stansbury, *Acta Met.*, **12**, 1033 (1964).

34. C. M. Libanati and N. E. De Reca, in *Diffusion dans les solides*, Y. Adda and J. Philibert, Paris, 1965.

35. N. E. De Reca, E. Walsoë and C. M. Libanati, *Acta Met.*, **16**, 1297 (1968).

36. J. Askill and G. B. Gibbs, *Phys. Stat. Solidi*, **11**, 557 (1965).

37. P. L. Gruzin, V. S. Emelyanov, G. G. Ryabova, G. B. Fedorov, 2nd Conf. Peaceful Use Atomic Energy, Geneva, 1958, p. 187.

38. W. S. Lyashenko, W. J. Bykov and N. W. Pavlinov, *Fiz. Met. i Metalloved.*, **7**, 362 (1959).

39. G. B. Fedorov and V. D. Gulyakin, *Met. i Met. Chist. Met. Sb. Nauchn. Rabot*, **1**, 170 (1959).

40. D. Volokoff, S. May and Y. Adda, *Compt. rend.*, **251**, 2341 (1960).

41. G. V. Kidson and J. McGurn, *Can. J. Phys.*, **39**, 1146 (1961).

42. J. I. Federer and T. S. Lundy, ORNL-3339 (1962) and *Trans. AIME*, **227**, 592 (1963).

43. G. V. Kidson, in *Diffusion in Body-Centred Cubic Metals*, American Society for Metals, Metals Park, Ohio, 1965, p. 329.

44. F. R. Winslow and T. S. Lundy, *Trans. AIME*, **233**, 1790 (1965).
45. A. N. Nesmeyanov, *Vapour Pressure of the Chemical Elements*, North Holland, Amsterdam, 1963.
46. C. Zener, *J. Appl. Phys.*, **22**, 372 (1951).
47. K. Compaan and Y. Haven, *Trans. Faraday Soc.*, **52**, 786 (1956).
48. A. Seeger, *Moderne Probleme der Metallphysik*, Springer, Berlin, 1965.
49. A. Seeger, G. Schottky and D. Schumacher, *Phys. Stat. Solidi*, **11**, 363 (1965).
50. D. Schumacher, A. Seeger and O. Härlin, *Phys. Stat. Solidi*, **25**, 359 (1968).
51. A. Seeger and H. Mehrer, *Phys. Stat. Solidi*, **29**, 231 (1968).
52. T. G. Stoebe and H. I. Dawson, *Phys. Rev.*, **166**, 621 (1968).
53. H. Mehrer and A. Seeger, paper presented at this conference.
54. P. Reimers, *Metall*, **22**, 577 (1968).
55. G. Schottky, *Phys. Letters*, **12**, 95 (1964).
56. H. Schultz, *Mater. Science Eng.*, **3**, 189 (1968/69).
57. Y. A. Kraftmakher, *Fiz. Tverd. Tela*, **5**, 950 (1963) and *ibid.*, **6**, 503 (1964).
58. Y. A. Kraftmakher, *Zhur. priklad. Mekh. Tekh. Fiz.*, **2**, 158 (1962).
59. A. A. Johnson, *J. Less Common Metals*, **2**, 241 (1960).
60. J. B. Conway and R. A. Hein, *Advances in Thermophysical Properties*, 1965.
61. J. P. Meakin, A. Lawley and R. C. Koo, *Lattice Defects in Quenched Metals*, New York, 1965.
62. Y. A. Kraftmakher and P. G. Strelkov, *Fiz. Tverd. Tela*, **4**, 2271 (1962).
63. H. Schultz, *Z. Naturforsch.*, **14a**, 361 (1959).
64. V. D. Shestopal, *Fiz. Tverd. Tela*, **7**, 3461 (1965).
65. O. M. Kamel and Y. A. Kraftmakher, *Fiz. Tverd. Tela*, **8**, 283 (1966).
66. G. M. Neumann and W. Hirschwald, *Z. Naturforsch.*, **21a**, 812 (1966).
67. G. M. Neumann, *Z. Phys. Chem.*, *N.F.* **51**, 165 (1966).
68. R. F. Peart and J. Askill, *Phys. Stat. Solidi*, **23**, 263 (1967).
69. A. C. Damask, G. J. Dienes and V. G. Weizer, *Phys. Rev.*, **113**, 781 (1959).
70. G. Schottky, A. Seeger and G. Schmid, *Phys. Stat. Solidi*, **4**, 419 (1964).
71. R. A. Johnson, *Phys. Rev.*, **152**, 629 (1966).
72. G. Schottky, *Z. Phys.*, **160**, 16 (1960).

Solute diffusion in α-Zr

G. M. HOOD

Materials Science Branch, Chalk River Nuclear Laboratories
Atomic Energy of Canada Limited, Chalk River, Ontario, Canada

ABSTRACT

Measurements of diffusion of ^{64}Cu in single crystals of α-Zr have been made in the temperature interval 720 to 835°C. They have shown diffusion rates for Cu that are about four orders of magnitude faster than self diffusion. This result suggests that α-Zr may be one of a number of metals capable of exhibiting extremely fast interstitial-like solute diffusion. In addition, single diffusion measurements of [65]Zn, [110]Ag, [113]Sn and [198]Au in α-Zr single crystals have been made between 821 and 840°C. The results of these experiments, in combination with the results for ^{64}Cu, indicate that, in α-Zr, solutes of low valence and small "size" diffuse more rapidly than solutes of high valence and large "size".

1   INTRODUCTION

Mass transport measurements may provide useful information with regard to the development and application of metals and alloys. Zirconium and its alloys find widespread utilization in the nuclear industry, yet comparatively little is known about the kinetics and mechanisms of atom movement in the low temperature phases of these materials. In an effort to expand information in this particular area, an exploratory programme of solute diffusion studies in alpha-zirconium has been initiated. This report gives an account of the work, which is in its preliminary stages, and the results to date. A review of previous work in this field has been given by Kidson[1].

Zirconium exists in two crystallographic forms. The low temperature alpha phase, which has an h.c.p. structure, transforms to a b.c.c. beta phase at

865°C. The alpha phase of Zr is the predominant one for most of its applications.

A useful reference temperature for diffusion experiments in metals is the melting point. Estimates of the effective melting temperature (T_m) of α-Zr, based on thermodynamic properties, indicate a figure of approximately 1600°C[2,3]. In order to measure diffusion coefficients characteristic of lattice diffusion the experiments should be performed in the temperature region $T/T_m > (0.5–0.7)$. Below this range, effects due to grain boundaries and dislocations may become important[4]. The upper limit of T/T_m for α-Zr is defined by the transformation temperature and has a value of 0.60, if the figure of 1600°C is adopted for the melting point. The probability, therefore, of making diffusion measurements in α-Zr, which are characteristic of lattice diffusion, is rather low unless specimens free of grain boundaries and of low dislocation densities are used.

The object of the present work was to make diffusion measurements of several solutes in α-Zr in an attempt to study valence and "size" effects. Because of the relatively low temperature range available the measurements have been mainly restricted to temperatures close to the transformation temperature.

2 EXPERIMENTAL

2.1 Specimen Preparation

Specimens 1, 2, 5, 7, 8 and 9 were single crystals cut from large-grained bars of α-Zr. The grains, 4 to 6 mm in diam., were developed by subjecting the bars to a strain-anneal process[5]. A typical impurity analysis is shown in Table 1. The remaining samples were cut from a single crystal of unspecified

Table 1   Typical α-Zr single crystal analysis

Element	P.P.M.
C	30
O	160
Al	25
Si	60
Fe	280
Zn	50
Hf	60

impurity content received from Oak Ridge National Laboratories (ORNL). The diffusion samples were cut to size by spark cutting or with the use of an acid saw or fine cut-off wheel. Surfaces for isotope deposition, flat to within one micron, were prepared by metallographic techniques. Prior to formation of the isotope tracer layers, all surfaces were briefly chemically polished on a metallographic cloth using a solution of 50 parts H_2O, 50 parts HNO_3 and 0.2 parts HF by volume.

2.2 Isotope Deposition and Diffusion Conditions

The tracer isotopes ^{64}Cu, ^{65}Zn, ^{110}Ag and ^{198}Au were prepared by neutron irradiation of "Specpure" materials. This was carried out in the NRX irradiation facilities of Chalk River Nuclear Laboratories (CRNL). ^{113}Sn was prepared by irradiation of Sn enriched to 80% in ^{112}Sn received from ORNL. Specific activities of 10–100 µC/mg were employed.

The active deposits were formed in most cases by evaporating the tracerrs from conical tungsten baskets in a vacuum of 10^{-6} mm Hg. The deposits were typically 500 Å thick. The deposition areas on the samples were defined by Al foil masks.

Thin films of ^{64}Cu were formed by ion implantation in a number of samples. This method was employed in an attempt to avoid effects thought to be due either to tracer hold-up at the surface oxide layers of the specimens or to limited solubility of the solute[6]. (The maximum solubility of Cu in α-Zr appears to be ~0.1 wt.%[7].) Implantation of ~10^9 40 keV ^{64}Cu atoms per specimen was carried out in the CRNL mass separator. This technique allows the use of extremely low solute concentrations and has the added advantage that these atoms are largely implanted below the surface oxide layers of the specimens. A similar technique has been successfully applied to tracer diffusion studies in aluminium[6,8]. The initial penetration of tracer on implantation was expected to be slight compared to the mean tracer penetration after diffusion. This point has not been experimentally verified. A maximum range of 1 micron is estimated for the implanted ^{64}Cu ions in the $\langle 0001 \rangle$ direction.

The Zn diffusion experiment was carried out by sealing off a 0.2 mg piece of radioactive Zn and a Zr specimen in a quartz tube under a vacuum of 10^{-6} mm Hg. Zn diffusion would initially take place from the vapour phase. All the measurable ^{65}Zn activity was absorbed by the sample during the diffusion anneal.

The experiment to determine the diffusion coefficients of [110]Ag and [113]Sn was performed on one specimen. The tracer layer was formed by evaporation of equal quantities of radioactive Sn and Ag on to one surface of sample 8.

Both this and the other diffusion anneals were done in a He atmosphere with the sample contained in a Zr capsule inside a quartz furnace tube. The specimens were isolated from the capsule sides by a tungsten coil. Sample temperatures were held constant to within $\pm 1\,^\circ$C during the course of a diffusion anneal and were measured by two chromel–alumel thermocouples in contact with the capsule.

2.3 Sectioning

Following the diffusion periods the samples were fixed to brass holders with Torrseal resin. The sides of the samples were removed with a fine file to avoid possible contributions from surface diffusion. The specimens were sectioned with a lathe or with a Jung Model K microtome. With the latter instrument, sections two microns thick were removed each time. Specimen alignment on the microtome prior to sectioning was by an optical method. A schematic diagram of the apparatus is shown in Fig. 1. A supported

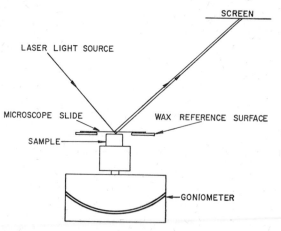

Figure 1 Schematic diagram of sample alignment procedure

reference surface of paraffin wax was formed around the sample in such a way that the latter could be oriented independently. The wax block, movable in the vertical direction, was cut by the microtome until a flat surface representing the cutting plane of the knife was obtained. A microscope slide was supported on this surface so that it also covered the adjacent specimen surface. A beam of laser light was then reflected off the slide and specimen surfaces on to a remote screen. The specimen orientation was adjusted until the images from both reflected beams appeared to coincide on the screen. Easy alignment of surfaces to within two parts in 10^4 was achieved in this manner.

The specimen sections were all weighed prior to analysis of their radio-activity. The weights were reproducible within ± 15 micrograms as determined on a Mettler microbalance. This corresponded to a weighing precision of about $\pm 5\%$ for individual two micron sections. An increase in accuracy corresponding to the number of sections included in each group resulted when several sections were analyzed together. Lathed sections 25 microns thick, or greater, were readily weighed to within a precision of 1%.

2.4 Activity Determinations

Radioactive assay of the gamma activity of individual sections or groups was performed with the use of a thallium-activated sodium iodide well-shaped crystal in conjunction with a 100 channel analyzer. The location of the desired gamma peaks was determined by counting radioactive standards. The activities were corrected for background contribution, normalized according to the weights of the samples and corrected for decay where necessary.

The beta activities of some of the ^{64}Cu-containing samples were counted in a Nuclear Unilux II liquid scintillation counter. For this procedure the samples were dissolved in 0.5 ml of a solution containing 50 parts H_2O, 50 parts HNO_3 and 10 parts HF by volume. The dissolution took place in a polythene counting vial. The total volume for each sample was made up to 20 ml by the addition of 4.5 ml of methanol and 15 ml of NE240 liquid scintillator. This latter method of counting was more efficient than gamma counting.

3 RESULTS

Generally, the penetration profiles appeared to follow a distribution according to the solution of the diffusion equation for an infinitely thin film of tracer diffusing into a semi-infinite cylinder. This solution is

$$C(x, t) = \frac{S_0}{\sqrt{\pi Dt}} \cdot \exp\left(-\frac{x^2}{4Dt}\right) \tag{1}$$

where $C(x, t)$ is the concentration of tracer at a depth, x, from the origin after diffusion for a period, t, S_0 is the original quantity of tracer at the surface and D is the tracer diffusion coefficient.

The penetration profiles for Zn, Ag and Au are shown in Fig. 2. These profiles appear to correspond to a distribution according to equation (1), although some uncertainty must exist in the cases of Ag and Au where the number of points is very limited.

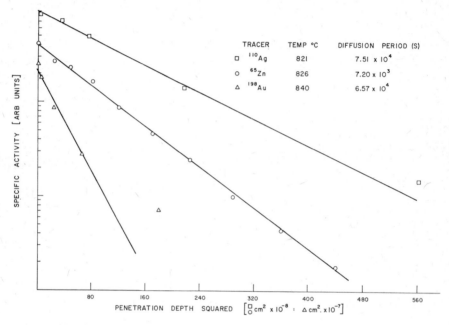

Figure 2   Tracer diffusion profiles in randomly oriented α-Zr single crystals

The Cu penetration profiles for specimens 3, 4 and 7, Figs. 3 and 4, do not obey equation (1). Possible origins of the curvature of those plots have been noted in section 2.2. Tracer loss during the diffusion anneals was slight, being never more than 10 % of the original deposit, and was not considered as a probable source of curvature. The diffusion coefficients calculated for specimens 3, 4 and 7 were based on the linear portion of the plots at deeper penetration. The D values obtained were consistent with the other Cu diffusion measurements. They could, however, be subject to appreciable error.

The penetration plots for specimens 5, Fig. 4, and 6, Fig. 5, where ion implantation was used, do not exhibit the extended curvature of samples 3,

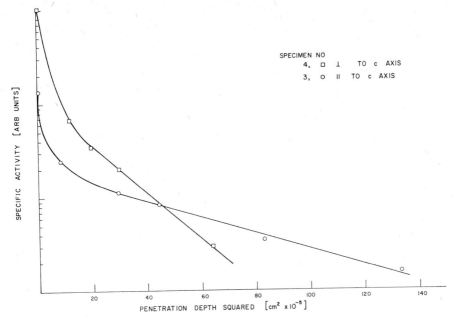

Figure 3 ⁶⁴Cu diffusion in specifically oriented α-Zr single crystals at 832 °C. Diffusion anneal interval = 3.6×10^3 s

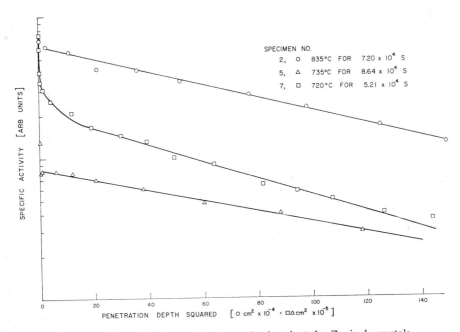

Figure 4 ⁶⁴Cu diffusion profiles in randomly oriented α-Zr single crystals

4 and 7. They do, however, exhibit initial high tracer concentrations. The experiment on specimen 6 was a composite one. The specimen was mounted at an angle on the target plate of the mass separator such that adjacent surfaces were presented at 45° to the ion beam. These surfaces were parallel and

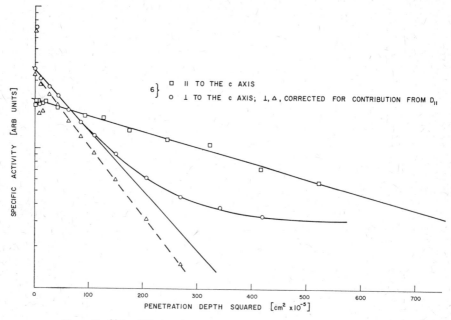

Figure 5 ^{64}Cu diffusion in a specifically oriented α-Zr single crystal at 735 °C. Diffusion anneal interval = 8.64×10^4 s

perpendicular respectively to the *c* axis. Due to the small size of the specimen it was not possible to mask off areas of each surface, hence all samples from the first series of sections, which were for diffusion perpendicular to the *c* axis, $D_\perp$, contained contributions from the adjacent surface. This almost certainly accounts for the curvature of the diffusion profile at deep penetration. By extrapolation of the curvature a constant contribution due to this overlapping effect was determined. From this estimated contribution the corrected profile for $D_\perp$, shown in Fig. 5, was obtained. The origin of the large oscillations in the first five points of the profile is not understood.

After completion of sectioning for $D_\perp$, the sample was remounted and sectioned to obtain the profile for diffusion parallel to the *c* axis, $D_\parallel$. It was considered that contributions to this latter profile from $D_\perp$ were not likely to be significant, since activity levels from this source after sectioning would be very low.

The activities of ^{110}Ag and ^{113}Sn from the sections of specimen number 8 were determined by analysis of their separate gamma peaks. No measurable penetration of Sn was found beyond the second section. This led to the estimated upper limit for Sn diffusion shown in Table 2.

Table 2 Results of tracer diffusion measurements in α-Zr single crystals

Solute	Zr specimen No.	Diffusion period (s)	Sample temperature (°C)	D (cm^2/s)
^{65}Zn	1	7.2×10^3	826	2.8×10^{-11}
^{64}Cu	2	7.2×10^4	835	3.3×10^{-8}
^{64}Cu$_\parallel$	3	3.6×10^3	832	3.4×10^{-8}
^{64}Cu$_\perp$	4			1.2×10^{-8}
^{64}Cu	5	8.64×10^4	735	3.4×10^{-9}
^{64}Cu$_\parallel$	6	8.64×10^4	735	1.13×10^{-8}
^{64}Cu$_\perp$		8.64×10^4	735	2.52×10^{-9}
^{64}Cu	7	5.21×10^4	720	3.5×10^{-9}
^{110}Ag	8	7.51×10^4	821	4×10^{-12}
^{113}Sn				$< 10^{-13}$
^{198}Au	9	6.57×10^4	840	1.3×10^{-11}

4 DISCUSSION

The results of the present measurements are summarized in Table 2 and Fig. 6. The representation of Zr self-diffusion is based mainly on the results of Flubacher[9].

The most significant feature of the results appears to be the very high rate of Cu diffusion compared to self-diffusion and the diffusion of other solutes. In this respect Zr may be compared with an increasing number of metals where diffusion of particular solutes, especially the noble metals, proceeds at anomalously high rates relative to self-diffusion. References to this work may be found in a recent paper by Dariel *et al.*[10].

In the present case, as in most other instances, an unrealistically high concentration of dislocations would be required to account for the fast diffusion rates in terms of dislocation enhancement. An alternative proposal, and one which seems to have been generally accepted, is that an interstitial contribution to diffusion may account for the observed phenomena. Kidson[11] has

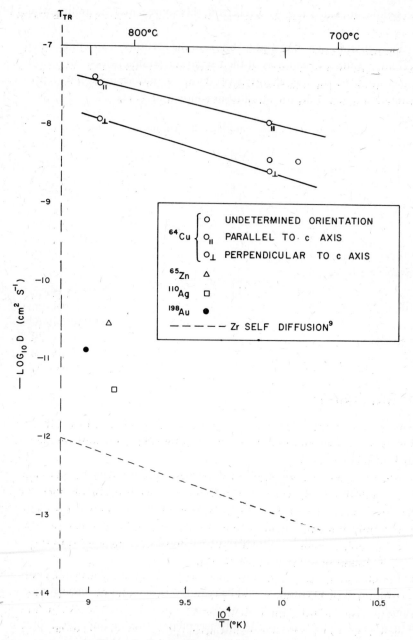

Figure 6   Tracer diffusion coefficients in α-Zr

given an account of the development of a diffusion equation which includes components due to both substitutional and interstitial diffusion.

The significance of atom and ion sizes with respect to interstitial formation has been emphasized by Dyson *et al.*[12] and Anthony and Turnbull[13]. In general the formation of interstitials appears to include the condition of little or no overlap between ion cores of solute and solvent. Estimates made in this way are liable to be somewhat crude in the absence of reliable figures for ion sizes in metals.

Such a calculation, based on the above condition, indicates the α-Zr lattice to be capable of accepting interstitial solutes of ionic radii less than 1.2 Å and 1.5 Å in the tetrahedral and octahedral sites respectively. The latter figures rest on the assumption that the effective Zr ion radius is 0.80 Å which corresponds to the ionic radius of Zr^{4+}. The corresponding radii for the Cu^{+1}, Ag^{+1} and Au^{+1} ions are 0.96, 1.26 and 1.37 Å respectively.

Thus, on the basis of size considerations, Cu might be expected to diffuse more readily via an interstitial mechanism than either Ag or Au. A further geometrical calculation indicates that to avoid ion core overlap at the saddle-point, the radii of the diffusing ions should be < 1.07 Å for interstitial jumps parallel to the *c* axis and < 1.05 Å for jumps with directional components perpendicular to the *c* axis. (The saddle-points are assumed to be the circumcentres of the triangular faces of the interstitial sites.) This property might be expected to lead to easier and probably faster diffusion parallel to the *c* axis than perpendicular to it, a speculation which is borne out by the experimental results on Cu. This anisotropy arises from the α-Zr lattice being somewhat "squashed" with a *c/a* ratio of 1.59.

The experimental observations on the diffusion of Cu, Ag and Au appear to be consistent with the above remarks. Apart from this evidence of an apparent "size" effect, a comparison of the slower diffusion of Zn compared to Cu, and Sn compared to Ag, suggests that valence is also an important factor in solute diffusion in α-Zr, although much more experimental work is required before any definite conclusions may be reached.

ACKNOWLEDGEMENTS

I would like to thank Mr. R. J. Schultz for his able assistance with much of the experimental work. The contributions of Dr. D. Santry and Mr. O. Westcott with regard to the ^{64}Cu implantation work and of Miss S. Collins for supplying x-ray photographs, are gratefully acknowledged. Finally thanks are due to Dr. T. A. Eastwood for a careful reading of the manuscript.

Diffusion in metals

REFERENCES

1. G.V.Kidson, *Electrochem. Tech.*, **4**, 193 (1966).
2. L.Kaufman, *Acta Met.*, **7**, 575 (1959).
3. A.Ardell, *Acta Met.*, **11**, 591 (1963).
4. P.J.Shewmon, *Diffusion in Solids* (McGraw-Hill Book Co., N.Y., 1963).
5. J.I.Dickson, University of Toronto, Ph.D. Thesis (1969).
6. G.M.Hood, *Phil. Mag.* (to be published).
7. C.E.Lundin, D.J.McPherson and M.Hansen, *J. Metals*, **5**, 273 (1953).
8. G.M.Hood, *Bull. Am. Phys. Soc.*, **13**, 487 (1968).
9. P.Flubacher, *Selbst-Diffusionsversuche in α-Zirkon*, E.I.R.-Bericht Nr. 49 (1963).
10. M.P.Dariel, G.Erez and G.Schmidt, *J. Appl. Phys.*, **40**, 2746 (1969).
11. G.V.Kidson, *Phil. Mag.*, **13**, 247 (1966).
12. B.F.Dyson, T.R.Anthony and D.Turnbull, *J. Appl. Phys.*, **37**, 2370 (1966).
13. T.R.Anthony and D.Turnbull, *Phys. Rev.*, **151**, 495 (1966).

Diffusion processes in partially ordered alloys

J. KUČERA and B. MILLION

Institute of Metallurgy of Czech. Acad. Sci. Brno, Žižkova 22

INTRODUCTION

In this paper we discuss the results concerning diffusion in Fe–Ni and Co–Ni systems from the point of view of the Cohen-Fine model[6] for short range order. The purpose of this discussion is to explain the self-diffusion behaviour of these alloys near the composition Ni_3Fe and Ni_3Co in the temperature region above T_C where non-zero short range order exists; T_C being the critical temperature of the order-disorder transformation.

RESULTS AND DISCUSSION

The self-diffusion of Ni in Fe–Ni systems has been measured by Walsoe de Reca and Pampillo[1] and that of Co in Ni–Co systems by us[2]. Some of the experimental results of these measurements are collected in Figs. 1 and 2 which show plots of log D vs. atomic concentration at different temperatures $(>T_C)$. For Ni_3Fe, $T_C = 776°K$. In the system Ni_3Co the order-disorder transformation has not yet been observed directly, however, there are many indirect indications that an ordered structure exists in this alloy too. These indications have been obtained from experiments involving positron annihilation measurements in Ni–Co alloys by Cizek et al.[3], tensimetric measurements of thermodynamic properties by Vrestal and Kučera[4] and calorimetric measurements by Velisek and Vrestal[5]; the last-named authors claim

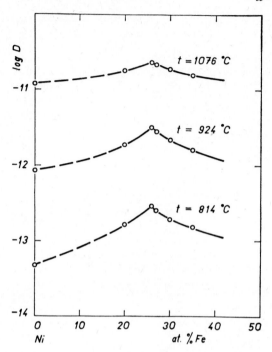

Figure 1   Concentration dependence of $\log D$ at $1076\,°C$, $924\,°C$, and $814\,°C$ in Ni–Fe alloys[1]

the critical temperature of the order–disorder transformation in Ni_3Co alloy to be $T_C \simeq 1033\,°K$.

Particularly noticeable in Fig. 1 and Fig. 2 are the Λ peaks. These occur at the AB_3 compositions and increase in size with decreasing temperature. In accord with the Λ-peaks in Figs. 1 and 2 is the appearance of minima or V-peaks near the composition AB_3 on the concentration dependence curve for ΔH and for $\log D_0$ as shown in the Fig. 3.

The existence of Λ and V-peaks can be explained on the basis of the Cohen-Fine assumptions[6] concerning alloy structures in which short range order exists and by using data obtained from the known results of Kuper, Lazarus and Manning[7] on diffusion in ordered CuZn alloys and Benzi and Gasparrini[8] on diffusion in ordered $AuCu_3$ alloys.

According to Cohen and Fine, short range order can be considered as a formation of ordered zones separated by a disordered matrix. The diffusion coefficient in the ordered zones varies (in the first approximation, if we neglect its long range order parameter dependence) according to an Arrhenius equation, the characteristics $D_{0_{(ord)}}$ and $\Delta H_{(ord)}$ of which are measured at temperatures $T < T_C$. In the disordered matrix, the temperature dependence of the diffusion coefficient is described by a second Arrhenius equation using the characteristics $D_{0_{(dis)}}$ and $\Delta H_{(dis)}$ which can be obtained by linear inter-

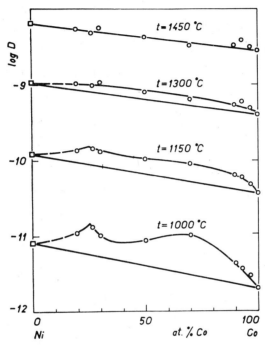

Figure 2   Concentration dependence of $\log D$ at $1450\,°C$, $1300\,°C$, $1150\,°C$ and $1000\,°C$ in Ni–Co alloys[2]

polation between the values characteristic of the components A or B resp. (Fig. 2). The accuracy of these values can be checked by measuring the concentration dependence of the diffusion coefficient D at sufficiently high temperatures for which $\log D$ is a linear function of concentration (see Fig. 2). At these temperatures $D_{(dis)} = D$ (Fig. 2). In the temperature range between T_C and the temperature at which only a disordered structure exists, the measured effective values of D also fit the Arrhenius dependence as well. An example of the temperature dependence of $D_{(ord)}$, $D_{(dis)}$ and D is given in Fig. 4, which has been drawn using Benci and his co-workers' data[9] for D,

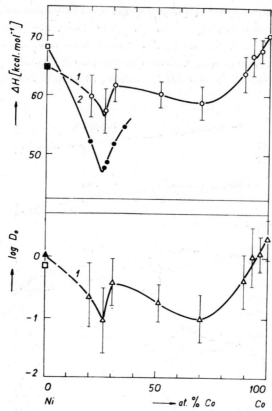

Figure 3 Concentration dependence of the activation enthalpy ΔH, and the frequency factor D_0 (1—Co in Ni–Co alloys, 2—Ni in Ni–Fe alloys[1]

Benci and Gasparrini's data[8] for $D_{(ord)}$ and Martin and co-workers'[10] and Kurtz and co-workers'[11] data for $D_{(dis)}$.

The relation between the quantities D, $D_{(ord)}$ and $D_{(dis)}$ can be obtained by using the first Fick law under the assumption that in an arbitrary specimen section perpendicular to the direction of diffusion flux I, the ratio of areas belonging to regions with ordered or disordered structure $S_{(ord)}/S_{(dis)}$ is constant. Then

$$S_{(ord)} \cdot I_{(ord)} = -S_{(ord)} \cdot D_{(ord)} \cdot \frac{\partial c}{\partial x} \qquad (1)$$

$$S_{(dis)} \cdot I_{(dis)} = -S_{(dis)} \cdot D_{(dis)} \cdot \frac{\partial c}{\partial x} \qquad (2)$$

Similarly the effective diffusion coefficient is given by the equation

$$S \cdot I = -S \cdot D \cdot \frac{\partial c}{\partial x} \tag{3}$$

where the total cross section of the specimen $S = S_{(ord)} + S_{(dis)}$. From the equations (1), (2), (3) we get

$$S \cdot D = S_{(ord)} \cdot D_{(ord)} + S_{(dis)} \cdot D_{(dis)}. \tag{4}$$

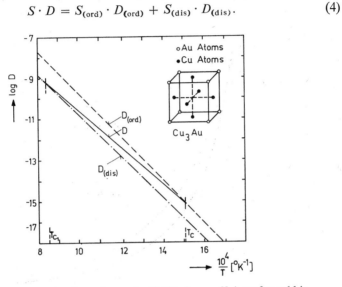

Figure 4 Temperature dependence of self-diffusion coefficients for gold in Cu₃Au alloys. $D = 0.0065 \exp(-38{,}250/RT)$; $D_{(ord)} = 3.15 \exp(-46{,}400/RT)$; $D_{(dis)} = 0.125 \exp(-45{,}200/RT)$

Expressing equation (4) in terms of the ordered and disordered phase volumes $V_{(ord)}$ and $V_{(dis)}$ we get

$$D = (V_{(ord)}/V) \cdot D_{(ord)} + (V_{(dis)}/V) \cdot D_{(dis)} \tag{5}$$

where the total specimen volume $V = V_{(ord)} + V_{(dis)}$. Introducing an order parameter by the relation $\sigma = V_{(ord)}/V$ which is quite analogous to the relation for the short range order parameter, we get

$$\sigma = \frac{D - D_{(dis)}}{D_{(ord)} - D_{(dis)}}. \tag{6}$$

The temperature dependence of σ for the Cu₃Au alloy is given in Fig. 5.

The value $\sigma = 1$ at the critical temperature is given by the assumption $D = D_{(ord)}$ at this temperature. This assumption is based on the results of measurements of self-diffusion coefficients in ordered and disordered β-brass solid solutions which have been carried out by A. B. Kuper and co-workers[7].

The temperature dependence of σ for the Ni$_3$Co alloy was calculated by using the very simplifying assumption: $\Delta H_{(ord)} = \Delta H_{(dis)}$ under the condition that $T_c = 1033\,°$K. The result of this calculation is also given in Fig. 5.

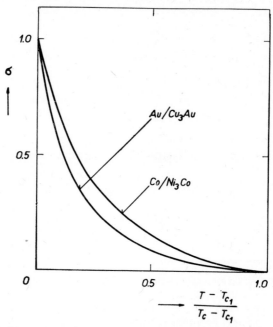

Figure 5   Temperature dependences of the order parameter calculated from diffusion characteristics for Cu$_3$Au and Ni$_3$Co alloys

On the basis of the mentioned supposition we are then able to explain the the existence of Λ-peaks near Ni$_3$Co and Ni$_3$Fe on the concentration dependence of $\log D$ as follows: if we measure the temperature dependence of the self-diffusion coefficient at the concentration corresponding to the composition AB_3 in the temperature range of partially ordered structure then the diffusion coefficient decreases with decreasing temperature more slowly than at any other concentration near AB_3. Therefore, at temperatures sufficiently close to T_c, a maximum on the concentration dependence of $\log D$ must appear at 75 at % Ni. The physical basis of this phenomenon relies on the maximum increase of the short range order parameter with decreasing temperature at this concentration. The appearance of a minimum in the concentration dependence of the activation enthalpy ΔH or of the frequency factor $\log D_0$, is naturally caused by the existence of a maximum in the concentration dependence of $\log D$.

REFERENCES

1. E. Walsoe de Reca and C. Pampillo, *Acta Met.*, **15**, 1263 (1967).
2. B. Million and J. Kučera, *Acta Met.*, **17**, 339 (1969).
3. A. Cizek, F. Parizek, J. Adam, I. Ja. Dechtjar and V. S. Michalenkov, *Czech. J. Phys.*, **B19**, 629 (1969).
4. J. Vrestal and J. Kučera, *Trans. AIME*, **245**, 1891 (1969).
5. J. Velisek and J. Vrestal, *Czech. J. Phys.*, **B16**, 937 (1966).
6. J. B. Cohen and M. E. Fine, *J. Phys. Radium, Paris*, **23**, 749 (1962).
7. A. B. Kuper, D. Lazarus, J. R. Manning and C. T. Tomizuka, *Phys. Rev.*, **104**, 1536 (1956).
8. S. Benci and G. Gasparrini, *J. Phys. Chem. Solids*, **27**, 1035 (1966).
9. S. Benci, G. Gasparrini, G. Germagnoli and G. Schianchi, *J. Phys. Chem. Solids*, **26**, 687 (1965).
10. A. B. Martin, R. D. Johnson and F. Asaro, *J. Appl. Phys.*, **25**, 364 (1954).
11. A. D. Kurtz, B. L. Averbach and M. Cohen, *Acta Met.*, **3**, 442 (1955).

4.7

On high surface diffusivities of solid metals
in the presence of adsorbed metallic vapours

J. HENRION and G. E. RHEAD

Université de Paris, Laboratoire de Chimie Appliquée, Paris, France

ABSTRACT

Self-diffusion coefficients as high as 10^{-1} to 1 cm^2 sec^{-1} have been measured for various metal surfaces in the presence of metallic vapours. Data are presented in particular for copper and gold with lead vapour, copper with bismuth and copper with thallium. Comparison is made with previous results for silver with adsorbed sulphur. The coefficients obtained are higher than those of the clean metal surfaces by as much as 10^4.

The diffusion coefficients (D_s) are obtained by measurements of twin boundary groove growth under conditions for which surface diffusion is the dominant growth mechanism. The temperature range of the experiments is $0.7T_m$ to T_m, the melting point temperature of the bulk substrate. The vapour is generated by a metal sample at a temperature lower than that of the diffusion specimen—in this way bulk condensation of the vapour is avoided. It is believed that the vapour, transported through an inert gas, is adsorbed in a single atomic layer which in some cases can form a mixed surface layer that is a true two-dimensional alloy.

Gjostein has shown that D_s values for several different metals can be normalized to the same curve if the diffusivities are plotted against T_m/T. At the melting point $D_s \cong 3 \times 10^{-4}$ cm^2 sec^{-1}. It is shown that the present results also fit this basic curve if the melting point is chosen to be not that of the metal substrate but that of a two-dimensional alloy with properties close to those of the corresponding three-dimensional alloy. D_s values higher than 3×10^{-4} cm^2 sec^{-1} can then be interpreted in terms of diffusion in a *two-dimensional liquid*.

INTRODUCTION

The study of surface self-diffusion began a decade ago when Mullins[1] formulated the phenomenological theory of mass transport along a solid surface. It then became possible, from measurements of changes in surface topography by means of interference microscopy, to deduce values of the surface self-diffusion coefficient.

The first results, as reviewed by Gjostein[2] and by Blakely[3], appeared to lack coherence. The data for different metals gave activation energies and pre-exponential factors covering very wide ranges. It was felt that many uncertainties arose from a lack of definition of the experimental conditions, in particular surface cleanliness.

Subsequent work has followed two different approaches: (i) studies of surfaces under ultrahigh vacuum conditions and with monitoring of surface cleanliness, for example by low energy electron diffraction (LEED), (ii) studies of the effect of a known adsorbate at a well defined chemical potential.

In the field of adsorption studies, investigations have been made of the effects of oxygen on copper[4], oxygen on silver[5], sulphur on copper[6] and sulphur on silver[7]. In the case of silver with adsorbed sulphur, the presence of the adsorbate was found to increase the diffusion coefficient by as much as 10^4. In order to extend this type of study to other adsorbates, we have undertaken a series of measurements of surface diffusion on f.c.c. metals in the presence of metallic vapours capable of producing adsorbed layers. We were guided in this direction by a study of Baily and Watkins[8] who showed from measurements of surface energy that an adsorbed monolayer of lead forms on copper in lead vapour. We have examined surface diffusion in this system and in the systems gold–lead, copper–bismuth and copper–thallium. We have found that these binary systems are characterized by very high diffusivities which are difficult to interpret in terms of current classical theories of the diffusion mechanism.

THEORY OF THE EXPERIMENTAL METHOD

A crystal at high temperature ($\sim 0.5T_m$ to T_m, the absolute melting point) tends to modify its surface topography in order to minimize the total surface energy. Changes in surface profile can be brought about by various processes: evaporation and condensation, volume diffusion and surface diffusion. The scaling laws due to Herring[9] show that if the transport distance is small enough then surface diffusion is always the dominant mechanism. In practice, it has been found that surface diffusion generally dominates for transport distances of several tens of microns.

Several reviews[10,2,3] describe in detail the theory of the various experimental methods that are available for determining the diffusion coefficient. We will outline the theory of grain boundary groove growth which forms the basis of our measurements.

At the junction of a grain boundary with a free surface the minimization of the total surface energy and grain boundary energy results in the formation of a groove with the necessary equilibrium dihedral angle at the root. The interferogram of Fig. 1 shows the shape of a groove formed by surface diffusion; it is characterized by two prominent humps on both sides of the groove. The groove width W (distance between the humps) is given theoretically[1,10] by the expression:

$$W = 4.6 \, (Bt)^{1/4} \qquad (1)$$

where

$$B = \frac{Ds \, (\gamma + \gamma'') \, \Omega^2 v}{kT} \qquad (2)$$

D_s, the surface self-diffusion coefficient,

γ, the surface energy,

γ'', the second derivative of the surface energy with respect to crystalline orientation,

Ω, the atomic volume,

v, the surface density of atoms,

kT, the thermal energy and

t, the duration of groove growth.

Groove growth by evaporation and condensation gives $t^{1/2}$ kinetics while volume diffusion gives $t^{1/3}$ kinetics[1,10]. It is thus possible to distinguish between the different growth processes. The measurements that we have made, together with calculations using known volume diffusion coefficients and vapour pressure data, show that in all the observations here reported surface diffusion was the dominant process for groove growth. The contributions from other mechanisms never amounted to more than the experimental error in W.

A difficulty encountered in this and in most studies of surface diffusion by mass transport techniques is the uncertainty of the values to be taken

Diffusion in metals

for γ and γ''. We have based our calculations on the best available data for γ and have neglected γ''. It is therefore possible that the results contain systematic errors that could amount to as much as a factor of two in each value of D_s.

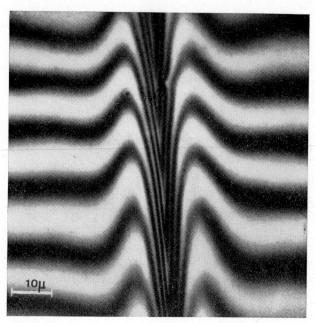

Figure 1 Interferogram of a typical grain boundary groove showing prominent humps due to the surface diffusion growth mechanism

EXPERIMENTAL

The polycrystalline specimens of copper (ASARCO 99.999 %) were cut to the dimensions $8 \times 5 \times 2$ mm approximately. After a mechanical polish with wet silicon carbide papers and electrolytic polishing in an orthophosphoric acid bath the specimens were recrystallized by annealing in purified hydrogen (palladium leak) at 900 °C for 60 hours. The grains thus obtained had an average diameter of about 1 mm. The specimens were again polished electrolytically until the smooth surface showed no trace of grain boundary grooves when examined under an interference microscope (Baker-Mirau type objective).

After this preparation the specimens were treated at a temperature between 700 and 1050°C in an atmosphere of hydrogen and metallic vapour at a total pressure slightly above atmospheric. This treatment was carried out under a constant flow of gas (~ 1 cm^3/sec) in a silica furnace tube with the specimen in a silica furnace boat. The metallic vapour was generated by a specimen of the pure metal (99.999% lead, bismuth or thallium from Johnson Matthey) in close proximity to the copper specimen. The silica furnace boats containing the two specimens were attached to silica rods ending in iron slugs in the cold part of the closed system. The positions of the specimens in the furnace could thus be adjusted by means of external magnets. The vapour generating specimen was always maintained at a temperature lower than that of the diffusion specimen in order to avoid bulk condensation.

For the experiments with gold, the specimens were cut to approximately $3 \times 3 \times 0.25$ mm from a foil of Johnson Matthey 99.999% gold. They were polished electrolytically in a cold chromium trioxide and acetic acid bath for 2 hours. After the preliminary annealing in hydrogen it was found impossible to remove the traces of grain boundary grooves by polishing. For this metal the diffusion coefficients were therefore evaluated from measurements of the difference in groove width before and after the treatment in lead vapour.

The treatments in metallic vapour were for periods from 10 minutes up to several days, depending on the specimen temperature and on the metal vapour pressure. At the end of the experiment the vapour generating specimen was rapidly displaced to a cold part of the system and the diffusion specimen then rapidly quenched. The surface was then examined under the interference microscope and groove widths were measured by means of an eyepiece graticule. A mean value was obtained from measurements of 20 grooves.

RESULTS

1 Copper in Lead Vapour

For this system a detailed study was made of the kinetics of groove growth at several temperatures. A certain anomaly appeared from the beginning of this study: the grooves at grain boundaries are generally much deeper and wider than those at coherent twin boundaries. It is thought that this difference is probably due to diffusion of lead into the grain boundary itself—an effect that may be reasonably excluded in the case of a coherent twin boundary. In the theoretical model adopted by Mullins[1,10], the boundary itself is considered to be inactive in the diffusion processes; we have therefore made measurements only of twin boundary groove growth throughout this work.

The results, presented in Fig. 2, were obtained with the lead specimen at a temperature only 3 to 5°C lower than the copper (the smallest difference found feasible without risking bulk condensation of lead). Under these con-

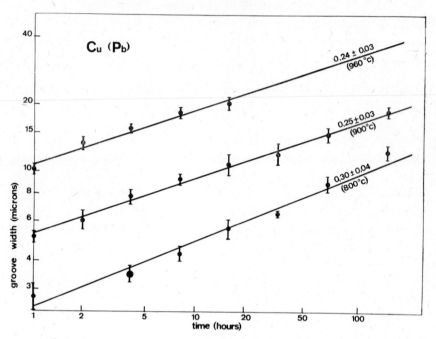

Figure 2 Logarithmic plot of twin boundary groove kinetics for copper in lead vapour. Least squares slopes

ditions the grooves exhibit the fastest growth rate: lowering the temperature of the lead decreases this rate. It is seen that the growth kinetics are satisfactorily close to the $t^{1/4}$ law of equation (1). The slightly higher exponent at 800°C can be attributed to a small contribution from volume diffusion. A significant contribution from transport of copper through the vapour phase can be excluded on the grounds that prominent humps were always observed in the groove profile: the mechanism of evaporation and condensation does not give humps[1].

In order to evaluate the surface diffusion coefficient from the groove width, it is necessary to know the surface energy. For temperatures below 900°C, we have referred to the results of Bailey and Watkins[8] for this system. These authors found values of γ which varied from 730 ergs cm^{-2} at 750°C to 810 ergs cm^{-2} at 900°C. To supplement this data, we have studied the variation of surface energy, over a wider range of temperature, by measuring the dihedral groove angles at twin boundaries. By using an analysis due to Mykura[11,12] it was then possible to find the ratio of the surface energy γ_S to the twin boundary energy γ_T. Between 700° and 900°C γ_S/γ_T increases with temperature. If one assumes a constant value of γ_T this behaviour corresponds to a surface with an adsorbed layer. Above about 920°C, the average

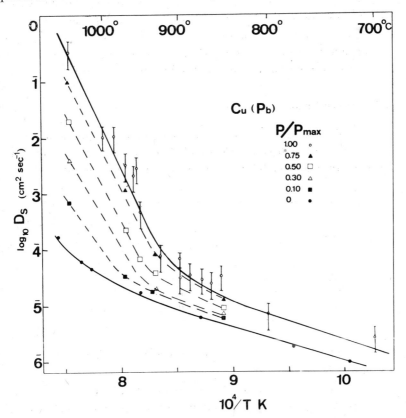

Figure 3 Temperature dependence of surface diffusion of copper at different values of lead vapour pressure. (Lower curve for copper in hydrogen due to Gjostein[13])

value of γ_S/γ_T was found to decrease from 250 to 140 which, when combined with Bailey and Watkins results[8], represents a change from 810 ergs cm^{-2} to about 500 ergs cm^{-2}. We have taken this lower value for temperatures above 900 °C.

The surface self-diffusion coefficients found for this system are plotted in Fig. 3. For conditions other than those represented in Fig. 2 the results were obtained from measurements of groove widths after a single anneal*. The ratio P/P_{max} is the ratio of the vapour pressure of lead under the experimental conditions—as determined by the temperature of the sample—to the maximum pressure when the lead was at (nearly) the same temperature as the copper. The lowest curve corresponds to the values obtained by Gjostein[13] for copper in hydrogen. For the surface energies at different values of P/P_{max} we have assumed a linear variation between 1670 ergs cm^{-2} for copper in hydrogen[13] and the values taken for $P/P_{max} = 1$. The results show that the presence of the vapour has a remarkable effect on surface diffusion.

2 Gold in Lead Vapour

A systematic study of the effect of lead vapour on gold was found to be very difficult. Some very high values for the diffusion coefficient were recorded (Table 1), but the reproducibility was poor. It is probable that the solubility

Table 1 Results for gold in lead vapour (maximum D_s values observed)

Temperature °C	P/P_{max}	D_s cm^2 sec^{-1} [a]	Pure gold[14] D_s cm^2 sec^{-1}
853	0.43	4.2×10^{-1}	9.0×10^{-6}
800	0.74	1.6×10^{-2}	2.2×10^{-6}
760	0.60	3.2×10^{-3}	9.0×10^{-7}
705	0.63	1.1×10^{-4}	1.8×10^{-7}

[a] These values are calculated on the assumption that $\gamma = 500$ ergs cm^{-2}

of lead in gold is the cause of these experimental difficulties. This solubility is high, $\sim 0.1 \%$, between 700 °C and 900 °C but practically zero below 400 °C. At the end of many experiments, especially those at high vapour

* This method was also adopted for the other systems.

pressures, a layer of lead covered the gold surfaces and made measurements impossible. We believe that this lead segregated to the surface from the bulk during quenching.

3   Copper in Bismuth Vapour

The results obtained for this system are shown in Fig. 4. A study was made, at 900°C, of the variation of D_s with the bismuth vapour pressure. A very rapid increase occurs.

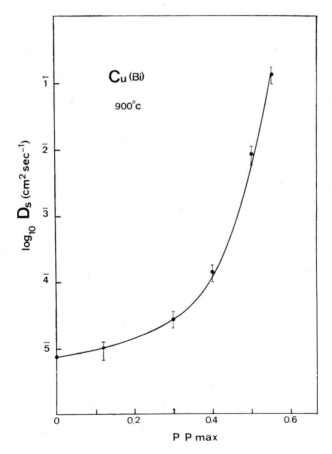

Figure 4   Surface diffusion of copper at 900°C as a function of the vapour pressure of bismuth

For high vapour pressures, $P/P_{max} > 0.55$, the results were difficult to reproduce: as in the case of gold with lead the bismuth frequently appeared in bulk on the surface at the end of the experiment.

No data is available for the surface energy of copper with adsorbed bismuth. For the calculations of D_s we have assumed a linear variation of 1670 to 500 ergs cm^{-2} between $P/P_{max} = 0$ to 0.5.

4 Copper in Thallium Vapour

For this system, the diffusion coefficients were measured for $P/P_{max} = 1$ (small temperature difference between the two samples). The results are shown in Fig. 5 together with Gjostein's results[13] for copper in hydrogen.

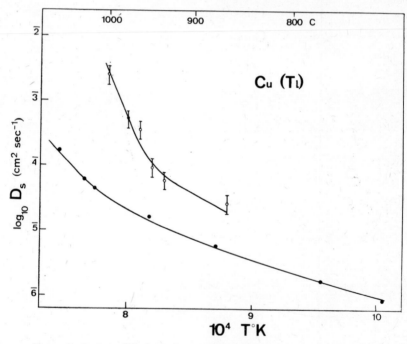

Figure 5 Surface diffusion of copper in thallium vapour ($P/P_{max} = 1$). (Lower curve for copper in hydrogen due to Gjostein[13])

As there are no available data on surface energy we have assumed a value of 500 ergs cm^{-2}. The results for this system are very similar to those for copper with lead.

DISCUSSION

A preliminary discussion of some of the results presented here has already been made[15,16]. We believe that the high values of D_s, some of which are even higher than diffusivities recorded for gases, are evidence of the existence of two-dimensional liquids. A brief review of current theories of surface diffusion is necessary in order to clarify this conclusion.

Gjostein[14] has made an important rationalization of some of the data for clean surfaces by plotting D_s as a function of T_m/T, T_m being the absolute melting point temperature and T the temperature of the experiment. The results obtained for several metals—Cu, Au, Ag, Ni, Fe—all fall close to the same curve. This curve is reproduced in Fig. 6. It is characterized by the value $D_s \simeq 3 \times 10^{-4}$ cm² sec⁻¹ at the melting point and the apparent activation energies $Q_s = 30 T_m$ cal/mole at high temperatures and $Q_s = 13 T_m$ cal/mole at low temperatures.

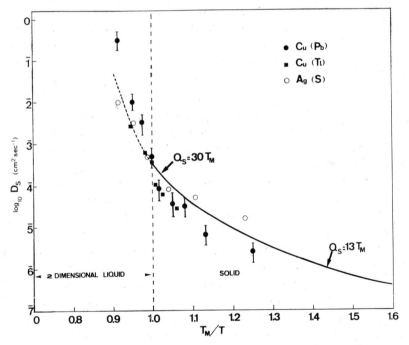

Figure 6 Surface diffusion data as a function of Tm/T. Full curve—base curve due to Gjostein[14]; Dashed curve—extension into the two dimensional liquid region. Results for surfaces with adsorbed layers. (For clarity the errors are omitted from two sets of points.)

Gjostein[14] has considered the interpretation of these two different activation energies by mechanisms of adatom (self-adsorbed surface atom) and surface vacancy migration respectively. Models have been set up to determine the concentrations of these diffusing defects as well as their energies of formation and migration. Later work[17,18] has suggested that the atom/vacancy hypothesis may not be satisfactory and that it is necessary to consider the effect of cooperative surface roughening as well as possible contributions from the migration of diadatoms and divacancies.

We believe that the high D_s values reported here cannot be interpreted in terms of the above models. Moreover, the current models do not explain an important feature of the data for clean surfaces: the constant value of D_s at the melting point. This value, 3×10^{-4} cm^2 sec^{-1}, can be understood in terms of the Lindemann theory of melting (see, for example, Gilvarry[19]). Melting is believed to occur when the amplitude of thermal vibration reaches a critical value. Thus, at the melting point is may be possible that each surface atom makes a diffusive jump for *each vibration*. Substitution into the random walk expression $D_s = a^2/4\tau$ of the value ~ 3 Å for the jump distance a, and a value of $1/\tau$ equal to a Debye frequency $\sim 10^{12}$ sec^{-1}, gives $D_s \sim 3 \times 10^{-4}$ cm^2 sec^{-1}. A fuller discussion of this point is given elsewhere[16].

It follows from the above argument that if we are to attempt to explain our results in terms of Gjostein's rationalization of the data, then D_s values above 3×10^{-4} cm^2 sec^{-1} must correspond to $T_m/T < 1$, i.e. to a liquid layer. The data of Fig. 3 (maximum values)* and Fig. 5 are replotted in Fig. 6 with the following melting points: Cu(Pb), 940°C, Cu(Tl), 960°C. The previously published data for silver with adsorbed sulphur (maximum D_s values)* are also plotted in Fig. 6 for a melting point of 800°C. When allowance is made for the experimental errors and the uncertainty in the position of the base curve it is seen that these three systems can fit into the same framework as that suggested for a clean metal.

The above melting points can be justified by assuming that in each case we are dealing with an adsorbed layer which is in fact a two-dimensional compound or alloy (adsorbate and adsorbent atoms in the same surface layer) and which has properties close to those of the corresponding three-dimensional substance.

---

* No attempt is made here to interpret the variation of D_s with the pressure of the adsorbate

For the case of silver with adsorbed sulphur, the measurements were carried out in a range of chemical potential of sulphur below that for bulk compound formation and for which Benard *et al.*[20] have shown the existence of an adsorbed monolayer. Moreover, studies of adsorbed layers of sulphur on various metals[21] by means of LEED have shown that there form mixed layers, or two-dimensional sulphides, that have crystalline structures close to those of the corresponding three dimensional sulphide. The melting point of 800°C that we have chosen is close to that of bulk Ag_2S: 825°C. Recent NMR studies of melting in very thin solid layers show that a monolayer can melt at a temperature slightly lower than the bulk melting point[22].

The surface energy measurements of Bailey and Watkins[8] have shown clear evidence for the existence of an adsorbed layer of lead on copper under experimental conditions very similar to those we have used. The possibility of diffusion in a thick liquid layer, in this system, can be excluded on account of the $t^{1/4}$ kinetics at 960°C: Robertson[23] has shown that grain boundary groove growth of copper immersed in liquid lead exhibits the $t^{1/3}$ kinetics characteristic of bulk diffusion of copper through the liquid; the groove widths in this case are markedly different from those observed for copper in lead vapour. The possibility of viscous flow can also be excluded since this mechanism would give groove growth proportional to time and groove profiles without humps[24]. We may therefore conclude that the observed mass transport occurred only by diffusion in the top surface layer. At the high temperatures of the diffusion experiments it is reasonable to suppose that the surface is a mixture of copper and lead atoms. The choice of a melting point at 940°C is compatible with the bulk monotectic temperature: 954°C. It should be noted that additional evidence of surface melting in this temperature range is shown by the decrease in surface energy found from measurements of twin boundary angles; this sudden decrease can be explained in terms of a surface transformation from the solid to the liquid state.

The results for copper in thallium vapour can be interpreted by analogy with the system copper–lead. The (bulk) phase diagrams for these two systems are very similar. The temperature chosen for melting of the surface (960°C) is close to the monotectic temperature: 968°C.

The results given in Table 1 for gold–lead are consistent with a surface melting point less than 730°C while the results for copper–bismuth (Fig. 4) indicate surface melting at less than 780°C. Both these temperatures are compatible with the bulk phase diagrams for these systems.

We conclude that above these surface melting points the mixed adsorbed layer, always in equilibrium with the underlying solid, has a liquid structure characterized by lack of translational order. In these conditions, the surfaces atoms are not fixed in sites of high coordination to the underlying substrate. Structures with surface atoms not in sites have been found to be quite common in LEED studies of ordered crystallized layers[21].

The data have been placed within the framework given by the base curve, but it still remains to be explained why the surface diffusivity for so many diverse systems can be rationalized in this way. We have already suggested a reason for the value of D_s at the melting point. It is surprising to note that there is no evidence of a discontinuity in passing from the solid to the two-dimensional liquid state. This fact suggests that the diffusion mechanisms, and also the surface structures, above and just below the melting point, are not appreciably different. Given that a very large fraction of the surface atoms are in continual diffusive motion at temperatures near the melting point it is difficult to envisage a mechanism described by unrelated movements of individual atoms. It becomes necessary to consider interactions between the moving atoms and possible cooperative motions. Such cooperative motions, involving perhaps very large numbers of atoms, offer a means of explaining the very high D_s values.

In these circumstances it would seem that characterising the data by an Arrhenius plot with Q_s and D_0 values is misleading and begs the question of migration by individual atomic jumps. In the high temperature region ($T_m/T < 1.2$) models based on ideas of diffusion in liquids may be more pertinent than those derived from the theory of the solid state.

ACKNOWLEDGEMENTS

We thank Professor J.Bénard for his considerable support and encouragement.

REFERENCES

1. W.W.Mullins, *J. Appl. Phys.*, **28**, 333 (1957).
2. N.A.Gjostein, *Metal Surfaces*, ASM-AIME Symposium, New York (1962), Chap. 4.
3. J.M.Blakely, *Prog. Mater. Sci.*, **10**, 395 (1963).
4. F.J.Bradshaw, R.H.Brandon and C.Wheeler, *Acta Met.*, **12**, 1057 (1964).
5. G.E.Rhead, *Acta Met.*, **13**, 223 (1965).
6. H.E.Collins and P.G.Shewmon, *Trans. Met. Soc. AIME*, **236**, 1354 (1966).
7. J.Perdereau and G.E.Rhead, *Surface Sci.*, **7**, 175 (1967).
8. G.L.J.Bailey and H.C.Watkins, *Proc. Phys. Soc. B*, **63**, 350 (1950).
9. C.Herring, *J. Appl. Phys.*, **21**, 301 (1950).
10. W.W.Mullins, *Metal Surfaces*, ASM-AIME Symposium, New York (1962), Chap. 2.

11. H. Mykura, *Acta Met.*, **5**, 346 (1957).
12. H. Mykura, *Acta Met.*, **9**, 570 (1961).
13. N. A. Gjostein, *Trans. Met. Soc. AIME*, **221**, 1039 (1961).
14. N. A. Gjostein, *Surfaces and Interfaces*, ed. J. J. Burke, N. L. Reed and V. Weiss (Syracuse Univ. Press), Vol. 1, p. 271 (1967).
15. J. Henrion and G. E. Rhead, *Compt. Rend.*, **267B**, 958 (1968).
16. G. E. Rhead, *Surface Sci.*, **15**, 353 (1969).
17. H. P. Bonzel and N. A. Gjostein, *Molecular Processes on Solid Surfaces*, Proc. of Batelle Conf., McGraw-Hill (1969).
18. P. Wynblatt and N. A. Gjostein, *Surface Sci.*, **12**, 109 (1968).
19. J. J. Gilvarry, *Phys. Rev.*, **102**, 308 (1956).
20. J. Bénard, J. Oudar and F. Cabané-Brouty, *Surface Sci.*, **3**, 359 (1965).
21. (i) J. Oudar and E. Margot; (ii) M. Perdereau, Colloque International sur la Structure et les Propriétés des Surfaces des Solides, Editions du C.N.R.S., Paris, 1970.
22. G. Karagounis, E. Papayannakis and C. I. Stassinopoulos, *Nature*, **221**, 655 (1969).
23. W. M. Robertson, *Trans. Met. Soc. AIME*, **233**, 1232 (1965).
24. W. W. Mullins, *J. Appl. Phys.*, **30**, 77 (1959).

DATE DUE

APR 1 8 2015			
			PRINTED IN U.S.A.